Contents

PRODUCT DESIGN:

Graphics with Materials Technology

2nd Edition

Lesley Cresswell
Jon Attwood
Alan Goodier
Barry Lambert

Edexcel
Success through qualifications

Heinemann Educational Publishers
Halley Court, Jordan Hill, Oxford OX2 8EJ
Part of Harcourt Education

Heinemann is a registered trademark of
Harcourt Education Limited

© Lesley Cresswell, Alan Goodier, Jon Attwood,
Barry Lambert 2004

First published 2002
Second edition 2004

08 07 06 05
10 9 8 7 6 5 4 3 2

British Library Cataloguing in Publication Data is
available from the British Library on request.

ISBN 0 435 75768 7

Designed by Wendi Watson
Typeset by 𝕋 Tek-Art, Croydon, Surrey

Original illustrations © Harcourt Education Limited,
2004

Illustrated by 𝕋 Tek-Art, Croydon, Surrey

Printed and bound in China by CTPS

Cover photo: © Power Stock

Picture research by Peter Morris

Acknowledgements
Every effort has been made to contact copyright
holders of material reproduced in this book. Any
omissions will be rectified in subsequent printings if
notice is given to the publishers.

Figure 1.18 on page 35 appears courtesy of the BSI.

Photo credits: Figure(s) 1.1 supplied by the
author; 1.2 Lampholder 2000; 1.3, 1.4, 1.5, 1.6,
1.7, 1.8, 1.9, Harcourt Education Ltd/Peter Morris;
1.16 supplied by the author; 1.17, 1.18, 1.19, 1.20
Harcourt Education Ltd/Peter Morris; 1.21
Harcourt Education/Chrissie Martin; 2.1 Harcourt
Education Ltd/Peter Morris; 2.2, 2.3, 2.4, 2.5, 2.6,
2.7, 2.8, 2.9, 2.10, 2.12, 2.13, 2.14, 2.15, 2.16, 2.17,
Harcourt Education Ltd/Tudor Photography;
3.1.18 EA Sports; 3.1.20, 3.1.21 Dyson; 3.1.25
Harcourt Education/Peter Morris; 3.1.27
BMW/Mini; 3.1.28 Corbis; 3.2.1 Rex Features;
3.2.2 Corbis; 3.2.3 Mary Evans Picture Library;
3.2.4 PSION; 3.2.5 Rex Features; 3.2.6 Science and
Society Photo Library; 3.2.7, 3.2.8 Harcourt
Education Ltd/Peter Morris; 3.2.9, 3.2.10 Corbis;
3.2.11 Christies; 3.2.12, 3.2.13 Bridgeman Art
library; 3.2.14 Corbis/Bettmann; 3.2.15 Apple
Corps; 3.2.16 Victoria and Albert Museum; 3.2.17
Harcourt Education/Peter Morris; 3.2.18 Baygen;
3.2.19 Aviation Images; 3.2.22 Victoria and Albert
Museum; 3.2.23 Harcourt Education Ltd/Peter
Morris; 3.2.24 Suede; 3.2.25 Christies; 3.2.26 Q8;
3.2.27 Victoria and Albert Museum; 3.2.29 SABA;
3.2.31 Ford; 3.3.4 Harcourt Education; 3.3.5 Giles
Chapman; 4.1.4 Shout/John Callan; 4.1.5
Harcourt Education/Peter Morris, Chris
Honeywell, Victoria and Albert Museum, Art
Directors and TRIP; 4.1.6 Science Photo Library;
4.1.8, 4.1.9 Science Photo Library; 4.1.10 Getty
News/AFP; 4.1.11 Epson; 4.1.14 Remarkable prod-
ucts; 4.1.16 Buro Happold; 4.2.2 Corbis; 4.2.3
Craft Space Touring; 4.2.6 Which?; 4.2.10
Shout/John Callan; 4.2.11 Nike; 4.2.12 Corbis;
4.2.13 Photodisc; 4.2.14 Corbis; 4.2.15 Envirowise;
4.2.17 AVAD; 4.2.19 Harcourt Education Ltd/Peter
Morris; 4.3.2 Denford; 4.3.11 Shout/John Callan;
4.3.13 Corbis; 4.3.17 Pro-Lok; 4.3.19 Corbis; 4.3.20
Alfa-Robo; 4.3.30 The Croc; 4.3.31, 4.3.32
Blazepoint; 5.1 Rex Features; 5.2, 5.3 Harcourt
Education Ltd/Tudor Photography; 5.4 AKG; 5.5
5.6, 5.7, 5.8, 5.9, 5.10, 5.11, 5.12, 5.13, 5.14
Harcourt Education Ltd/Tudor Photography; 6.3,
6.4, 6.6 Harcourt Education Ltd.

Student work appears courtesy of A Welch, T Still,
A Tkacz; R Asksey, H Capper and R Keyworth.

Part 1

Part 1

Introduction

This Advanced Design and Technology book is designed to support the Edexcel Advanced Subsidiary (AS) and Advanced GCE Specification for Product Design: Graphics with Materials Technology.

The book follows the structure of the Specification and is intended to support you through the course. The content of the book will provide you with a great deal of knowledge and understanding and will help you prepare for assessment. As with any Advanced level course you are advised to read around the subject to broaden your knowledge and understanding.

How to use this book

Part 1

Part 1 provides advice on how to use the book, explains how the course is structured and offers guidance on how to manage your own learning during the course. This includes advice on planning, organising and managing your work. You should read this section before starting your course.

Parts 2 and 3

Parts 2 and 3 provide unit-by-unit guidance on each of the AS Units 1–3 (Part 2) and the Advanced GCE Units 4–6 (Part 3). These sections will provide you with knowledge and understanding to help you through your course. They will also help you prepare for internal and external assessment and increase your chances of success. You should therefore refer to these sections of the book for guidance on the subject content and to help you understand how each unit is assessed.

Parts 2 and 3 also look at the structure, subject content and assessment requirements of the Edexcel Specification. Each of the three AS and three A2 units use similar headings and subheadings to those found in the Specification, so you know that you are covering the content of the course. Each unit is structured as shown in Table 1.1.

Tasks, questions and information appear throughout the text as follows:

- The tasks:
  - help you to understand issues such as industrial practices
  - give you practice in some aspect of designing or manufacturing
  - or help you practise specific skills, such as how to do market research or test the suitability of materials.
- 'Factfile' boxes contain information which may explain:
  - technical terms
  - or illustrate points in the text.
- 'Think about this!' boxes explain different issues, such as industrial practices, or 'values issues' which may influence your design decisions.
- 'To be successful you will…' boxes appear at the end of each section in the coursework units. They contain the assessment criteria that you will need to meet in order to be successful.

Table 1.1 *Structure of book's AS/A2 units*

Summary of expectations for the Unit	Unit content	Student checklist/practice exam questions
The first page of the unit summarises: • what you are required to do • what you will learn • how the unit is assessed	This covers the subject content in detail. It: • explains what you will learn in each unit • helps you understand the assessment requirements • provides tasks, questions and information to guide you through the unit	The last page of the unit provides checklists and practice exam questions to help you: • check the progress of your coursework • revise for and prepare for the exams • be as successful as possible

- In A2 Units 4–6 you may find 'Signposts' to the AS units. These refer you back to information or topics that have been discussed in the AS course.
- The coursework Units 2 and 5 will provide many opportunities for you to generate evidence for your Key Skills portfolio. You may find it helpful to check out Key Skills requirements for Communication, Application of Number and Information Technology at Level 3.
- Technical terms are in bold when they first occur and are explained in context in the text. They also appear in a glossary on pages 307–319.

How the course is structured

There are three units at Advanced Subsidiary (AS) and three units at Advanced GCE (A2).

The AS units

The three AS units combine to make the AS course. The AS units:

- build on the knowledge, understanding and skills you developed through the study of GCSE Design and Technology
- provide a discrete course leading to an AS qualification, or
- provide the first half of the Advanced GCE course. The AS units contribute 50 per cent of the Specification content. You must follow the AS course before progressing to A2.

The A2 units

The A2 units combine with the three AS units to make the Advanced GCE course. The A2 units:

- build on the knowledge, understanding and skills developed in the AS course, to achieve the full Advanced GCE standard
- provide the other 50 per cent of the Specification content
- enable you to achieve a greater level of sophistication and more in-depth knowledge and understanding.

A summary of the AS and A2 units is provided in Table 1.2.

How the AS units are assessed

Units 1 and 3 are externally assessed by examination. Unit 2 is the coursework unit. This is assessed by internal marking and external moderation by the Edexcel Moderator (see the summary in Table 1.3).

How the A2 units are assessed

Units 4 and 6 are externally assessed by examination. Unit 5 is the coursework unit. This is assessed by internal marking and external moderation by the Edexcel Moderator (see the summary in Table 1.4).

Table 1.2 Summary of the AS and A2 units

AS 50% of the Specification content		A2 50% of the Specification content	
Unit 1	Industrial and commercial products and practices	Unit 4	Further study of materials, components and systems with options
Unit 2	Product development I	Unit 5	Product development II
Unit 3	Materials, components and systems with options	Unit 6	Design and technology capability

Table 1.3 Assessment of Units 1–3

AS 50% of the Specification content		Assessed by	% of AS course	% of A2 course
Unit 1	Industrial and commercial products and practices	External assessment 1½-hour examination	30%	15%
Unit 2	Product development I	Internal assessment Coursework project	40%	20%
Unit 3	Materials, components and systems with options	External assessment 1½-hour examination	30%	15%

Table 1.4 *Assessment of Units 4–6*

A2 50% of the Specification content		Assessed by	% of A2 course
Unit 4	Further study of materials, components and systems with options	External assessment 1½-hour examination	15%
Unit 5	Product development II	Internal assessment Coursework project	20%
Unit 6	Design and technology capability	External assessment 3-hour examination	15%

Unit guidance

The following section guides you through the three AS units and the three A2 units.

Unit 1 Industrial and commercial products and practices

This unit enables you to develop an understanding of industrial and commercial practices through product analysis. Throughout the unit you should investigate the design, manufacture, use and disposal of a range of products. The products you investigate should include both 2D and 3D elements, such as a product and its packaging. You will find information about 2D/3D elements in the sections describing Units 2 and 5 below. The areas of study for Unit 1 include:

a Basic product specification: develop a product specification for a range of products
b Working characteristics of a range of materials and components
c Scale of production
d Manufacturing processes
e Quality
f Health and safety
g Product appeal

Throughout the unit you should undertake a variety of tasks to enable you to understand the subject content. For example, you could:

- work collaboratively with others on some investigative activities, e.g. when analysing a range of products
- work individually on some tasks, e.g. when developing creative communications skills to record the investigation of products (communications skills can include writing, drawing, sketching, graphics, charts, flow diagrams, systems diagrams, computer-aided design (CAD), modelling, etc.)
- work individually on the detailed analysis of products, to gain a personal understanding of product development and manufacture.

Assessment (written exam)

Unit 1 is assessed through a 1½-hour product analysis exam, which assesses your understanding of the unit subject content. The style of assessment remains the same each time the unit is assessed, but the product to be analysed will be different.

Unit 2 Product development I

Product development I is a full coursework project. Unit 2 builds on the knowledge, understanding and skills you gained during your GCSE coursework. At AS level you are expected to take a more commercial approach to designing and making a product, to meet needs that are wider than your own. This could mean designing and making a one-off product for a specified user or client, or designing and making a product that could be batch or mass-produced for users in a target market group.

An AS Graphics with Materials Technology project should:

- reflect a study of the 'technologies' involved in the AS Graphics with Materials Technology Specification
- include a 2D element developed from traditional and modern graphics media
- include a 3D model using at least one resistant material listed in Unit 3A (Classification of materials)
- include a coursework project folder that demonstrates good quality graphic communication.

When choosing an AS coursework project you should ensure that the project enables you to meet all the assessment criteria. Before you start your coursework you should refer to the section on Unit 2, which explains the assessment criteria in detail. As you work through your AS coursework project, refer to this section as and when you need.

Assessment (coursework project)

Your AS project should comprise a product and a coursework project folder. The project will be marked by your teacher or tutor and moderated by the Edexcel Moderator.

Unit 3 Materials, components and systems with options

This unit has two sections, both of which must be studied. Section A includes subject content related to materials, components and systems. Section B has two options, each with subject content. You must study one option from Section B. A summary of Unit 3 is given in Table 1.5.

During the unit you should undertake a variety of tasks to enable you to understand the subject content. For example, you could:

- work individually on some activities, e.g. when using a database to research materials and components
- work individually on some tasks, e.g. when undertaking practical tasks to develop understanding of working properties and processes
- work collaboratively with others on some activities, e.g. when testing materials.

Assessment (written exam)

Unit 3 is assessed through a $1\frac{1}{2}$-hour written exam which assesses your understanding of the unit subject content. You are required to answer two sections:

- Section A: Materials, components and systems
  Section A consists of short-answer, knowledge-based questions, worth a total of 30 marks. Examiners will look for a description, explanation or annotated sketch which show your understanding of the topic or process. You are advised to spend approximately 45 minutes on Section A.
- Section B: Options
  Section B consists of two compulsory questions, each worth 15 marks. You are expected to demonstrate understanding of the technology associated with the option studied. You are advised to spend approximately 45 minutes on Section B.

Unit 4 Further study of materials, components and systems with options

Unit 4 has two sections, both of which must be studied. Section A includes further subject content related to materials, components and systems. Section B has two options, each with subject content. You must study the same option from Section B that you studied in Unit 3. A summary of Unit 4 is given in Table 1.6.

During the unit you should undertake a variety of tasks to enable you to understand the subject content. For example, you could:

- work individually on some tasks, e.g. when undertaking practical tasks to develop understanding of the relationship between properties and the selection of materials
- work individually on some activities, e.g. when using the Internet to research information about new technologies and new materials
- work collaboratively with others on some activities, e.g. when developing understanding of the impact that 'values issues' have on the design, development, use and disposal of a range of products.

Table 1.5 Unit 3 areas of study

Unit	Level	Components	Areas of study
3	AS	Section A: Materials, components and systems	• Classification of materials and components • Working properties and processes • Testing materials
		Section B: Options	• Design and technology in society • CAD/CAM

Table 1.6 Unit 4 areas of study

Unit	Level	Components	Areas of study
4	A2	Section A: Further study of material, components and systems	• Selection of materials • New technologies and the creation of new materials • Value issues
		Section B: Options	• Design and technology in society • CAD/CAM

Assessment (written exam)

Unit 4 is assessed through a $1\frac{1}{2}$-hour written exam which assesses your understanding of the unit subject content. You are required to answer two sections:

- Section A: Materials, components and systems
  Section A is similar in style to Unit 3, consisting of short-answer knowledge-based questions, worth 30 marks. Examiners will look for a more in-depth response when describing or explaining topics or processes. You are advised to spend approximately 45 minutes on Section A.

- Section B: Options
  Section B is similar in style to Unit 3, consisting of two compulsory questions, each worth 15 marks. You are expected to demonstrate understanding of the technology associated with the option studied. You are advised to spend approximately 45 minutes on Section B.

Unit 5 Product development II

Product development II is a full coursework project. Unit 5 builds on the knowledge, understanding and skills that you gained during your AS coursework. At A2, you will need to work more independently, which may involve using a wider range of people to support you in your work.

At A2 you should take a commercial approach to designing and manufacturing, to meet needs that are wider than your own. This could mean designing and making a one-off product for a specified user or client, or designing and manufacturing a product that could be batch or mass-produced for users in a target market group. You will need to develop your A2 project in collaboration with potential users or with a client (such as a local business or organisation). You should make use of feedback from your user, client or target market group in order to access the full range of marks.

An A2 Graphics with Materials Technology project should:

- reflect a study of the 'technologies' involved in the A2 Graphics with Materials Technology Specification
- include a 2D element developed from traditional and modern graphics media
- include a 3D model using at least one resistant material listed in Unit 3A (Classification of materials)
- include a coursework project folder that demonstrates good quality graphic communication

- include an increased emphasis on industrial applications and commercial working practices.

Your A2 project should demonstrate clear progression from the standard achieved at AS level. This can be achieved through demonstrating a higher level of 'design thinking':

- undertake research that targets more closely the problem/design brief
- select and use relevant research information
- make closer connections between research and the development of ideas
- use a higher level of understanding about materials, processes and manufacturing techniques
- demonstrate greater understanding of relevant technical terminology
- use higher level communication and presentation skills
- demonstrate appropriate use of ICT; including finding a balance between computer-generated images and those that are hand drawn.

Assessment (coursework project)

Your A2 project should comprise a product and a coursework project folder. The project will be marked by your teacher or tutor and moderated by the Edexcel Moderator, using the same assessment criteria as the AS project. Before you start your coursework you should refer to Unit 5, which explains the assessment criteria in detail. As you work through your A2 coursework project, refer to this section as and when you need.

Unit 6 Design and technology capability

This unit is called Design and technology capability because it assesses the knowledge, understanding and skills you have gained during your Advanced GCE course.

Unit 6 focuses on the designing and making process including knowledge and understanding of product development and manufacture. Since this knowledge and understanding is taught throughout the whole Advanced GCE course, no new learning is expected during Unit 6. However, it is essential that you review and revise what you have already learned and undertake exam practice in order to prepare fully for this exam.

Assessment (Design Paper)

- Your centre will be sent a Design Research Paper in March of the year of the exam. This will give you a context for design, together with bullet points that give you direction about what to research.

- In the three-hour Design Exam there will be one compulsory design question that is based on the research context. You will be asked to produce a design solution and describe how your solution can be manufactured.
- You may take all your research materials into the exam and use them as reference throughout, but the research material is not submitted for assessment.

Managing your own learning during the course

The purpose of this section is to help you take more responsibility for planning and managing your own work. This is an essential feature of any course that you will be undertaking at AS and Advanced GCE level. The ability to manage your own learning is an essential skill in higher education and is highly valued by employers.

In order that you may take responsibility for your work, you need to be very clear about what is expected of you during the course. This book aims to provide you with such information.

- Read the whole of Part 1 before you start the course, so you understand the course structure and the assessment requirements. It will also give you an overview of the requirements of each unit.
- Get to grips with the coursework projects that you need to produce and with the deadlines that you are required to meet. Investigate the coursework assessment and mark scheme.
- Before you start a unit, read the relevant 'Summary of expectations' in Parts 2 or 3. This will give you an understanding of the unit requirements and provide information about how each unit is assessed.

Taking responsibility for your own learning

Once you have a clear understanding of the course requirements, you can plan your time and your work. Taking responsibility for this will enable you to be more independent and take more responsibility for your own decisions. Being independent will require you to manage your own learning, develop project management skills and use a wide range of support beyond your teacher or tutor.

Project management

Learning to manage your work is called project management. This is an essential skill in higher education and in employment. Project management means knowing:

- what is required

- planning and setting targets
- managing time, resources, budgets and people
- monitoring progress and 'getting it right'.

In order that you manage your own projects successfully, you will need to develop the following:

- research skills
- communication skills
- an understanding of industrial and commercial practice
- ICT skills.

Research skills

You will need to use both primary and secondary research to help you design, develop and manufacture your product.

You can use primary research to identify:

- user preferences – such as buying behaviour, taste and lifestyle of target market groups
- market trends – such as style, design and colour trends
- existing products – using product analysis to investigate product design and manufacture.

You can use secondary research, including information from books, magazines, catalogues, databases, CD-ROMs or the Internet, to identify:

- existing products and price ranges
- information about materials, components, systems, processes, technology, production, quality and safety issues
- the work of other designers.

Communications skills

During your Design and Technology course you will need to apply a range of communications skills, such as the following:

- talking and listening to others
- reading, analysing and recording research information
- developing an understanding of form, function and design language
- drawing and sketching
- using professional practice – such as writing reports and presenting ideas to peers or clients.

Understanding industrial and commercial practice

You will be expected to develop an understanding of industrial approaches to design and manufacture during your course and to evidence them in your coursework. You will learn about industrial practices in Unit 1, but there are other approaches you could use to enhance this understanding. For example, you could:

- investigate work-related materials produced by a business
- use the expertise of visitors from business
- make an off-site visit to see a business at work
- use work experience to inform your understanding of how organisations work
- use modern contexts to develop products for real/imaginary companies, e.g. that project a brand image appropriate to a specific company or retailer
- use ICT to research information about products, e.g. finding out about materials, components, systems, processes, the way products are marketed, about company values.

You should evidence your understanding of industrial practices in AS and A2 coursework through the designing and manufacturing activities that you use to develop products. These may include:

- developing design briefs and specifications
- undertaking market research and product analysis
- generating, developing and evaluating ideas
- modelling and prototyping prior to manufacture

- producing a production plan
- testing against specifications.

In order to evidence industrial practices you should also use industrial-type terminology. To learn about this you should investigate the glossary on pages 307–19 and refer to the two coursework units in Part 2.

ICT skills

You should investigate how you could use ICT in your coursework projects and use it *where appropriate and available*. The use of ICT is assessed in the AS and A2 coursework projects, when developing and communicating design proposals; in planning manufacture and in product manufacture.

Currently, you will not be penalised for non-use of ICT. Where CAD is available, however, you should aim to find a balance between computer-generated images and those that are hand drawn. It will be essential to make use of your drawing skills in some of the exams, in particular in Unit 1 and Unit 6.

In order to enhance your Design and Technology capability you could investigate how you could use:

- ICT for research and communications – such as using the Internet, e-mail, video conferencing, digital cameras or scanners
- word processing, databases or spreadsheets for planning, recording, handling and analysing data
- CAD software to model, prototype, test and modify design proposals in 2D/3D
- computer-aided manufacture (CAM) for computer control, using CNC machines.

Part 2
Advanced Subsidiary (AS)

Industrial and commercial products and practices (G1)

Summary of expectations

1. What to expect
This unit will develop your understanding of industrial and commercial practices through the investigation of a range of manufactured products.

2. How will it be assessed?
The work that you do in this unit will be externally assessed through a $1\frac{1}{2}$-hour Product Analysis examination. The style of questions will remain the same in each exam, so you will know what to expect. The product will either be a one-off, batch or mass-produced product. The exam questions are set out in Table 1.1.

Table 1.1 Assessment criteria for Unit 1

Assessment criteria	Marks
a) Outline the product design specification for this product.	7
b) Justify the use of: i) material/component ii) material/component.	3 3
c) Give **four** reasons why the product is one-off/batch/mass continuously produced.	4
d) Describe the stages of production for this product. Include references to industrial manufacturing methods in your answer.	16
e) Discuss quality issues for the product.	8
f) Discuss the health and safety issues for the product.	8
g) Discuss the appeal of the product.	8
h) Quality of written communication.	3
Total marks	60

3. What will be assessed?
The following list summarises the topics covered in Unit 1 and what will be examined:

a **Basic product specification** – Develop product design specifications for a range of products, under the following headings:
  - purpose/function
  - performance
  - market
  - aesthetics/characteristics
  - quality standards
  - safety.

b **Working characteristics of a range of materials and components** – Understand that properties and working characteristics influence the choice of materials and components used in a range of products.

c **Scale of production** – Understand that products are manufactured by different manufacturing systems.

d **Manufacturing processes** – Understand how one-off, batch and high-volume manufacture of products is achieved using:
  - preparation
  - processing
  - assembly
  - finish.

e **Quality** – Understand the importance of quality of design and manufacture:
  - quality control in production
  - quality standards.

f **Health and safety** – Understand the principles of health and safety legislation and good manufacturing practice:
  - safe use of product
  - safe procedures in production.

g **Product appeal** – Influences on the design, production and sale of products.

4. How to be successful in this unit
To be successful in this unit you will need to:
- analyse a range of manufactured products using the assessment criteria in Table 1.1
- apply your knowledge and understanding to a given product in the exam
- give clear and concise answers, using specialist vocabulary
- use clear, annotated diagrams where they could make the answers clearer.

5. How much is the unit worth?
This unit is worth 30 per cent of your AS qualification. If you go on to complete the whole course, then this unit accounts for 15 per cent of the full Advanced GCE.

Unit 1	Weighting
AS level	30%
A2 level (full GCE)	15%

Understanding commercial products and industrial practices

This unit will help develop your understanding of products and the industrial and commercial practices by which they are manufactured. It will provide you with useful information about materials, manufacturing processes, product assembly, quality and safety. In industry, product analysis includes the collection of data from catalogues and trade literature and the practical 'hands on' analysis of existing products, which may be own company products or those of a commercial rival. Product analysis enables designers to:

- evaluate the properties of the materials used in a product
- assess the product's fitness-for-purpose for a specific end-use
- see how well the product's characteristics and price meet user requirements
- evaluate the processes used to manufacture the product
- work out how the product was assembled
- examine the product's quality of design and manufacture
- find out why an own company product is selling or not selling as predicted
- see how and why a competitor's product is successful
- develop a product design specification and design ideas for a new product.

Developing a strategy for undertaking product analysis

Think about the last time that you bought a new mobile phone or a portable CD player. When you shop for products such as these, you evaluate the function, style and value for money that they provide. You also make value judgements about the product's quality of design and manufacture. This means that you are already well practised in judging the worth of a product from a user's point of view. You now need to apply your product analysis skills in a more formal and objective way. This is best achieved by practice.

Developing product analysis skills

The key to developing product analysis skills is to look, examine and question. Looking at products will involve activities that enable you to examine them by eye and by taking them apart. Asking questions about products should target what you need to find out. It is important to investigate different types of products that have been manufactured by a range of production

processes, different levels of production and from a variety of materials. This will involve analysing a selection of one-off, batch-produced and high-volume products. Remember to include actual physical products that you can handle as well as products from catalogues. It is important to undertake the following exam practice:

- practice developing design specifications for different types of products
- undertake short tasks and timed tasks based on each of the assessment criteria
- practice doing timed product analysis encompassing all of the assessment criteria.

Using the product analysis assessment criteria

The sections of this unit are organised around the structure of the assessment criteria (see Table 1.1). The sections will take you through each criterion, explaining what each involves, what you are expected to do and how many marks are awarded, so you know what to expect in the exam. Each section provides back up knowledge and understanding, so you can be sure that you are covering the required material about commercial products and industrial practices. It is worth remembering that in the exam, the assessment criteria (the questions you are asked) remain the same each time and it is only the product that is different. Your responses to all the assessment criteria will depend on the scale of production of the product.

Figure 1.1 *In the product analysis exam you will be given photographs and details of a Graphics with Materials Technology product*

a) Outline the product design specification for the product (7 marks)

In the exam you are asked to address at least four of the following bulleted headings:

- purpose/function
- performance
- market
- aesthetics/characteristics
- quality standards
- safety.

When you undertake product analysis during your course it is important to address all of the above specification headings. This will give you practice in developing meaningful product specification criteria. When you undertake product analysis during the end-of-unit exam you should use the given headings to help structure your answers. You will have the option of addressing all the specification criteria or concentrating on just four of them. The key to achieving all seven marks is to provide seven appropriate points, even if you are addressing only four of the headings. Each of your specification points needs to be justified, which means providing a fully explained answer. For example for the purpose/function 'The boxed puzzle should not be too big' needs more explanation, such as 'The wooden blocks should be a suitable size to easily fit into a child's hand'.

Factfile

In industry, a product design specification is an essential document that sets out the criteria for the development of a product. It is also used as a checklist against which the performance of the product can be measured. All product design specifications vary according to the type of product and its end-use. For example, the specification for breakfast cereal packaging with a promotional toy would be different from that of a wooden boxed puzzle.

Purpose/function

In the exam paper you will be given photographs and details of a Graphics with Materials Technology product. You will need to take a few minutes to carefully study this information so you are clear about what you are being asked to respond to in the assessment questions. The first product specification criteria

asks you to define the function and purpose of the product:

- the aim or end-use of the product, e.g. 'A co-ordinated stationery gift set aimed at cat lovers. The clear cover enables the buyer to see the design of the writing paper and envelopes inside.'
- how the product will be used or what it should do, e.g. 'A stationery gift set with a cover that protects the contents and keeps the stationery organised.'

It is important that you fully explain (justify) the function and purpose of the product in order to gain the available marks.

Task

Choose a product with which you are familiar, such as a product that you use at home. Use the following questions to develop a specification for its purpose/function.

- What is the aim of this product?
- What is the need it provides for?
- What is the end-use of this product?
- Where, when and how will it be used?
- What benefits will it bring to the users?
- How does it fulfil its purpose?

Performance

In this product specification criterion you are asked to explain the 'performance requirements of the product, materials and components'. Make sure that you read again the information given in the exam paper about the product and the materials and components it is made from. This will enable you to explain:

- how the product should perform, e.g. 'The child's book should be free-standing and be light enough for a young child to hold.'
- how the materials/components should perform, e.g. 'The laminated card pages should be able to withstand fair wear and tear.'

Market

In this next product specification criteria you need to think about two different aspects of the term 'market': the retail market for the product and the target market group. The size of the market usually determines the level of production for a product.

The retail market for the product has a considerable influence on how it is designed and manufactured. For example, a hand-made boxed jigsaw with an Arts and Crafts-style 'garden' design might be batch produced for sale in a museum shop. On the other hand, a menu holder for a global fast food chain might be produced in high volume, although the menu itself might be batch produced in different languages. When you analyse a product it is therefore important to look for design clues that can help you decide the product retail market and therefore the kind of user who might buy it. For example, the retail market for a child's free-standing 'car' book may be a specialist toy shop and the target market will be adults looking for a value-for-money learning aid.

The second aspect of the term 'market' is the target market group (TMG), i.e. the users of the product. All products are designed with users in mind. Many manufacturers use extensive market research to establish their TMG, without which it would be almost impossible to design a product. Establishing the TMG means knowing who you are designing for and what their needs are; are the users male or female, young or old, able-bodied or disabled, expert or novice users of the product? The type of end-user will help determine the characteristics of a product to enable it to meet user needs and therefore have a good sales potential. When you think about the TMG you may need to make references to ergonomics. This is the science of designing products for human use, matching the product to the user. The child's free-standing 'car' book, for example, will be ergonomically designed to fit the size of the child's hand and have high strength to weight so the child can easily push it along. When you analyse a product you will need to decide if the product is made in a specific size to meet the needs of children, women or men, and if it needs to be made in specific dimensions to make the product more usable or to improve the users' interaction with it.

Task

Describe the retail market and the needs of the target market group users for each of the following products:

- a perfume bottle and packaging designed by Jean Paul Gaultier (Figure 2.1 page 41)
- a menu and menu holder for a global fast food chain
- a point-of-sale display unit for stationery to be sold in a museum shop.

Aesthetics/characteristics

Aesthetics play a very important role in the design and marketing of products as the main reason for buying a product is often how it looks and the image it can give the user. For many products appearance and design characteristics are based on the manufacturer's perception of user needs. When you address the aesthetics/characteristics of the product in the exam, you need to focus on the aesthetic properties and characteristics that the product *should* have in order to meet the TMG requirements. You will gain few marks for simply describing the product. For example the aesthetics and characteristics of a boxed wooden block picture puzzle might be that it should have:

- a design that is based on a children's television cartoon character
- eye-catching text/graphics appropriate to the TV cartoon character
- an attractive contrasting colour scheme
- a design that is suitable for the age range
- a smooth/tactile, easy-to-clean finish.

In the Product Analysis Exam you would not be expected to produce a long list of points about the aesthetics/characteristics of a product, because the maximum marks available for any of the specification headings is four. To achieve four marks for aesthetics/characteristics, you would therefore need to provide four different, fully explained and appropriate specification points.

Quality standards

When you explain how the product can meet quality standards you need to refer size, dimensions and the use of tolerances. You should also refer to the relationship between quality, cost and scale of production. For example, a boxed hand-made jigsaw may use more expensive materials than a high volume thermoplastic menu holder, although both levels of production would make use of a quality system of some kind.

All products need to be manufactured to the correct size and dimensions to ensure that the component parts fit together well and the product is capable of performing its function. All the component parts therefore need to be designed and manufactured to a high standard with precision and accuracy. This is where British Standards (BS) come in. There is an enormous range of published BS, which are used by manufacturers (not BS) to help them produce a quality product. Although you are not expected to quote a BS number, it is always preferable to explain the relevance of a standard, such as a BS, to ensure that a brass paperweight has a scratch-free surface finish or that the packaging is durable and does not fall apart and the print quality is consistent.

Successful manufacturers incorporate the use of a quality management system (QMS) in design and manufacture in order to achieve a quality product. You could therefore refer to the use of a QMS or total quality management (TQM). One aspect of TQM is the use of quality control (QC) checks which use agreed tolerances to check the accuracy of dimensions and sizes in order to manufacture identical products. You could also quote the use of ISO 9000, which is an internationally agreed set of standards for a QMS. ISO 9000 ensures that a manufacturing customer receives the product that has been agreed. For example, an ISO-registered manufacturer of a child's 'car' book with thermoplastic wheels would only buy in the wheels from a supplier that was also ISO registered, because they would know that the thermoplastic wheels would meet an agreed BS. Any manufacturer that has a QMS in place can apply for a Kitemark or offer a warranty or guarantee of product quality.

Task

A one-off product such as a hand-made boxed jigsaw, made from plywood in a laminated board presentation box, is likely to be made to a high quality and be relatively expensive to buy. On the other hand, a mass produced board game from a high street store may be made from laminated board in a carton board box and be a fraction of the cost of the hand-made product, although it would have been produced to meet BS through the use of the manufacturer's quality assurance system.

a) Describe how a designer-maker could ensure that the hand-made boxed jigsaw was a high-quality product.
b) Describe how the manufacturer of a mass produced board game could achieve a quality product.

Safety

Safety is an essential feature of all manufactured products, many of which have to comply with safety legislation before they can be sold. Safety standards are set by British Standards (BS) to ensure the safety of the user.

Products such as games or cosmetic containers and their packaging, are rigorously tested to ensure that they will not cause injury through normal use or misuse. This includes the use of non-toxic materials, minimising the risk of cuts from sharp edges or paper and eliminating the risk of choking on loose parts (especially for products used by young children).

A product that displays a Kitemark is independently tested at regular intervals to make sure it complies with a relevant safety standard. The Kitemark symbol shows potential customers that the product is safe and reliable. When a product is sold across the European Union (EU) it has to meet safety standards set by the EU. Manufacturers who claim to meet these standards can use the CE mark, but if the product is found not to comply with the safety standard it can be seized and the manufacturer, importer or supplier prosecuted.

In this section you will need to make and explain points that are specific to the product in question, rather than making general statements. For example a board game with small component parts may require labelling with a legible graphical symbol, stating that the game is not for children under the age of thirty-six months. You will not be expected to quote a BS number, but must make sure that what you explain is appropriate to the product. For example, a child's 'car' book would need to use non-toxic printing inks, have no sharp corners and have thermoplastic wheels large enough not to be a choking hazard. Other points about safety need explaining, e.g. rather than stating that the product needs to have a Kitemark, you could explain that the product needs to meet a relevant safety standard and pass regular tests in manufacture to be awarded a Kitemark.

Task

Describe the safety features of a cast brass paperweight in a presentation cardboard box, with reference to the following criteria:
- compliance with safety standards
- ergonomics
- safety when opening the box
- finish
- safety of use and disposal.

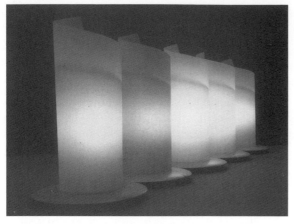

Figure 1.2 *Batch-produced lights*

Task

The batch-produced lights shown in Figure 1.2 are made from thermoplastic, paper and wood. Each light can be fitted with a different coloured filter on its polymer-laminated paper shade to change the lighting mood. The wooden base has a lamp holder for a low energy 10-watt fluorescent lamp.

Draw up a product design specification for the light.

Think about this!

In the Product Analysis examination paper the product specification headings will be given so you do not have to commit them to memory.

Remember to address at least four of the headings so you have the opportunity to achieve the seven marks available. You can still address all seven headings to gain seven marks, but must explain each point.

The specification statements require well justified points about how the product should be designed. Remember that the product specification is not a description of the given product, but criteria which define what the product should be like. The given scale of production is important, because it influences many aspects of the product specification.

b) Justify the use of materials and components (6 marks)

In this section you will be asked to justify the use of materials and components used in the product given in the exam paper. There are different parts to the question with marks allocated to each part.

You will be asked to justify the use of different materials, or different components, or the use of materials and components. In your response you will need to give reasons why the working properties and characteristics of the given materials or components make them suitable for the product and its function. Simply listing the properties of the given material is not enough to gain the available marks. When you undertake product analysis as part of your course, keep a reference book handy so you can look up materials properties to help you decide why the material is suitable for the product.

Working characteristics of a range of materials and components

Physical and mechanical properties

In order to explain the suitability of a material for the product given in the Product Analysis exam, you will need to understand the material's physical and mechanical properties. For example:

- clear PVC is suitable for the cover of a stationery box, because it is readily available in sheet form, is rigid, tough and easy to cut and fold
- medium density fibreboard (MDF) is suitable for the pieces of a jigsaw, because it is durable, easily cut and shaped, paints well and is readily available in large sized sheets, making it economical for mass production.

The physical and mechanical properties of materials make them suitable for different types of manufacturing processes and different types of products.

- Physical properties describe the handling characteristics of a material, including its feel.
- Mechanical properties describe how the material reacts to forces acting upon it.

Paper, plastics, woods and metals may have different mechanical properties but all of them can be deformed. Laminating wood, for example, increases its strength, a mechanical property.

Laminating paper or card increases its rigidity, strength and stability. Other materials such as plastics and metals have the ability to change into a molten or plasticised state at certain temperatures. It is important to differentiate between physical and mechanical properties and to be able to apply them to paper, card and board, woods, metals and plastics. Table 1.2 lists common physical and mechanical properties.

Aesthetics

Aesthetics refers to all the ways in which we use and interact with products, whether by sight or touch. These days the decision to buy a product is often based on its appearance, rather than its technology, because many products have similar functional characteristics. The success of one product over another is therefore often dependent on its aesthetic characteristics and the user's subjective experience of it. For example, owning a certain type of vacuum cleaner or toaster can say something about you. When you refer to the aesthetic properties of materials you will need to explain how their working properties impact on the shape and form of the product, using terms such as those shown in Table 1.3.

The properties of paper, card and board

Paper, card and board are used for modelling and commercial packaging. Their properties and working characteristics can be classified by size, weight and thickness and by their ability to be cut, shaped, formed and joined. Most paper, card and board can be recycled, depending on the applied surface finish. For example card coated or laminated with a polymer makes it difficult for the card to decompose or be recycled.

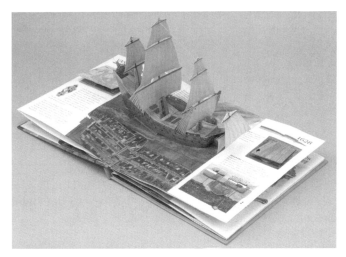

Figure 1.3 *This pop-up book is made from card, with a laminated cover. Some of the pop-ups have supports made from thin strips of rigid polystyrene (HDPS)*

Size, weight and thickness
- Paper and boards are available in standard sizes and standard packs.
- Papers and thin cards come on rolls, which make them suitable for continuous printing processes.
- Paper and boards are specified by weight (grams per square metre – gsm) or thickness (micrometers – microns).
- Paper becomes board at around 220 gsm.

Cutting, shaping, forming and joining paper and card
The end-uses of paper, card and board depend on their ability to be cut, scored, creased, folded, moulded, pinned, stapled, bound or glued.

Table 1.4 shows the properties and end-uses of different types of paper, card and board.

Table 1.2 *Physical and mechanical properties*

Physical	Mechanical
Appearance, corrosion resistant, chemical resistant, density, ease of application, electrical conductivity/insulation, fusibility, optical properties (how easily light passes through), thermal conductivity/insulation	Brittleness, ductility, durability, elasticity/flexibility, hardness, malleability, plasticity, stability, stiffness, strength, toughness

Table 1.3

Words related to texture	Words related to colour	Shape and form of the product
Hard, soft, abrasive, smooth, textured	Warm/cool, mellow/dull, bright/vivid, neutral/colourless, pastel/light, dark/sombre, eye-catching Appropriate for images, e.g. a red fire engine	Functional, engineered, stylish, sleek, modern, traditional

Table 1.4 The properties and end-uses of different types of paper, card and board

Type	Properties	End-use
Bank	45–63 gsm White or tinted, thin, light, uncoated	Used to make carbon copies
Bond	63–120 gsm Strong durable with a matt, even surface	Outline sketches of page layouts Sketching, developing ideas, writing, printing
Tracing paper	60–90 gsm Pale grey, thin, very smooth and transparent, heavier weights have a high cost	Outline sketches of page layouts, sketching and developing ideas, accurate working drawings
Cartridge paper	90–150 gsm Creamy white, fine textured and totally opaque, fairly expensive but cost depends on weight and quality	General purpose drawing and painting, booklets, brochures, printing
Layout paper	50 gsm Off white, very thin, smooth, and translucent, fairly expensive	Outline sketches of page layouts, sketching and developing ideas
Card and board	Over 220 gsm Strong and rigid in a large range of colours, thicknesses, surface finish, texture and quality; can be scored, creased, bent and coated or laminated to improve quality or performance; can have good printing surface; card is cost-effective and easy to mass produce; boards are often built from two or more layers of differing quality to lower costs; relatively expensive compared to paper, but depends on its weight, finish and quality	Thin card used for photocopying and printing, card used for 3D modelling and packaging mock-ups, card and board used for packaging boxes and cartons
Corrugated card	Fluted paper sandwiched between paper layers (liners), available in single, double or triple boards; durable, good strength to weight ratio, impact, puncture and tear resistant; good quality liners give good printing surface; low cost, can be made using recycled materials	Thinner corrugated card with fine flutes, used for perfume bottles, larger fluted corrugated card used for protective packaging

The properties of plastics

Plastics are also known as 'polymers' because of their special structure of interlinked chains of molecules. Most early plastics were so called because they were 'plastic', meaning soft and pliable. The very plastic nature of plastics, and the suitability of plastic materials for vacuum forming, injection moulding and blow moulding, enables the development of an enormous range of different product shapes. Plastics can be divided into two main categories: thermoplastics and thermosetting plastics.

- Thermoplastics are the most common type of plastic, made from long tangled chains of molecules which have very few cross links. When thermoplastics are heated they become soft, allowing them to be formed into different shapes and then return to their original shape on reheating. This is known as plastic memory. Thermoplastics generally soften at low temperatures (as low as 100 °C), making thermoplastic products unsuitable for use in high temperatures.

- Thermosetting plastics are made up from long chains of molecules, which are cross linked, resulting in a very rigid molecular structure. Thermosetting plastics can only be heated up and re-formed once, after which they are permanently set in a shape. It is for this reason that they can be used in temperatures in excess of 400 °C.

Thermoplastics are used extensively in packaging because they are:

- lightweight and versatile
- strong, tough, rigid, durable, impact resistant and water resistant
- transparent or available in a wide range of colours
- readily available in granule, tube or sheet form
- easily shaped, formed and joined by cutting, folding, drilling, moulding, welding or gluing
- easily printed
- low cost and recyclable.

Tables 1.5 and 1.6 show the properties and end-uses of thermoplastics and thermosetting plastics.

Table 1.5 *The properties and end-uses of thermoplastics*

Thermoplastic	Properties	End-use
Acrylic	Rigid, hard, brittle, durable, glossy, easily scratched, polishes well, weather resistant, non-toxic, good electrical insulator, easily thermoformed, easily joined using Tensol cement, available in sheet form including transparent, translucent and bright colours, can be recycled	Signage, point of sale displays, leaflet holders, transparent boxes/lids, containers or closures for perfume products
Rigid polystyrene (HDPS)	Light, hard, rigid, non-toxic, water resistant, brittle with low impact strength, often transparent, toughened HDPS is impact resistant and can be in range of colours, can be recycled	Disposable cups cutlery and plates, bubble packs, CD/cassette cases, yoghurt pots, children's toys
Expanded polystyrene (LDPS)	Durable, lightweight, buoyant, impact and water resistant, good heat insulator, easily cut with hot wire cutters, can be recycled	Take-away packaging, egg cartons, hot drinks cups, fruit, vegetable and meat trays, packaging for electrical and fragile products
PVC	Rigid or flexible, hard, tough, lightweight, good chemical and weather resistance, available as a film or sheet	Shrink wrapping and cling film blister packaging, packaging for toiletries, medicines, food, confectionery, water and fruit juices
PET	Very tough with high tensile strength, impact resistant, good chemical/temperature resistance, alcohol and oil resistant, transparent with good optical properties, can be recycled	Mineral water and soft drinks bottles, food containers, microwaveable food trays
High density polyethylene (polythene) (HDPE)	Surface has waxy feel, waterproof, resistant to chemicals, can be sterilised, can be recycled	Bleach and medicine bottles, containers for washing up liquid and cosmetics, thin sheet packaging

Table 1.6 *The properties and end-uses of thermosetting plastics*

Thermosetting plastic	Properties	End-use
Epoxy resin	Clear, rigid, very strong, good resistance to wear and chemicals, good adhesive	Adhesive for unlike materials, e.g. metals to plastics, PCBs
Urea formaldehyde	Stiff, hard, strong and brittle, good electrical insulator	Electrical plugs and fittings, adhesive

Tasks

1 Using Table 1.5 to help you, list the physical and mechanical properties of acrylic, HDPE, PVC, PET and polystyrene.
2 State the type of plastic suitable for use in each of the following products. Give three reasons why each plastic is suitable for the product:
   • shop sign
   • drinks bottle
   • medicine bottle.

The properties of woods

As a natural resource, hardwoods and softwoods produce timber with a number of disadvantages, such as knots and irregularity of grain. Timber can also suffer from changes in moisture content, leading to warping and twisting if not properly seasoned. Wood is also prone to decay and fungal and pest attacks. Timber is relatively expensive to buy and does not come in large sizes, although it can be machined and joined in many ways.

Wood can be produced in standard stock sizes and sections, suitable for manufacturing into products. The use of standard sizes is cost-effective for product manufacture because it reduces the need for initial cutting to a workable size.

• Hardwood is a close-grained, strong and tough timber from deciduous trees that carry their seeds in fruit. It is slow growing (60–100 years to mature) and therefore expensive.
• Softwood is a more open-grained timber from cone-bearing coniferous trees that are evergreen with needles. Although it is fast growing (20–30 years to mature) and ideal for growing commercially, it produces weaker

timber that can split easily. Softwood is less expensive to buy because it can produce long straight lengths with little waste on cutting.

The properties and characteristics of timber can be categorised by their grain, colour, texture, hardness and elasticity. Hardwoods generally have a greater mechanical strength than softwoods. Table 1.7 shows the properties and end-uses of hardwoods and softwoods.

The properties of manufactured boards

Manufactured boards are available in large, wide sheets up to 3 metres × 2 metres. They are relatively inexpensive and, because there is no grain running through them, they are more stable. Manufactured board can be finished by painting and staining, and can be veneered or covered with plastic laminates that give the impression of a solid wood finish. Table 1.8 shows the properties and end-use of plywood and MDF.

The properties of metals and alloys

Metals can be divided into three main categories: ferrous, non-ferrous and alloys.

- Ferrous metals contain iron. They are almost all magnetic and unless treated corrode very easily.
- Ferrous alloys such as mild steel and stainless steel are made from a mixture of iron and carbon. Iron is soft and ductile and carbon is hard and brittle. By varying the amount of carbon used, steels with different hardness can be produced.
- Non-ferrous metals such as copper, aluminium and zinc are pure metals that contain no iron.
- Non-ferrous alloys include brass, which is made from a mixture of copper and zinc.

Metals are ideally suited to manufacturing processes such as casting using aluminium, brass, copper and zinc alloys. Table 1.9 shows the physical and mechanical properties and end-uses of a range of metals.

Task

Compare a child's toy made from softwood, one made from hardwood and one made from manufactured board.

a) Investigate the properties of the softwood, hardwood and manufactured board and explain why they are suitable for their end-use.

b) Draw up a product design specification for one of these products.

Task

Explore two different lighting products that incorporate the use of metals.

a) Investigate the properties of the different metals and explain why they are suitable for their end-use.

b) Draw up a product design specification for one of the lighting products.

Table 1.7 The properties and end-uses of hardwoods and softwoods

Wood	Type	Properties	End-use
Beech	Hardwood	Close-grained, hard, tough and strong, works and finishes well but prone to warping, high cost	Toys, stools
Jelutong	Hardwood	Uniform grain with few knots, pale cream, soft, fine, shapes easily, high cost	Pattern making, modelling and moulds
Pine	Softwood	Straight-grained, fairly strong, but knotty and prone to warping, light brown or yellow, easy to work, readily available and inexpensive	Vacuum forming moulds, product and architectural models, high-quality packaging

Table 1.8 The properties and end-uses of plywood and MDF

Manufactured board	Properties	End-use
Plywood	Made from odd number of veneers glued with grain in alternate directions, two outside layers in same direction, generally made from birch, strong and fairly inexpensive	Jigsaw puzzles, games, architectural modelling
MDF	Very stable, dense, no grain with smooth faces, easily cut and shaped, can be joined with wide range of knock down fittings, takes range of surface finishes, paints well	Children's toys and games, moulds for vacuum forming blister packaging, 3D prototype products, e.g. mobile phones

Figure 1.4 *This paperweight is made from brass, which is a good choice because it is weighty enough to hold papers in place, does not corrode, is durable, non toxic, will not break if dropped and is a readily available material. It has a shiny finish and attractive golden colour; it looks like gold but is much cheaper*

Finishes, properties and quality

Surface finishes and coatings are used to improve a product's functional and aesthetic properties, to improve its useful life and to improve the overall quality of the product. For example paper and thin card can be laminated with plastic sleeves to provide a protective, attractive and long-lasting finish.

Surface finishes

Surface finishes protect materials from oxidation, liquids, scratching or soiling and provide improved aesthetics.

- Tinplate finishes on steel reduce corrosion, e.g. food cans.
- Anodising gives aluminium and its alloys a consistently high quality finish. Coloured dyes can be added before final lacquering.
- Paints are inexpensive, durable and waterproof and are available in a wide range of colours. Polyurethane paints are tough and scratch resistant.
- Varnish can not only enhance the look of wood, but can give a clear, glossy, tough, waterproof and heatproof protective finish that can be wiped clean, e.g. on a child's toy.
- Varnish can also be applied during the printing process to provide a gloss finish to paper and board. Lacquer can also be applied to a printed surface, by machine rollers and dried using ultra violet lamps, to provide a very high gloss finish. Most paperbacks and many cartons have this lacquered finish.

Table 1.9 *The physical and mechanical properties and end-uses of a range of metals*

Material	Physical properties	Mechanical properties	End-use
Steel (ferrous alloy)	Poor resistance to corrosion (rusts)	Tough, ductile and malleable, good tensile strength, easily joined by welding or brazing, cannot be hardened and tempered	Tin-plated steel used for aerosols, food tins, drinks cans, bottle tops, biscuit and paint tins, trays, and tin toys
Tin (non-ferrous, pure metal)	Good corrosion resistance	Soft, low tensile strength, ductile and malleable, low melting point, easily joined	Coating for steel to make tin-plate *(see end-uses for steel)*
Stainless steel (ferrous alloy)	Resists wear and corrosion	Hard, strong, durable and tough, lightweight, difficult to cut and file	Kettles, lighting, ball bearing type games
Copper (non-ferrous, pure metal)	Good conductor of heat and electricity	Malleable and ductile, easily joined, polishes well, expensive	Wire, PCBs
Aluminium (non-ferrous, pure metal)	Corrosion resistant, good conductor of heat and electricity, good fusibility	High strength/weight ratio, lightweight, malleable and ductile, difficult to join, polishes well	Aerosols, drinks cans, cosmetic and toiletry products, screw caps, foil laminates for cartons
Zinc (non-ferrous, pure metal)	Good corrosion resistance	Ductile and easily worked	Coating for steel
Brass (non-ferrous alloy of copper and zinc)	Corrosion resistant, good conductor of heat and electricity, good fusibility	Harder than copper, casts and machines well, polishes well, cheaper than copper	Screws, electric plug pins, candlesticks, clocks

Laminating paper and card with clear polypropylene film, increases rigidity, strength, durability and stability. Laminating protects and enhances the surface finish of food cartons. Paper and board can also be laminated with aluminium foil to provide a barrier to moisture in drinks cartons. A grey board core can be laminated with colour printed facings, e.g. to prolong the life of a young child's book, making it less likely to tear.

- Embossing paper or board creates raised, depressed or rough surfaced areas on business cards, letterheads, brochure covers and packaging.
- Polishing machines finish polymers, such as acrylic, to an attractive glossy finish.

Surface coating, decoration and self-finishing

- Colour can be added to paper, card and board by dyeing during manufacture or by printing. Colour can create an image or style, make a product look heavy or light and give a sense of quality.
- Surface coating of papers on one or both sides, with a mineral coating, improves their smoothness, strength and stability.
- Surface decoration can include the use of gloss and emulsion paints, enamel spray paint, varnish, wood stain, acrylic paints and vinyl transfer techniques, to provide improved aesthetic qualities.
- Plastics are said to be self-finished because they generally do not require edge or surface finishing due to the use of high quality moulds. Plastics can be given rough and textured surfaces from the designs introduced onto the mould surface.

In the product analysis exam, it is therefore important when justifying the use of a specific material to think about the relationship between materials properties, any applied surface finish or

Figure 1.5 *The menu and holder comprise a laminated, colour printed card menu and an acrylic menu holder*

coating and the required level of quality of the product. For example, specialist boards are expensive and can be used for packaging high-quality cosmetics. Recycled papers and boards may cost less, but their quality is generally lower. Recycled packaging is often used for niche markets, e.g. for an 'eco' aware market.

Task
Look at Figure 1.5 and justify the use of:

a) laminated, colour printed card for the menu.
b) acrylic for the holder.

c) Give four reasons why the product is one-off, batch or continually mass produced (4 marks)

In this section you will be asked to give reasons why the given product is manufactured either as a one-off, or is batch or continually mass produced. Four marks are available, so you will need to explain four appropriate reasons for the given scale of production. Your responses should be applied to the product in question, rather than being general statements about

one-off, batch or mass production. For example a product such as the batch produced stationery gift set shown in Figure 1.6 makes use of modern technology, which enables a fast turn around in design graphics, so different styles of gift sets can be produced according to demand. If the current style does not sell as expected, production can quickly be changed

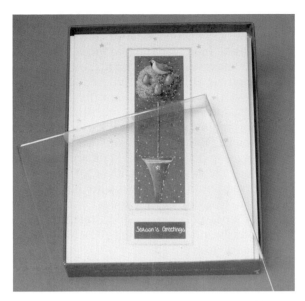

Figure 1.6 *This batch-produced stationery gift set makes use of CAD to enable easy and fast changes in design graphics, and just in time (JIT) production for quick response to demand. The use of CAM in offset lithography enables fast, cost-effective production and reduces labour costs. JIT reduces the need for storage and cuts costs*

over to meet an emerging design trend. The use of modern technology makes use of unskilled labour and reduces labour and storage costs, good reasons why the gift set is batch produced.

Scale of production

The scale of production is one of the most important decisions to be made when developing a product, because it has an impact on all design and manufacturing decisions, including:

- the number of products manufactured
- the choice of materials and components
- the manufacturing processes, speed of production and availability of machinery and operators
- production planning, the use of just in time (JIT) and stock control including the use of ICT
- production costs – including the benefits of bulk buying, the use of standard components and the eventual retail price.

One-off production

One-off production includes single, often high-cost products, such as a light made from laminated paper and stainless steel, manufactured to a client's specification. This kind of one-off product is often very high cost because a premium has to be paid for any unique features, more expensive or exclusive materials, time consuming hand production and the finishing processes involved.

- One-off products can include prototypes that are made before full commercial production commences. For example, a child's jigsaw made from plywood or MDF, with a different image on each side, packaged in a card box, would be meticulously planned so the product is cost-effective to produce. The prototype would be manufactured as a one-off, but production planning would be for batch or high volume.
- Bespoke production is often achieved by one designer-maker and quality is checked as the work progresses. The tools and equipment may be less automated than in high-volume production, but the end product is often more individual and high quality. For example, a bespoke upmarket domino game in a presentation box might be hand made using a PC, a laser printer and a coping saw.
- Architectural models, shop and exhibition displays and signage for commercial vehicles or independent retailers are often one-off products (see Figure 1.7).

Tasks

1 Explain the benefits to the client of having interior lighting manufactured as a bespoke product.
2 Give four reasons why commercial vehicle signage is sometimes manufactured as a one-off product.

Figure 1.7 *Vinyl signage is often produced as a one-off for a small company. The vinyl letters/graphic images can be applied to magnetic sheeting, which is then attached to the vehicle sides*

Batch production

Batch production involves the manufacture of identical products in specified, pre-determined 'batches', which can vary from tens to thousands. Stationery, business cards, brochures, point-of-sale displays and promotional packaging can all be batch produced.

The tooling, machinery and workforce must be flexible to enable a fast turn around so production can be easily adapted to another product manufacture, depending on demand. Batch production often makes use of flexible manufacturing systems (FMS) to enable companies to be competitive and efficient. The use of computer integrated manufacturing systems (CIM), CAD/CAM and automated materials handling, processing and assembly, enable production downtime to be kept to a minimum.

Batch production results in a lower unit cost than one-off production, because economies of scale in materials buying enables cost savings and identical batches of consistently high-quality products are manufactured at a competitive price. (See Figure 1.6.)

High-volume production

The high volume production of most consumer products makes use of faster, more automated manufacturing processes and a largely unskilled workforce. High volume products are designed to follow mass market trends, so the product appeals to a wide national and international target market. Production planning and quality control (QC) in production enables the manufacture of identical products. Production costs are kept as low as possible so the product will provide value for money. For example, the mass production of a free-standing 'car' book with injection moulded wheels, might make use of:

- bulk buying of thermoplastic granules or wheels to reduce production costs
- standard sizes of laminated card and a standard process such as injection moulding, to ensure cost-effective and efficient manufacture
- CNC manufacturing equipment for offset litho printing of the card
- fast printing, cutting and gluing of the pages and assembly of the wheels, using a dedicated production line and automated processes to reduce labour costs
- quality control to enable manufacture of consistently identical books that meet specifications.

Continuous production

Continuous production is used to manufacture standardised high volume products that meet everyday mass market demand. This type of production is highly automated and uses machines that can run continuously for processes such as forming or printing. The machines might only be stopped for routine maintenance. For example, soft drinks bottles and cans, soap powder boxes, standard medicine bottles or toiletry packaging are often produced continuously, until the company decides to make changes to the form of the product or the design graphics. Coca Cola bottles and cans have been produced continuously in their present form for a number of years.

Tasks

1 The menu and holder shown in Figure 1.5 comprise:
   - a laminated, colour printed card menu
   - an acrylic menu holder.
   Give four reasons why this menu and menu holder are mass produced.
2 Analyse the stages of manufacture of the following cans made by continuous production:
   - a three-piece welded (tinplate) can used for processed foods
   - a two-piece drawn and wall ironed aluminium can used for soft drinks.

Factfile

In-line production and assembly makes use of low-cost unskilled labour. Most assembly lines operate on a 'just in time' (JIT) basis in which ICT is used to plan the ordering of materials and components so they arrive just in time for production. JIT is often used in quick response manufacturing (QRM), where products are produced quickly in the exact quantities needed to fulfil demand. JIT and QRM reduce the need and the costs of keeping stock.

Computer integrated manufacturing (CIM) systems enable manufacturers to manage the whole design and manufacturing process, including communications, information handling, stock control and production planning, resulting in increased productivity and competitive manufacture.

Task

Remember that in the Product Analysis exam you need to provide four different, fully explained points about the scale of production in order to achieve the available marks. Use the following questions about the scale of production to help you:

- Is the product designed to follow mass market trends?
- Does the product appeal to a wide target market or a single user?
- Is the product high cost or is it sold at a low retail price? Is the product value for money?
- Does the production make use of standardised materials, components and processes?
- Does it use specialist tools, equipment, moulds or patterns?
- Is the product fast or time consuming to manufacture? What impact does this have on costs?

- Can the materials and components be bought in ready manufactured?
- Does the product make use of production planning and quality control (QC) in production to make identical products? How is this achieved?
- Is the product made by quick response (QR) manufacture? How is this achieved?
- Are just in time (JIT) and stock control features of the product manufacture? How are they achieved?
- Is the product made in large quantities, with a dedicated production line, using automation?
- Is the product batch produced in specified quantities, using CAD/CAM to enable a fast turn around to adapt production to another product manufacture?
- Is the product made for a specific client, using specialist materials and processes?

d) Describe, using notes and sketches, the stages of manufacture for the product, including references to industrial manufacturing methods (16 marks)

This section is the most demanding on the Product Analysis paper, the one that carries the most marks and the one where you should spend the most time. In this section it is even more important that you plan your responses fully enough to ensure that sufficient relevant points are identified and explained. Note where the most marks are awarded and allow enough time to achieve the available marks. You are directed to use the following headings in your answer:

- preparation (of tools, equipment and components)
- processing
- assembly
- finish.

Using these headings will enable you to structure your responses. Remember that one word or generalised comments will gain few if any marks. Examiners are looking for knowledge and understanding of the stages of manufacture of the *given* product, so the use of technical terms about industrial manufacture is essential. You do not need to include detailed references to quality control and safety testing in this section.

Preparation (of tools, equipment and components)

Preparation deals with the preparation for the manufacture of a given product, *not* the lead up research, design and prototyping stages. In your response to the given product, you will need to list the preparation of its different components, referring to the use of any specialist tools or equipment. For example to achieve four marks for the preparation of the cardboard presentation box shown in Figure 1.8 you could include any four of the following processes:

- buying in card of an appropriate weight/thickness, using just in time (JIT)
- consider buying in pre-manufactured boxes
- producing lay plans of multiple nets to ensure efficient use of the card
- buying in printing inks and foam for the insert (JIT)
- preparing dies for the boxes
- selecting/finalising the method of printing (offset litho for card printing)
- preparing CAD graphics prior to printing
- producing printing plates for each colour

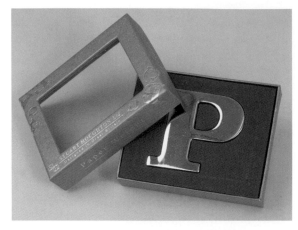

Figure 1.8 *This presentation cardboard box was designed to hold a brass paperweight in a foam insert*

- setting up the proofing press for a test print run/hot foil blocking
- sizing the card prior to printing
- setting up machinery for the production cutting of the box net
- setting up machinery for die cutting the foam.

The presentation box shown in Figure 1.8 was designed to hold a brass paperweight (see Figure 1.4). Preparation of the paperweight could include selecting a cost-effective supply of good quality brass, JIT buying in of the brass, the production of dies/moulds for casting and finalising the method of casting. Alternatively the box manufacturer could buy in paperweights if a suitable supplier could be found.

Selection and buying in of materials

In all levels of production, preparation for manufacture incorporates the selection and buying in of the most suitable materials. The best material for the job is not necessarily the most expensive because in many cases there needs to be a compromise between quality and cost. In high-volume manufacture in particular, there needs to be a continuous supply of cost-effective materials of known performance and of an acceptable quality (i.e. the chosen material must be fit for its purpose). Buying in materials for different levels of production depends on the following factors:

- the required aesthetic, physical and mechanical properties
- the scale of production and related cost/quality requirement. Is the product custom-made signage or a soft drinks can made by continuous production?

- the availability of a supply of reliable, cost-effective materials for mass production
- the availability of specific high-quality materials for one-off production
- the suitability of the material for the manufacturing process. For example, thermoplastics are suitable for injection moulding
- the required dimensional accuracy of the finished product. Can the material be precision cut?
- surface finishes appropriate to the required product quality, e.g. does the material require laminating or embossing?

Task
Using the materials buying in factors listed above, explain your choice of material for a prototype mould for a new perfume bottle, a batch-produced lightweight pendant light and a mass-produced boxed chess set.

Buying in pre-manufactured standard components

It is generally more economically viable for manufacturers to bulk buy standard components using catalogues from specialist suppliers. The use of standard components can often provide reliable, available and cost-effective solutions to manufacturing needs. For example a manufacturer of a child's 'car' book might 'buy-in' the injection moulded wheels and the Velcro™ used to fasten the book's card clasp. The book manufacturer's business is to design and manufacture children's books and whilst their volume of production is likely to be high, the company is not necessarily in the business of producing individual components. These are often made in standard forms and sizes by specialist companies who will manufacture and supply components in very high volume at a lower unit cost than would be possible by the book manufacturer.

Stock control and just in time

Computer integrated manufacturing (CIM) systems enable manufacturers to manage the whole design and manufacturing process, including communications, information handling, stock control and just in time (JIT) manufacture, resulting in increased productivity and competitively priced products. JIT uses a computer-aided stock management system, which enables materials to be delivered from suppliers just before each manufacturing stage, reducing the cost of keeping materials in stock.

Figure 1.9 *Hot foil blocking adds a high-quality visual effect to packaging*

Print planning

Print planning involves choosing the most appropriate printing process, such as offset lithography, screen printing or letterpress for black and white or full-colour printing. Print planning involves the buying in of inks, the preparation of CAD graphics prior to printing, the production of printing plates for each colour, the setting up of presses for a test print run or hot foil blocking, and sizing the paper prior to printing. Hot foil blocking uses a heated die to press a foil coating on to the surface of paper/card.

Task
Describe the preparation stage of manufacture for the mass-produced menu and menu holder shown in Figure 1.5.

Processing

In the processing section examiners are looking for correct use of terminology when naming graphics materials and processes appropriate to a given product, i.e. 'printing the pages in full colour using offset litho'. You are *not* expected to give detailed descriptions of manufacturing processes or to provide detailed diagrams. You should have an overview of the processes appropriate for named woods, metals or plastics. For example you will need to correctly name a *specific* plastic such as HDPE and state an appropriate forming process. In the processing section for the brass paperweight and its presentation box shown in Figures 1.4 and 1.8 you could gain marks for listing any of the following manufacturing processes:

- casting the brass letters
- filing or fettling to remove metal flashing from the letters
- barrelling to clean the casting and improve its quality
- colour printing the box using offset litho and hot foil blocking to add a visual effect

- cutting out the net and die cutting the corners using press knives
- scoring, creasing and folding the box net
- applying adhesive to box net
- die cutting the foam to fit the box and accept the paperweight
- collecting scrap materials for recycling.

The chosen method of processing for a particular product has an enormous influence on the product design. You will therefore need to be familiar with a number of common processes, such as printing, thermoforming plastics, casting, cutting and abrading. Computer control is a feature of many of these automated processes. The use of making aids such as jigs, patterns, standard components and quality assurance enable accurate production.

Printing processes
Offset lithography
Offset lithography uses a rotary method of printing, using printing plates made from strong, thin, sheet aluminium, which is etched with the required image. The printing plate is wrapped around a revolving cylinder. The printed image is first printed on a rubber blanket (offset), then transferred from the blanket to the paper. The flexible rubber blanket makes it possible to print on a wide variety of surfaces, including metal cans, which are heat treated after printing to give a rub- and scratch-resistant surface. Offset lithography is the most

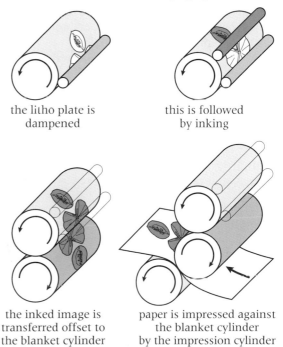

the litho plate is dampened

this is followed by inking

the inked image is transferred offset to the blanket cylinder

paper is impressed against the blanket cylinder by the impression cylinder

Figure 1.10 *Offset lithography*

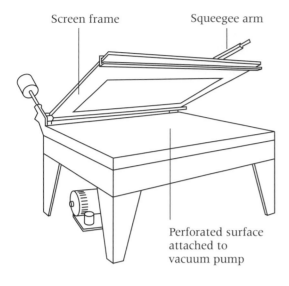

Figure 1.11 *A screen printing press*

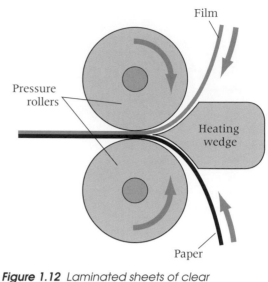

Figure 1.12 *Laminated sheets of clear polypropylene are applied after printing*

widely used, economical and versatile process because it prints good quality images on papers for business cards, stationery, menus, brochures, posters, magazines and newspapers.

Screen printing

Commercial screens are prepared using light sensitive emulsion that is exposed to a film positive of the required image. Some fully automatic presses feed and deliver the printing medium (substrate) automatically, holding the substrate and squeegee still, while the screen moves. These presses can print up to 6000 copies per hour, but are also economical for short runs, of less than 100 copies.

Screen printing uses a different screen to print each colour and leaves a thick film of ink, so it is not suitable for printing fine text. However it can print on virtually any material, such as paper, card, plastics, wood or metal. It can also print white on black, and metallic and fluorescent colours, better than any other process. Screen printing is used for posters, plastic and metal signage, and point-of-sale displays.

Letterpress

Letterpress combines four processes: inking, printing, feeding and delivering the printed image.

Although it is used for printing high-quality books, letterheads and business cards, or for one-off print jobs, it has mainly been superseded by the use of offset lithography.

Full colour printing

Colour images have to be separated, using a laser scanner or CAD, into the four process colours, cyan (blue), magenta (red), yellow and black (CMYK). Separate printing plates are made for each colour. In offset litho the four-colour printing process uses dots of CMYK to make up the full-colour image. The dots are printed in one operation, with each colour printed a fraction of a second after the previous one.

> ## Task
> Investigate the manufacturing processes required for printing a metal drinks can, a full-colour poster, a plastic sign, a menu and a high-quality colour book.

Applied finishes

> ## SIGNPOST
> 'Finishes, properties and quality' Unit 1 page 23

Packaging is laminated using polymer films or aluminium foil after the printing process. Varnish is applied after printing to achieve a glossy, protective surface. Varnish can be printed on the same printing machine or by 'spirit varnishing' on a special machine. Very high gloss is achieved using lacquer, which is dried (cured) using ultra-violet (UV) light. Most paperbacks and cartons have this finish.

Forming processes

Injection moulding

The injection moulding process makes use of a high cost mould that is injected with liquid polymer made by heating thermoplastic granules. Once the polymer cools and solidifies the formed product is ejected. Injection moulding is suitable for complex shapes with holes, screw fixings and integral hinges, formed by thinning the plastic.

Blow moulding

In the blow moulding process a hollow thermoplastic tube (the parison) is extruded between a split mould and gripped at both ends. Hot air is blown in to the parison, which expands to take on the shape of the mould, including any relief details, such as threads and surface decoration. Once the polymer cools and solidifies the product is ejected. Blow moulded containers need not be symmetrical and can integrate handles, screw threads and undercut features. The distinctive Jif lemon design, unchanged in over forty years, was one of the first examples of blow moulded polythene.

Vacuum forming

In vacuum forming a thermoplastic sheet is clamped and heated, blown and stretched. Air is sucked out to force the softened sheet over a mould pushed up from below. Once the plastic cools and solidifies, cold air is blown up from below to release the formed product.

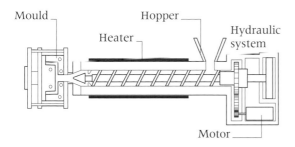

Figure 1.13 *Injection moulding is used in high-volume, continuous production of products such as toy wheels, snap-on mobile phone covers, sweet dispensers, storage and packaging*

Task

a) Examine the shapes of a yoghurt pot and a shampoo bottle.
b) Explain how and why the properties of the different materials used in these products influence their manufacturing processes.

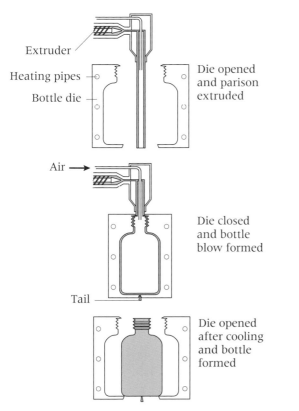

Figure 1.14 *The stages of blow moulding, used in the high-volume, continual production of hollow products such as bottles for soft drinks and shampoo*

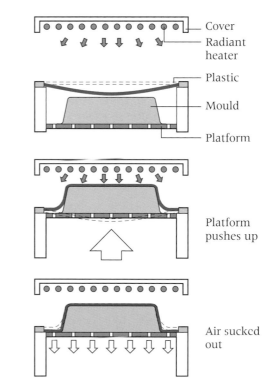

Figure 1.15 *Vacuum forming is used to batch produce food packaging such as yoghurt pots, chocolate box trays or blister packs*

Casting

Casting involves the use of a molten metal such as aluminium, copper, steel or brass solidifying in a mould, which is then opened to remove the casting. Flashing sometimes occurs when the metal leaks out between the sections of the mould. Die casting produces high quality castings with a smooth surface finish and fine detail clearly reproduced. For example, pencil sharpeners cast from zinc based alloys do not need surface finishing and will not rust.

Methods of cutting and abrading
Die cutting (cutting and creasing)

Press knives (cutting dies) are used to cut or press repeat shapes from sheets of card or cartonboard. The dies are sharp metal blades with synthetic sponge on either side, manufactured in specific shapes to match the required card cut-outs, e.g. for pop-up books or for the corners of playing cards. The dies are fixed in a press (a 'cutting and creasing' machine) and forced down through the sheets of card. Blunt dies are used for creasing and sharp dies for cutting.

Folding

Folding is used to produce small brochures as four pages (single fold) or eight pages (accordion, gate or roll-over styles). The paper is fed at the required distance into a buckle folding machine and scored at the fold line by two inward-facing nip rollers.

Abrading

Abrading using disc, belt and orbital sanders or lathes is a wasting process, used to clean up and improve the finish of wood, metal and plastic components prior to processing. For example a cast brass paperweight may be polished with emery cloth, wire wool or abrasive powder on a buffing wheel.

Using jigs, patterns, standard components, making aids and appropriate processes for accurate production

The use of making aids, such as jigs, patterns and standard components, together with quality control (QC) ensures accurate and consistent production. Standard components bought in from ISO 9000 registered suppliers are of a consistent form and size. Making aids ensure the flow of products through the factory is smooth and uninterrupted, enabling efficient manufacture. For example, cutting dies enable the accurate production of pop-up books and packaging. In injection moulding, the accuracy of dies is essential to ensure the quality of finish of products such as mobile phone covers.

The preparation of manufacturing aids is very costly and therefore mainly associated with high-volume production. For example the automated blow moulding process is extremely expensive to set up as the moulds need to have a very high-quality finish and can cost tens of thousands of pounds. Sample formed products are generally made and tested before commencing full mass production, to ensure that an acceptable quality product is made within the specified tolerance.

Computer control of manufacturing processes

The manufacture of graphics with materials products may include plotting, printing and cutting, using computer-controlled equipment.

- CNC plotter-cutters can cut shapes in card, self-adhesive vinyl and thin polymers for use in signage, promotional products and advertising. The cutting blades are easily controlled by a CAD file and can be adjusted for depth of cut and pressure. This enables fine control of scoring for folding and cutting.
- Magazines, brochures and catalogues are printed in high volumes using computer control at every stage of production to ensure proper print register and the highest quality end product. Web-fed offset litho presses can produce up to 40,000 copies per hour in full colour. The paper is supplied in 15-kilometre reels and the press automatically stops if there is a paper break.
- Computerised lay planning and cutting are fully automated processes controlled by digital data sent directly from the computer screen to the CNC cutting machine control unit.
- Laser cutting is computer controlled to enable the fast and accurate cutting of metals in complex shapes.
- Computer controlled engraving machines can cut the X, Y and Z axes allowing the engraving of 3D surfaces, curves and lettering, useful for jewellery, drinks coasters and sign making.

Task
a) Examine the shapes of a simple metal pencil sharpener and a pop-up book.
b) Explain how and why the properties of the different materials used in these products influence their manufacturing processes.

Assembly

In the Product Analysis Exam you will be asked to produce an appropriate order of assembly for a given one-off, batch or high volume product and refer to the assembly joining processes. You could use a list or flow diagram to describe the order of assembly, but remember to use full sentences and to match your responses to the available marks. For example the order of assembly for a child's 'car' book (see Figure 1.16) could include any of the following processes:

- assemble the pages in the correct sequence
- apply the hinge to the edge of the outside page using a contact adhesive
- glue the card clasp to the edge of the book using a clear adhesive
- glue the Velcro™ to the book and clasp using a contact adhesive
- fit the four thermoplastic wheels through the pages in pairs.

Design and assembly

Considerations about the assembly of a product need to take place early on in the design process, as the product design needs to take into account the materials and manufacturing processes to be used. For example, the initial design will need to take into account how the component parts align and the order in which they need to be assembled. In high volume production, designing for manufacture is directly related to designing for cost. The main aim of this is to minimise assembly costs and to enable high-quality products to be made.

Joining processes

You need to be aware of the following joining processes used in assembly:

- mechanical
- adhesives
- binding.

Mechanical assembly processes include semi-permanent and permanent fastenings.

- Semi-permanent or temporary fastenings, including nuts, bolts and screws, are the most common methods of joining material where the joint might have to be occasionally dismantled. Self-tapping screws are used when one side of the work is inaccessible or when joining dissimilar materials. They are used to secure sheet metal, plastics and composite materials. Washers and locking devices prevent the screws from working loose.
- Permanent fastenings are those in which one or more of the components have to be destroyed to separate the joints. This includes rivets, which are made of ductile materials and rely on deformation to fasten and hold components in place.

Adhesives

The choice of adhesives depends on the material, the required strength of the join and the speed of drying. Sheet materials that have slots and flaps may need an adhesive to make them permanent or, if carefully designed, can be made to be taken apart.

- PVA is a strong adhesive for papers and woods.
- Solvent welding cements are used to join thermoplastics such as polystyrene and acrylic.
- Thermoplastic and thermosetting plastic contact adhesives are used extensively in industry. Thermoplastic adhesives, used to join wood, metal and plastics, soften when heated and have good resistance to moisture. Thermosetting adhesives (epoxy adhesives) cannot be softened and provide high-strength joints, but they have low impact strength and tend to be brittle. Epoxy adhesives are used to join expanded polystyrene to most materials.

Binding

Binding fixes a set of sheets together to form a single document, which may be a brochure or a book. The main methods of holding sheets together include the following:

- perfect binding is a simple method used on magazines and paperbacks. The folded sections have the back fold trimmed off and glue is spread down the spine before the cover is glued on
- saddle stitching is where a booklet is opened over a 'saddle' and stapled along the back fold before final trimming
- sewn binding is used for hardback books, which are sewn with threads down each section.

Tasks

1 Explain the difference between permanent and semi-permanent mechanical joints and give an example of a product where each might be used.
2 Name and give reasons for your choice of adhesive to join papers, woods, metal and plastic.

Finish

SIGNPOST
'Finishes, properties and quality' Unit 1
page 23

The last part of section d) deals with the finish given to a product, whether it be surface coating, applied finishes, self-finishing or surface decoration. The analysis of a range of manufactured products made using paper, card and board, woods, metals and plastics, will help you understand different types of finishes and the reasons for applying them.

Finishing is the final process in the manufacture of a product and is a very important aspect of quality. In high volume and batch production, random quality checks are made on complete assembled products as part of the on-going quality assurance system. Although finishes vary according to the product and materials used, the general reasons for their use include protection against degradation, resistance to surface damage, to provide surface decoration to improve the appearance, or to add style or brand logos to the product. Some finishes may be added to provide information about the product, such as bar codes or age warnings on children's toys. Finishes should:

- be colourfast with a uniform colour and texture
- not run, peel or blister
- resist environmental and operating conditions (corrosion or mistreatment)
- resist physical damage (wear or scratching)
- resist staining or discoloration

- be easy to keep clean.

The completed products are finally packaged to protect the product and enable efficient handling, so products are ready for dispatch and distribution to retailers. Products can be shrink-wrapped on special machines, or stacked on pallets and the whole pallet shrink-wrapped for protection.

Tasks

1 Identify and compare six examples of products and analyse the surface finish of each. Use the following questions to help you:

- What kind of material is the product made from?
- Is the finish coated, laminated or anodised?
- Does the surface finish result from any manufacturing process?
- Was the finish applied before or after assembly?
- How was the finish applied?

2 Select a range of products such as a point-of-sale display, a pop-up book, a promotional gift or a board game and packaging. Describe, using notes and sketches, the stages of manufacturing for each under the following headings:

- preparation
- processing
- assembly
- finish.

e) Discuss the importance of quality in the manufacture of the product (8 marks)

In this section you will be asked to demonstrate your understanding of the importance of quality in design and manufacture. You will need to explain each point you make. For example, stating that the quality of colour should be checked at certain stages during printing demonstrates a lack of technical knowledge. You would need to make reference to a *specific stage* of printing, *how* the colour should be checked and to a *specific* printing process. Remember to avoid discussing safety in this section.

Quality control in production
Quality assurance (QA)

There is no such thing as absolute quality, because the concept of quality changes over time. The attitude of companies towards quality must therefore be constantly reviewed if they are to satisfy consumer expectations, which are influenced by:

- the product design, its appearance and the image it gives the user
- its build quality, performance in use and value for money.

Customer satisfaction, in which cost and quality are in harmony, is an aspect of quality assurance (QA), which covers every area of product development from design to delivery to the customer. The use of QA ensures that identical products are manufactured on time, to specification and budget. Manufacturers of batch and high-volume products that demonstrate the use of a total quality management (TQM) system throughout the company can apply for ISO 9000, the international standard of quality. ISO 9000 approved companies only buy in standard components from other approved suppliers. TQM gives them control over raw materials, a record at every production stage to aid product/process improvement, a reduction in waste/reworking and customer satisfaction. TQM means that quality is built in and monitored at every stage of production, to enable the product to be made 'right first time'. The use of a TQM system means that the company's reputation is enhanced, so it will get repeat orders and increase its profitability.

Although a system like TQM is not likely to be used in the manufacture of a one-off product made by a designer-maker, it would be used in the production of a one-off prototype, such as a child's MDF jigsaw, packaged in a card box. TQM would ensure that the whole design process incorporates both the design of the jigsaw and box and the manufacturing process, so that quality is built into the product.

Quality control (QC)

Quality control is the practical means of achieving quality assurance. It is concerned with monitoring the accuracy of production for conformance to specification, at critical control points (CCPs) in the product's manufacture. QC ensures the manufacture of identical products with no faults and provides feedback to the quality assurance system, to make sure that it is working properly. QC makes use of inspection and testing against manufacturing specifications and against agreed quality standards.

- Specifications provide clear details about materials, dimensions and tolerances, processes and assembly.
- Tolerance is the numerical difference between the limits in size that the design will tolerate and still function as required. For example, if the tolerance limit for a 5.00mm thick acrylic sheet is set at +/– 0.02mm, all sheets between 4.98mm and 5.02mm would be acceptable, but anything outside the tolerance would be rejected. Packaging nets need accurate geometrical drawing, with minimal tolerance, to ensure that when folded and glued they fit together correctly and so meet the quality specification.
- Inspection examines the product to determine if the specified tolerance limits are being met. It takes place during preparation, processing, assembly and finish. Inspection includes the correct setting up of machinery and monitoring processes, such as letterpresses, varnishing, casting or die cutting. Final inspection monitors the quality of the materials, dimensional accuracy, appearance and surface finish. For example, a presentation box would need to be checked for accuracy of cutting, folding and gluing. Random sampling involves the inspection of a small batch of the finished products. The overall product quality is then determined using statistics.

Quality control during the final print run

Print quality is maintained by various quality control checks, which include the use of standard colour bars, crop marks, registration marks and greyscale.

- Standard colour bars on four-colour printing proofs are matched to the ink density on the printing press. Ink density is then measured by a colour reflection densitometer, to make sure that there is no colour variation through the print run.
- Crop marks enable accurate cropping of text, photographs or illustrations to meet production requirements.
- Registration marks are used on sets of overlays, artwork, film or printing plates, so that when they are superimposed during printing, the work registers (matches) correctly. Specialist computer software automatically generates register marks outside the page area.
- Greyscale is a tonal scale printed from no colour to black. It is used in quality control to check consistency in colour and black and white photographic processing.
- QC is increasingly making use of sensors and probes to monitor correct image registration, colour balance and repeat design graphics. Digital QC is continuous and accurate, providing instant 'real time' feedback on production, which reduces waste and costs.

Task

Analyse a 3D product made using paper/card and one resistant material, together with its packaging.

a) Write a report of the quality assurance system that could be used by the manufacturer, identifying and describing the key factors of the system, including:
  • critical control points
  • quality control techniques used in printing and other processing
  • the role of inspection.
b) Justify the use of a quality assurance system in one-off production and high-volume manufacture.

Figure 1.16 *A child's 'car' book*

Quality standards

Quality assurance (QA) involves the use of quality standards (QS), which enable manufacturers to meet the requirements of users. The aim of QA is customer satisfaction, the relationship between performance, price and aesthetic appeal. All manufacturers want to produce saleable products. A product like a bicycle that was considered state of the art 50 years ago would not be a saleable product today, which is one of the reasons why QS are continually developing, to keep up with technological change.

Quality standards are agreed by the customer (such as a retailer of 'own brand' products) or laid down by organisations such as the British Standards Institute (BSI), European standards (EN) or the International Organisation for Standards (ISO). Agreed QS and those laid down by organisations are incorporated into the product specification.

Testing against external quality standards

Testing is concerned with the performance, durability and life expectancy of the product. In quality control, quality indicators (QIs) are used at critical control points (CCPs) to check for conformity to specification. QIs are variables or attributes that are capable of being monitored.

• Tolerance is a variable factor because it is concerned with dimensions that can be measured, such as length, weight or performance. Test results are compared with the specification.
• Attributes cannot be measured: they are either right or wrong. Visual inspection of a boxed cast paperweight, for example, checks if the

box is the correct colour or if the cast paperweight is fault free with no blow holes. If it is faulty it is rejected; if it has no surface faults it is said to conform to its specification.

Tasks

1 Figure 1.16 shows a child's 'car' book, which comprises six laminated, colour printed card pages, two pairs of injection moulded wheels and a folded card clasp with a Velcro™ seal. Quality control for the car book might include checks on the standard of card lamination, print quality or that the wheels are correctly aligned. Describe four other quality checks that could be made on this product.
2 Discuss the importance of quality in the manufacture of a cast paperweight and presentation box; a kite made from paper, polymer and metal; and vinyl signage. Use the following headings:
  • quality control in production
  • quality standards.

f) Discuss the health and safety issues associated with the product and its production (8 marks)

In this section you are asked to demonstrate your understanding of the principles of health and safety (H&S) legislation and good manufacturing practice. Make sure that you match your responses to the marks available, explaining each point you make.

Safe use of the product

Product safety and usability are of prime importance in today's market, where users regard safety as a basic requirement of a product's performance. For example in the 1950s American cars were sold on looks alone, whereas today's cars include airbags and side impact bars as standard. Product performance also extends to safety of the environment. Increasingly manufacturers are liable to environmental legislation, which encourages design for recycling and the safe disposal of waste. Although most plastic packaging does not degrade in landfill, there are biodegradable polymers on the market such as Biopol and a new type of polythene that degrade to CO_2, water and biomass.

Guarantee of product reliability, performance and safety

Manufacturers today are expected to guarantee a product's reliability, performance and safety. Special requirements apply to toys for children under the age of 36 months (see Figure 1.17).

Figure 1.17 *Special labelling requirements apply to toys for children under the age of 36 months*

The toys must be made from non-toxic materials that cause no harm if put in the child's mouth, have no small parts that may present a choking hazard and have no sharp edges. Any flexible polymer sheeting used for toys should be thicker than 0.038mm to avoid asphyxiation caused by the polymer sheet covering the face of a child. Toy bags made of impermeable material with an opening circumference greater than 380mm should not have a drawstring or cord closure. Packaging must also be made from non-hazardous materials and have no sharp edges or corners. Food packaging must comply with BSI safety standards, e.g. 'BS EN 646 Paper and board intended to come into contact with foodstuffs'.

Labelling requirements

Most labelling requirements relate to labels, signs, symbols and warnings about safety, found on packaging or the product itself.

- The CE mark (see Figure 1.18) shown on a product sold across the EU, means that the product has met the required legal, technical and safety standards. If the product is tested and found to fail the standard, it can be withdrawn from sale and the manufacturer prosecuted.
- The Kitemark (see Figure 1.18) is awarded if a product meets a given standard and the manufacturer has a quality system in place to ensure that every product is made to the same standard. Independent testing houses, approved by the BSI, test products at regular intervals and confirm that they comply with the relevant standard. If you see a Kitemark symbol on a product, you will know that it is safe and reliable.
- Special labelling requirements apply to toys for children under the age of 36 months (see Figure 1.17). BS EN 71 describes how toys must be accompanied by appropriate clear and legible warnings to reduce risks in their use,

Figure 1.18 *The BSI Kitemark, the European CE mark and the age warning symbol*

such as suitability for the age of the child, about a choking hazard, or the use of flexible polymer sheeting or toy bags.

> ### Task
> Identify how the design of a product such as a child's toy has safety factors designed into it, making reference to the following:
> * materials used
> * size of component parts
> * the use of flexible polymer sheeting
> * packaging.

Safety procedures in production
Principles of health and safety legislation

The Health and Safety at Work Act (1974) provides a body of law dealing with the health and safety (H&S) of people at work and the general public who may be affected by the work activity. The legislation exists to ensure the provision of safe working conditions, which reduce the risks of accidents. This legislation also includes:

* The Factories Act (1961)
* COSHH (Control of Substances Hazardous to Health)
* The Management of Health & Safety at Work Regulations (1992).

Principles of health and safety at work

The 1974 act states that risk assessment is a legal requirement, which means that all manufacturers must have an H&S system in place. Not only are employers *and* employees responsible for ensuring safety at work, but so are manufacturers of equipment and all suppliers of materials and components. Employers are responsible for the H&S of their employees and employees must follow the safe working practices set out by the employer, with guidance from the Health and Safety Executive (HSE).

Making risk assessments

Risk assessment means identifying hazards that may cause potential harm to employees and users. The risk that a hazard may occur must be evaluated, and the chance of injury and potential damage that could occur must be eliminated or controlled. Risk assessment used in manufacture could include the use of:

* H&S regulations and risk assessments procedures
* H&S notices around the workplace
* staff training in the safe use of machines, tools and materials

* guards and emergency stop buttons on machines
* machine servicing and up-to-date log books
* the correct use of personal protective equipment (PPE), such as clothing, goggles or gloves
* appropriate ventilation, heating, lighting and noise levels, e.g. ventilation when casting brass
* COSHH data, first aid instructions and fire safety signs.

> ### Task
> Draw up the key stages of production for a toothpaste tube and its outer packaging. Identify the risk assessment procedures necessary to manufacture it as safely as possible.

> ### Task
> Figure 1.19 shows a blister pack for a Tipp-Ex Pocket Mouse. Discuss the health and safety issues associated with this product and its production, under the following headings:
> * safe use of the blister pack for Tipp-Ex Pocket Mouse
> * safety procedures in the production of the blister pack for Tipp-Ex Pocket Mouse.

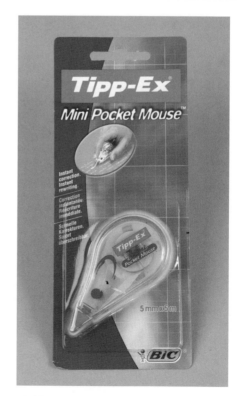

Figure 1.19 A blister pack for a Tipp-Ex Pocket Mouse

g) Discuss the appeal of the product (8 marks)

This last section asks you to discuss the appeal of the given product, using headings such as 'form', 'function', 'trends/styles' and 'cultural'. The headings may vary in each exam, depending on the product in question, so you will need to match your responses to the available marks. Remember to use the headings to structure your answer. You should approach this section from the point of view of the buyer, so ask yourself 'Why would the user buy this product?' Remember that this section is *not* a repeat of a), so do *not* write out another product design specification which will gain few if any marks.

Product appeal

Trends and styles are very important criteria in product development these days. There is also a need to use new technology, to understand the market and to continually re-innovate to prevent products from becoming out of date. Even for technical products like mobiles or hand-held computers and their packaging, aesthetics, market appeal and function are vital to the product's form and use. The economic success of most manufacturers depends on their ability to identify target market group (TMG) needs and to create products that meet these needs. These days the TMG may be global and cultural differences are becoming increasingly important. Environmental issues are also set to become a market driver as legislation clicks in. In the exam you will need to discuss the following:

- form and function, which relates to the product's design features and technical performance. How do they meet TMG requirements? Who are the TMG users? Are they adults or children, male or female? Why is the product suitable for them? What are its specific characteristics that make it more appealing than any other product?

- trends/styles, which relate to the aesthetic appeal of the product. In what way does the style of the product appeal to the TMG? How does if fit in with their lifestyle? What kind of image does it give to the user? Does the product's style and colour follow current market trends? Is this a selling point?

- cultural, which can include other values issues that influence the design of the product. Is it recyclable? Does it have a Kitemark or a warranty? Is it designed to meet British, European or International legislation? What kind of retail outlet would sell it? Is it cheap and cheerful or high quality? Is it expensive or value for money?

Tasks

1 Identify a current best-selling product and evaluate its appeal.
- State the TMG for the product and the image it gives the users.
- Explain the characteristics that make it more appealing than any other similar product on the market.
2 Compare the key design features of two perfume containers and their packaging, produced by different manufacturers.
- List and compare the aesthetic and performance characteristics of each product.
- Draw up a list of TMG needs and explain how each product meets them.
- Explain which product is the best buy and why.

Figure 1.20 *The distinctive Jif lemon design, unchanged in over forty years, was one of the first examples of blow moulded polythene. Explain the lasting appeal of this product*

Quality of written communication (3 marks)

Remember that three marks are available for the 'Quality of written communication'. If you always use the headings given in the exam paper to structure your answer, you will be more likely to organise and present your responses clearly and logically. When you work through this unit keep a technical notebook where you write definitions of technical terms, so you use them appropriately in the exam. Three marks can make a difference between grades.

Exam preparation

Practice exam question

Figure 1.21 *A stainless steel bottle opener with a presentation cardboard box and foam insert*

a) Outline the product design specification for the bottle opener and its packaging. Address at least **four** of the following headings:
- purpose/function
- performance
- market
- aesthetics/characteristics
- quality standards
- safety. (7)

b) Justify the use of:
- stainless steel for the bottle opener (3)
- cardboard for the box. (3)

c) Give **four** reasons why the bottle opener is batch produced. (4)

d) Describe, using notes and sketches, the stages of manufacture of the bottle opener and its packaging under the following headings:
- preparation (of tools, equipment and processes) (4)
- processing (6)
- assembly (4)
- finish. (2)

Include references to industrial manufacturing methods in your answer. Do **not** include details of the production of stainless steel, foam or cardboard in your answer. Do **not** include detailed references to quality control and safety testing in this section.

e) Discuss the importance of quality in the manufacture of the bottle opener and its packaging, using the following headings:
- quality control in production (3)
- quality standards. (5)

f) Discuss the health and safety issues associated with the bottle opener and its packaging under the following headings:
- safe use of the product (2)
- safety procedures in production. (6)

g) Discuss the appeal of the bottle opener and its packaging under the following headings:
- form (2)
- function (2)
- trends/styles (2)
- cultural. (2)

h) Quality of written communication (3)

Total: 60 marks

UNIT 2 Product development I (G2)

Summary of expectations

1. What to expect

You are required to submit one coursework project at AS. This project enables you to build on the knowledge, understanding and skills you gained during your GCSE course. At AS level you are expected to take more responsibility for planning your work. Your AS project should comprise a product and a coursework project folder. It is important to undertake a project that is appropriate and of a manageable size, so that you are able to finish it in the time available.

2. What is a Graphics with Materials Technology project?

An AS Graphics with Materials Technology project should:

* reflect study of the 'technologies' involved in the AS Graphics with Materials Technology Specification
* include a 2D element developed from traditional and modern graphics media
* include a 3D model using at least one resistant material listed in Unit 3A (Classification of materials)
* include a coursework project folder that demonstrates good quality graphic communication.

For example an AS Graphics with Materials Technology project could include the use of a 2D element, such as a drawing of a building in its environment, linked to a 3D element, such as an architectural/interior design model. The 3D model must include the use of wood, metal or plastics. The 3D model should be considered a prototype that is too 'big' to be constructed in the final product version (whereas a Resistant Materials Technology project could be constructed).

3. How will it be assessed?

The AS project covers the skills related to designing and making. It is assessed using the criteria given in Table 2.1.

You must attempt to cover all the Assessment Criteria A–G. The G criterion has been included to reflect that your project meets all the requirements of the Specification (see 'What is a Graphics with Materials Technology project?'). It is very important to check the appropriateness of your project with your teacher or tutor *at the start of the project*. If you take account of how the marks are awarded when planning your work, you will be able to spend an appropriate amount of time on each part the project. This will give you more chance of finishing in the time available and increase your chance of gaining the best possible marks. Your AS project will be marked by your teacher and the coursework project folder will be sent to Edexcel for the Moderator to assess the level at which you are working. It may be that after moderation your marks will go up or down.

Table 2.1 *AS coursework project assessment criteria*

Assessment criteria		Marks
A	Exploring problems and clarifying tasks	10
B	Generating ideas	15
C	Developing and communicating design proposals	15
D	Planning manufacture	10
E	Product manufacture	40
F	Testing and evaluating	10
G	Appropriate project	10
	Total marks	110

4. Choosing a suitable project

As a student designer your choice of AS coursework project is very important, because you need to balance designing and making a successful product with meeting the needs of the project assessment criteria.

Remember that the level at which you should be working is one year on from GCSE. Do not be too ambitious in your choice of project, as testing and evaluation cannot properly take place on incomplete work!

At AS level you are expected to take a more commercial approach by designing and making a product that meet needs that are wider than your own. This could mean designing and making:

- a one-off product for a specified user
- *or* designing and making a product that could be batch or mass produced for users in a target market group.

The key to success is to choose a project that is both enjoyable and challenging, so that you will feel inspired to finish it in the time available.

If you identify a realistic user need and a problem to be solved, you can start to focus on the kind of product that you could design and make to solve the problem. The needs of others are therefore a good starting point for design. It will enable you to ask questions about the purpose and potential for a product and how it will benefit the user(s).

Whatever kind of product you design will need to be planned and you will need to include details of how this single product will be manufactured in your school or college workshop. Even if you are designing for batch or mass production, you will still only be making *one* of them. You are *not* required to manufacture the product in quantity, although you may need to produce identical components for use in the product. You will, however, need to detail in your production plan the changes necessary in the manufacture of your one single product, if it was to be made in quantity.

5. The coursework project folder

The coursework project folder should be concise and include only the information that is relevant to your project. It is essential to plan and analyse your research and be very selective about what to include in your folder. This will help you make decisions about what you intend to design and make.

Do not be tempted to waste your valuable time finding out information that has no relevance to your project, because you will not gain any marks for it.

Your coursework folder should include a contents page and a numbering system to help its organisation. The folder should comprise around 20–26 pages of A3/A2 paper. A title page, the contents page and a bibliography should be included as extra pages.

Table 2.2 gives an approximate guideline for the page breakdown of your coursework project folder.

Table 2.2 *Coursework project folder contents*

Suggested contents	Suggested page breakdown
Title page with Specification name and number, candidate name and number, centre name and number, title of project and date	extra page
Contents page	extra page
Exploring problems and clarifying tasks	4–5
Generating ideas	3–4
Developing and communicating design proposals	5–6
Planning manufacture	3–4
Product manufacture	2–3
Testing and evaluating	3–4
Bibliography	extra page
Total	20–26

In the section on 'Product manufacture' it is essential to include clear photographs of the actual manufacture of your product. This will provide photographic evidence of modelling and prototyping, any specialist processes you have used including the use of CAD/CAM, and show the stages of manufacture.

Please note, however, that the guideline for the page breakdown of your coursework folder is only a suggestion. You may find that your folder contents vary slightly from the guideline because of the type of project that you have chosen.

6. How much is it worth?

The coursework project is worth 40 per cent of your AS qualification. If you go on to complete the whole course, then this unit accounts for 20 per cent of the full Advanced GCE.

Unit 2	Weighting
AS level	40%
A2 level (full GCE)	20%

A Exploring problems and clarifying tasks (10 marks)

1. Identify, explore and analyse a wide range of problems and user needs

Your choice of AS coursework project is very important, because you are expected to take a more commercial approach to designing and making a product that meets need wider than your own. This could mean designing and making a one-off product for a single specified user or client, or designing and making a product that could be batch or mass-produced continually for users in a **target market group**. Whichever type of product you design and make, you will need to explore and identify a realistic need or problem through investigating and analysing the needs of people in different contexts.

2D and 3D elements

Your choice of project is important because it must include both 2D and 3D elements. The 2D/3D elements should be linked by the theme or context for design, so that one supports and underpins the other. For example, if your theme is 'brand image' the 3D element could be a promotional gift, supported by the 2D element. This could be point-of-sale advertising or marketing materials, both related to the 3D promotional gift.

For your coursework project you are, therefore, asked to produce the following:

- a 3D model or prototype product, incorporating at least one resistant material from Unit 3A, such as wood, MDF, plastic or metal – the 3D outcome should be semi-functioning
- a 2D element developed from traditional and modern graphics media – the 2D element should be linked to and support the 3D outcome
- a coursework project folder that summarises the development of the 2D/3D elements – the folder should demonstrate good quality graphic communication.

A simple example of a coursework project would be the development of a 3D prototype product such as a perfume bottle or calculator, together with its packaging. Although the perfume bottle or calculator would not have to 'work' as a functioning product, it would still need to be a good quality concept model, testable against the specification criteria. The 2D graphic element – the packaging – would be developed from graphics media as a separate outcome from the coursework project folder.

Don't forget to consult with others, including your teacher or tutor, to ensure that you choose a suitable project.

Exploring and identifying the needs of users

The needs of users change according to **demographics**, expectations and lifestyle. In other words, the kind of products that people want to use change according to their age, and their expectations change according to the kind of lifestyle they aspire to. Many manufacturers and retailers target potential customers and attempt to match their lifestyle needs with products. They often do this by using **market research** to identify the **buying behaviour**, taste and lifestyle of potential customers. This can establish the amount of money they have to spend, their age group and the types of products they like to buy. New products can then be developed to meet customer needs. (Consider the questions in Figure 2.1.)

There are many starting points for investigating the needs of users. You may, for example, decide to investigate trends in leisure pursuits, lifestyles or demographics. This might require you to find out where people shop, their gender and age group, their level of disposable income, their **brand loyalty**, the newspaper and magazines they read, the music they like, the TV programmes they watch, their leisure interests and where they eat out. This kind of in-depth

Figure 2.1 *What problem does this product solve? What is the target market for this product? What are its key design features? How does this product meet the needs of its users?*

investigation can provide numerous contexts for design – perhaps you could even find a 'gap in the market' for a new product and devise a marketing campaign for it. Maybe there is the need for a new leisure centre or the traffic flow in your town centre needs re-designing – could you produce an architectural model and interior/exterior **presentation drawings**?

Task

Finding out about the needs of users

The needs of your users, or target market group, can provide a good starting point for design. Manufacturers often build up **customer profiles** to help establish the characteristics of products that customers want. You can use a similar practice. Write a customer profile to describe target market groups for the following products and their packaging. In doing this you could think about customers' lifestyles, values and tastes.

a) chess set   **c)** tube of toothpaste
b) computer game   **d)** WAP phone

Identifying and analysing a realistic need or problem

The characteristics of many products need to change gradually over time if they are to fulfil the developing needs of users and sell into a market at a profit. Once the requirements of users are known, they can be analysed and problems which give rise to the need for new products identified. Figure 2.2 shows how one student analysed and identified a need. Often a new product may be developed from an existing one by modifying its aesthetic or functional characteristics to meet changing user needs. This kind of product development often involves redesigning to add value to a product – to improve its performance, function or appeal.

Updating existing products through colour and styling to create a more modern or fashionable look is sometimes used to increase the market share of products that have been on the market for some time. The redesign of products with a strong brand image, such as sweets or DIY tools, needs to be carefully managed. Any changes must be gradual so that the recognisable brand image is maintained.

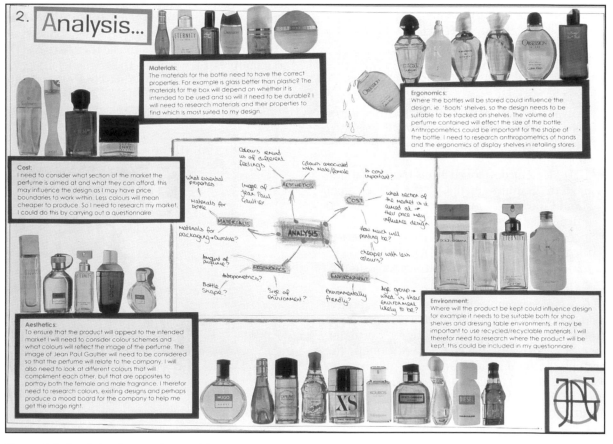

Figure 2.2 Analysis of a need or problem

One of the most important areas of product development is the creation of new brands that target specific market groups. The young sports market is a prime target for this type of **marketing**. There are many views about the morality of marketing brands because some young people can be pulled in to a cycle of needing to be seen to be wearing the latest 'logo' or **brand name**. What do you think? Are young people under pressure to be seen in the latest styles? Is this kind of brand marketing ethical?

Clarifying the task

Once you have identified user needs and a problem to be solved, you can start to focus the problem by asking questions about the kind of product that you could design and make to solve the problem. You can think about the purpose and potential for the product: How will it benefit users? Will the product be a prototype product that could be mass produced or a one-off specialist product like a scale model? Are there any products currently on the market that fulfil a similar need? What kind of materials do they use? What kind of materials and processes could you use? Asking questions like these will help you clarify what you are trying to achieve and enable you to write a design brief.

> *To be successful you will:*
> - Clearly identify a realistic need or problem.
> - Focus the problem through analysis that covers relevant factors in depth.

2. Develop a design brief

Developing a design brief will focus your mind on what you want to do. The design brief should develop from your exploration and analysis of user needs and problems and from your identification of a potential product that is feasible for you to make.

Your design brief needs to be holistic – in other words, it needs to encompass the development of both the 2D and 3D outcomes. It should also be simple, concise and should explain what needs to be done, but it should not include the solution to the problem. It is intended to give you direction but not to be so precise and specific that you are not left with any room for development.

> *Task*
>
> **Developing a design brief**
>
> Your design brief should be simple, concise and explain what needs to be done, without going into too much detail. For example: 'Design and make a children's board game and its packaging.'
>
> a) For the design brief above identify the type of product to be designed, the purpose of its packaging and the range of potential users.
> b) Choose three different contexts. Write a design brief for each, identifying the 2D/3D elements, their purpose and the target market group.
>
> Try not to be too ambitious when you write your design brief because even the simplest idea has a habit of becoming more complex as it develops! Use your design brief to help plan your research, so that you can target what you need to find out.

> *To be successful you will:*
> - Write a clear design brief.

3. Carry out imaginative research and demonstrate a high degree of selectivity of information

Research

Your research needs to be targeted and it needs planning in order to find out useful information that will help you make decisions about what you intend to design and make (see Table 2.3). Read your design brief and your analysis of the problem – this will help your planning. For example the design brief: 'Design and make a children's board game and its packaging' would lead to research into the needs of children of different ages and the buying behaviour of the purchasers (adults or children?). It would also include the investigation of existing board games and packaging, including research into their design, materials, components and manufacturing processes, and the quality, cost and safety needs of users.

Task

Market research

For a product of your choice, produce a market research report that identifies the buying behaviour, taste and lifestyle of potential consumers. Find out the size of the target market group, their preferred brands and the competition from existing products.

You should undertake both primary and secondary research using a range of sources, such as the ones listed in Table 2.3. Use these research ideas and your research planning to target the research you need to undertake. This will depend on the type of product you intend to design and make – you are not expected to research everything listed in Table 2.3.

Writing a bibliography

Reference all your secondary sources of information in a bibliography and include:

- sources of any information found in textbooks, newspapers and magazines
- sources of information from CD-ROMs, the Internet, etc.
- sources of scanned, photocopied or digitised images.

Product analysis

In industry product analysis is often used to obtain information about the design and manufacture of competitors' products. Analysis can include data research, using catalogues or trade literature and analysis of the products themselves, either by eye or using physical analysis of the product's component parts (see Figure 2.3). You can analyse

Table 2.3 *Research ideas and where to find information*

Research ideas	Information sources
Market trends, design and style trends, **niche markets**, new product ideas, user requirements, buying behaviour, lifestyle, demographics	Market research, window shopping, shop surveys, visits, the work of other designers, art galleries/museums, user surveys, questionnaires, product test reports, consumer reports, people, the Internet
Materials, components, corporate image, advertising trends, systems, processes, scale of production, product performance requirements, product quality, product price ranges, value for money	Analysis of existing products, materials testing, books, newspapers, electronic media, CD-ROMs, databases, the Internet, exhibitions, reports, magazines, catalogues, libraries, local colleges, industry, people
Quality control and safety procedures	Books, safety reports, *Design and Technology Standards for Schools*, information via the Internet, industry
Values issues, e.g. cultural, social and environmental	Keep up with the latest news, events, films, exhibitions, cultural and social issues, use the Internet to find out about company values and environmental issues

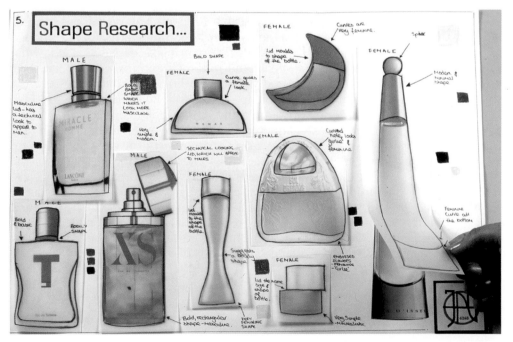

Figure 2.3
Student research into the shape of existing perfume bottles

commercial products that are similar to the 2D and 3D elements of the product you intend to design and make. This can provide information about design, style, advertising and marketing trends, corporate image, packaging, materials, components, processes and assembly. It will also develop your understanding of quality of design and manufacture and of the concept of value for money.

At this stage of your project you may find that some activities need to be carried out simultaneously. For example, research and the analysis of the problem are often bound together – some analysis may be necessary in order to focus some of the research, while analysis of the research is necessary if the gathered information is to be useful when making decisions about what to design.

Analysing research

You will need to analyse your research and select *useful* information that is *relevant* to the design of your intended product. Do not be tempted to include pages of irrelevant information to 'pad out' your project folder, as this will gain you no extra marks! The Moderator needs to see evidence of your ability to be selective and demonstrate how your research has linked into the problem drawn out from your analysis. Your research analysis should, therefore, provide you with a good understanding of the problem and the desired aesthetic, functional and performance requirements of the product you intend to design and make. This will enable you to move on to the next stage in the development of your product – writing a design specification.

> ### To be successful you will:
> - Carry out a wide range of imaginative research, selecting information that will help you make decisions about what to design and make.

4. Develop a design specification, taking into account designing for manufacture

> ### Factfile
> A design specification is detailed information that guides a designer's thinking about what is to be designed. It is used to help generate, test and evaluate design ideas and to help develop a **manufacturing specification**.

In the same way that analysis and research are often bound together, the design brief, the research process and the development of the product design specification are all linked and may be developed simultaneously. Your product design specification should include measurable characteristics that will help you design with manufacture in mind. You will need to include enough detail to develop feasible design ideas, but leave room for creative thinking. Use the following checklist to help develop your design specification:

- the product purpose, function and aesthetics
- market and user requirements
- the expected performance requirements of the product, materials and components
- the kind of manufacturing processes, technology and scale of production you may use
- any values issues that may influence your design ideas, such as cultural, social or environmental

Specification

'Must have large area of solar panels' This is so that the design is efficient and that there is no wasted space.

'The development must be space efficient' This is so no large habitat is destroyed, so as not to use up valuable space which is also expensive to purchase.

'Must have use of modern materials' So as to make it an appealing, fashionable place.

'Must consider environmental impact of structure' There is no point building a structure to help save the environment that damages the environment through it's construction.

'Must have back up power' Solar panels do not produce electricity through the night.

'Must be interesting for visitors' Visitors are not going to see the benefits of solar power and become 'converted' if the plant is so boring and uninteresting that they don't want to come to see what happens there.

'Must be pleasant environment to work in' It is possible that the location will be isolated and for the workers who'll probably spend a long time there on site between breaks, it needs to be pleasant so boredom and complacency doesn't set in. In addition, it maintains morale when staff have a structure they are proud of.

'Must not have too much visual impact on the surroundings' Could be a tourist location or residential area, they do not want it ruining their outlook.

'Must not have a large impact on habitats/protected species'

Most importantly… profits depend on electricity production, therefore… **'Must capture as much light as possible through design of the structure throughout all daylight hours'.**

Figure 2.4 Part of a student specification

- any quality control and safety procedures that will constrain your design
- time, resource and cost constraints you will have to meet.

You are expected to identify an appropriate scale of production, such as a one-off product for a specified user, or a prototype product that could be batch mass or continually produced for a target market group. The scale of production will impact upon all your design and manufacturing decisions. For example, a one-off architectural scale model would need to meet the requirements of the client. This might be an architect, a local authority planning department or a model-making consultancy. This kind of model would require accuracy of scale and manufacture and could be simple or complex. On the other hand, a prototype educational toy that could be batch or mass produced would need to meet the requirements of a range of children and their parents, but may require processes to be simplified for ease of manufacture. You would need to take these requirements into account when producing your prototype. In both of these scenarios, you would be working in a similar way to an industrial designer or architectural model maker.

To support either of these scenarios you would also need to develop the 2D graphical outcome that is separate from the coursework project folder. For the architectural model you could develop presentation drawings that show the building in situ. The presentation drawings may need to be reproduced for display elsewhere (for example in an exhibition or a library) or used in architectural publications, in which case you would need to consider reproduction methods. For the prototype educational toy, you may have decided to produce marketing materials, which would need to be reproduced in high volume.

Your design specification should guide all your design thinking and provide you with a basis for generating design ideas. A design specification, which may start out in outline, can change and develop as research is carried out, until a final specification is reached. This is used as a check when testing and evaluating design ideas and will provide information to help monitor a product's quality of design. The design specification is, therefore, an essential document that sets up the criteria for the design and development of your product. Specification criteria can also be used later on to guide your thinking when developing a manufacturing specification.

B Generating ideas (15 marks)

1. Use a range of design strategies to generate a wide range of imaginative ideas that show evidence of ingenuity and flair

You are expected to generate a range of feasible design ideas, based on the criteria that you have set up in your product design specification. Sketches need to be lively and include annotation to explain your thinking. Remember to practise sketching and annotating ideas, as this will help you in your coursework and in the Unit 6 Design examination. Your design specification should be a great help when designing, as it is much easier to develop ideas from a starting point than starting from nothing!

Imagine starting with a blank sheet of paper and being able to design anything you like. Where do you start? How do you know what is required? No designers work like this because, firstly, there is not enough time to work in this way and, secondly, products are designed for people who have definite likes and dislikes and lifestyles that demand products with specific aesthetic, functional and performance characteristics.

Many designers say that designing becomes easier when there are limitations under which to work because these limitations provide a framework within which to design.

Inspiration for design ideas
Many sources can be used for inspiration – your design specification and research, of course, but you may also decide to base some aspects of your ideas on a theme. Try using some of the following to inspire your work:

- natural forms
- an artist or the work of other designers, e.g. Brody
- an art or design movement, e.g. Art Nouveau
- a theme, e.g. 'miniature' or 'techno'
- influences from exhibitions, music or films
- new technology, e.g. using new materials or processes.

Many product designs in an industrial context are modifications rather than original ideas. Inspiration for this type of designing can come through product analysis. In industry it may involve:

- adapting branded products to produce 'own labels'
- developing products to appeal to different target market groups
- developing products through following new legal guidelines that are related to environmental or safety issues
- adapting products through the use of new or different materials or processes.

You can use product analysis to inspire your design work. It is a good way of collecting information about a product type – about materials, components, manufacturing processes, style, colour, quality and price. If you work out the design specification for an existing product this can often help you develop a specification for your own product. However, you must ensure that any product you develop in this way is truly original and not simply a copy of an existing product.

Generating ideas

It is often a good idea to keep a notebook for quick sketches – these can be pasted, or scanned and pasted, into your coursework project folder, rather than be redrawn (your project folder

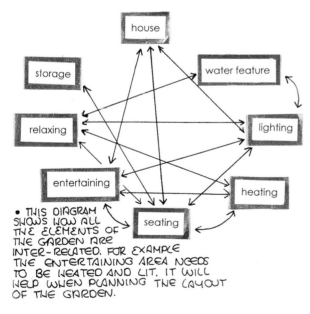

- THIS DIAGRAM SHOWS HOW ALL THE ELEMENTS OF THE GARDEN ARE INTER-RELATED. FOR EXAMPLE THE ENTERTAINING AREA NEEDS TO BE HEATED AND LIT. IT WILL HELP WHEN PLANNING THE LAYOUT OF THE GARDEN.

Figure 2.5 *This student used annotation to evidence her design thinking*

should show evidence of creative thinking rather than stilted copied out work). This may lead to a certain untidiness, but you will find it more inspiring if you approach your initial design work in a similar way to using a sketch book – use hand-drawn sketches, colour ideas and written notes to show your ideas, which can be developed and modified later. Use a medium that you feel comfortable with and that is easy to work with, for example pencil, ballpoint pen or fine markers are ideal for quick line drawings and for simple shading. At this stage the examiner is looking for evidence of your design thinking and later on to see how you develop these ideas. It is not always necessary to develop a wide range of totally different ideas, although you should always try to produce variations related to what the product looks like, with alternative ideas of how it will function, especially if there are moving parts.

2. Use knowledge and understanding gained through research to develop and refine alternative designs and/or design detail

Evidencing the influence of research on your design ideas should not be a problem, as this influence should come through naturally from your design specification, which is based on your research and analysis. Also, you should be using your research and specifications as inspiration for ideas – so be sure to make this explicit when you are developing your initial ideas by adding brief notes to explain your thoughts.

As you experiment with first ideas and gradually refine your thinking, you may find that you start to think about the possible materials or processes you could use, to work out if your ideas are feasible. You may also start to play around with combinations of ideas or work on the fine detail of some of your ideas. Making use of the information you acquired during your research phase should enable you to develop and refine alternative ideas to the stage where you can select one (or possibly two) as the most promising.

The type of 2D and 3D outcomes being designed will influence the research you need to do and the kind of ideas that you produce. For example, ideas for a prototype board game may be concerned with its aesthetic and functional aspects – you may have to consider the production of playing pieces as well as storage of component parts. However, ideas for the design of point-of-sale advertising may be concerned with colour, style and font size in relation to the image you are trying to promote about a prototype product.

> *To be successful you will:*
> • Demonstrate effective use of appropriate research.

3. Evaluate and test the feasibility of ideas against specification criteria

As you sketch your first ideas you should refer back to your design specification to see if you are going in the right direction, to see if there is any aspect that you have missed, or to look for more inspiration. Make sure that you use your design specification as a tool for measuring the appropriateness of your design ideas. Always make evaluative comments on decisions made. Make sure you demonstrate your technical knowledge and use technical terminology, rather than making general references such as 'made from plastic' or 'I don't like the look of it'. At AS level you are required to justify the rejection or selection of ideas. It is always helpful to talk to others at this stage, as talking over your ideas often provides the step forward you need. There are a number of options here – discussing your ideas with a teacher or tutor, presenting your ideas to a small group of fellow students, or talking to potential users of your product. In industry this would be a normal part of product design and development, in order to determine the feasibility of ideas, ease of manufacture and market potential. You should evaluate your ideas against the design specification so that you are

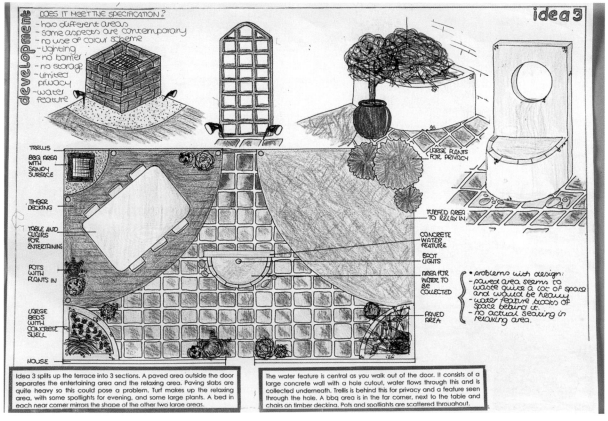

Figure 2.6 *Generating ideas*

3.

Mood Board...

It is very important in design to ensure that the image of the product being designed relates to that of the company. I have therefor gathered together a collection of images related to Jean Paul Gaultier and his work. This will enable me to portray the right image.

FIRST THOUGHTS ON JPG

French
Modern & original designs
A lot of red & black
very artistic
squares
Stripes & geometric patterns/shapes
Sexy feel

Figure 2.7 It is important to justify why your chosen idea is worth developing. One way of doing this is to refer back to the product image you were trying to achieve.

able to justify, using written notes, why your chosen design is worth developing.

During the course of your design development you may find that there are some aspects that you need to find out more information about. You may need to modify your design specification in some way, if, for example, you find after evaluating ideas against the design specification there are some aspects that are not feasible.

> *To be successful you will:*
> * Objectively evaluate and test ideas against the specification criteria.

C Developing and communicating design proposals (15 marks)

1. Develop, model and refine design proposals, using feedback to help make decisions

Your aim should be to develop your ideas until you produce the best possible solution to your problem. This can be done by modelling, **prototyping** and testing. Modelling your design proposal in 2D and 3D will provide you with information about its feasibility. You can also get feedback by finding out what potential users think about your ideas. Ask them a range of questions, based on the requirements of the design brief and specification. The feedback you receive should help you make decisions about the aesthetic and functional characteristics of your design proposal. You can then refine your ideas, if necessary, until you find the best possible solution.

> *To be successful you will:*
> * Develop, model and refine the design proposal, with effective use of feedback.

2. Demonstrate a wide variety of communication skills, including ICT for designing, modelling and communicating

You should use a variety of 2D and 3D communication skills to develop, model and refine your design proposals (see Figures 2.8 and 2.9). Communication skills can include writing, drawing or using ICT for word processing or CAD. Modelling will enable you to see how your design proposal will perform, or what it will look like, before you make any final decisions. The modelling method you use will depend on the type of product you are designing and the aspect of its design that is to be modelled. You should use materials that are quick and easy to cut, shape and join; alternatively you could use construction kits or computer modelling. The use of computers for 2D and 3D modelling is an essential part of industrial design.

> ## *Factfile*
> **Modelling in industry**
> Modelling is a key industrial process, because it enables the testing of the product design before manafacture. Even though a number of development models may be made, in the long run it saves time. Costly mistakes, such as the product not functioning properly or not meeting user requirements, are thus avoided.

2D modelling

2D modelling techniques can include, for example, drawing layouts, using crating, pictorial or perspective drawings, sections, exploded views, using cut-and-paste or computer-aided design (CAD) software. Using CAD you can scan in your image and try out different variations, or use parametric design software to trial 'what if' scenarios. However, you should only use CAD if it is appropriate; you will not be penalised for its non-use. Although most graphic designers use CAD, most will have been trained in drawing techniques. Sometimes it is more appropriate to use hand-drawn techniques, e.g. when thinking through the development of an idea. You should try to develop both hand and computer skills.

> ## *Factfile*
> **Prototyping in industry**
> As a product idea is modelled it gradually becomes more detailed until it becomes a 'prototype' 3D product. Prototyping is a key industrial process, because it enables the product *and* its manufacturing process to be tested. This avoids costly mistakes, such as the product not meeting the specification or customer requirements.

3D modelling

You should use 3D modelling and prototyping techniques to test your design proposal before manufacture. Modelling by hand to make 3D working mock-ups can involve the use of different materials, such as:

- laminated card to model curved structures
- clay, plasticene or Styrofoam to build 3D models
- wire, strips of wood, laminated card, drinking straws or even spaghetti to develop wire frame models.

Figure 2.8 *Modelling ideas can help you to make decisions about the feasibility of your design*

Figure 2.9 *Modelling prior to manufacture*

If you are working on a scale model (e.g. for theatre, film or a building), or on the development of a prototype product, you may need to make further sketch models to explore different aspects of the design. These can help you plan:

- the most appropriate manufacturing processes
- how long different processes might take
- the materials, components, equipment and tools you need
- the order of assembly of component parts
- how easy the product will be to manufacture in the time available – do you need to simplify anything?
- any forward ordering of materials, so they are ready when you need them
- estimated costs of materials and manufacture
- where and how you will check the quality of your product.

As your design proposals develop, your ideas will become more refined until you reach the best possible solution – your final design proposal.

> **To be successful you will:**
> - Use high-level communication skills with appropriate use of ICT.

3. Demonstrate understanding of a range of materials/components/systems, equipment, processes and commercial manufacturing requirements

You are expected to demonstrate an understanding of the materials, components and/or systems that are appropriate to the manufacture of your product. Modelling and prototyping should enable you to do this because they enable you to test and trial materials and components. Selecting the most suitable materials is vital to the success of your project. They should be appropriate to your chosen scale of production. For example, special building materials may be required for the design of an interior. What are the properties of these materials and how can you replicate them in a scale model? What kind of materials are appropriate for a child's toy that is to be produced in high volume?

Selecting materials

Technologists have difficult decisions to make when selecting the right material for the job,

especially when new materials are appearing all the time! Deciding which materials to use in a product design is not easy. Making good choices requires understanding of a wide range of materials. When choosing materials you need to take into account the following:

1. Materials availability
   - How rare or common is the material?
   - How simple, easy or safe is its manufacturing process?
   - Does the material come in standard forms and preferred standard sizes?
2. Moral and social issues
   - Is the material harmful to work with or use, even though it has some very useful properties?
   - Should you take the risk and use the material or not?
   - Does the production of the material exploit a labour market in any country?
3. Cost
   - What are the processing and delivery costs of the material?
   - How much will it cost to store the material? Is it possible to use just-in-time ordering of materials to reduce storage costs?
   - What are costs of waste materials? Can these be recycled?
4. Method of production
   - Are the material's properties appropriate to the chosen method of production?
   - How many products need to be made?
   - Will the combination of materials and the production method produce a high-quality product?

When choosing materials you should also consider influences on the product styling:

- What is the product's purpose? What is it for? What should be its durability and life span?
- What is the product's function? How is it required to work? What does it need to do?

Other considerations include the aesthetic requirements of the product. You should consider:

- form, such as the 3D shape and size
- visual properties, such as colour, reflection, transparency and surface pattern
- textural properties, such as how rough or smooth, including surface detail and the finish used.

The final choice of materials is often a compromise, which means that it partly satisfies all the product requirements, but does not completely satisfy them all! In some cases some requirements are more important than others, such as cost being more important than function. It is only when all the information is collected that a decision can be made. There is rarely one correct answer or absolute best material. Usually, there are several materials that will do the job.

Illustrating the final design proposal

You should clearly illustrate your final design proposal to show what your 2D and 3D outcomes will look like. You should also explain how the 2D outcome will be achieved and how the 3D outcome will be manufactured.

You should use an appropriate graphic style, which could be hand- or computer-generated. You will not lose marks if you choose to use hand techniques as opposed to using ICT. However, you could enhance your design and technology capability by using a range of computer techniques, such as word processing or CAD software (see Figure 2.10). You are not required to know how to use a specific language, computer or software package, but should use what is appropriate and available.

Factfile

Illustrating the final design proposal

Professional product designers aim to illustrate the final design proposal in the most convincing way possible. To do this they use a variety of techniques such as:

- 2D and 3D presentation drawings
- schematics
- detailed drawings of the design
- small-scale and large-scale models
- written reports.

The technique chosen to illustrate the final proposal needs to meet the client's budget, be suitable for the type of information to be communicated and for the character of the product. For example, the design of a pop-up book or its point-of-sale advertising could be presented in a simple, colourful style. A prototype perfume bottle, on the other hand, would require a more sophisticated presentation because it is aimed at a more exclusive target market.

You should annotate your final design proposal, using appropriate technical terminology. Identify the materials, components and processes required to manufacture your product. This may require you to produce front or back views of your product, or to produce exploded views to explain any design detail. Your annotation should demonstrate an understanding of the working characteristics of your chosen materials, components and processes. This information should be related to your chosen scale of production. For example, if you design a one-off product for an individual user, it may require the use of special materials, processes, tools or equipment. If you design a product intended for mass or continuous production, it may require the use of easily available materials, or processes that reduce waste or that need to be simplified for ease of manufacture. Remember that if you are designing for mass production you will still only be making one product in your workshop.

To be successful you will:
- Demonstrate a clear understanding of a wide range of resources, equipment, processes and commercial manufacturing requirements.

---

4. Evaluate design proposals against specification criteria, testing for accuracy, quality, ease of manufacture and market potential

You should test your final design proposal against the design specification to evaluate how it will meet its quality requirements, how easy it will be to manufacture and how it will meet user requirements. The use of feedback from an individual user or users in a target market group will give you some idea of your product's market potential. Remember to include evaluative comments about your product in your design folder. This evaluation should enable you to justify why this is the best solution for the problem, by explaining:

- how your design proposal meets the specification
- how it will meet the aesthetic, functional, cost and quality requirements of users
- how easy it will be to manufacture in the time available
- the availability of any specialist equipment and your skill level in using it.

To be successful you will:
- Objectively evaluate and test your design proposals against the specification criteria.

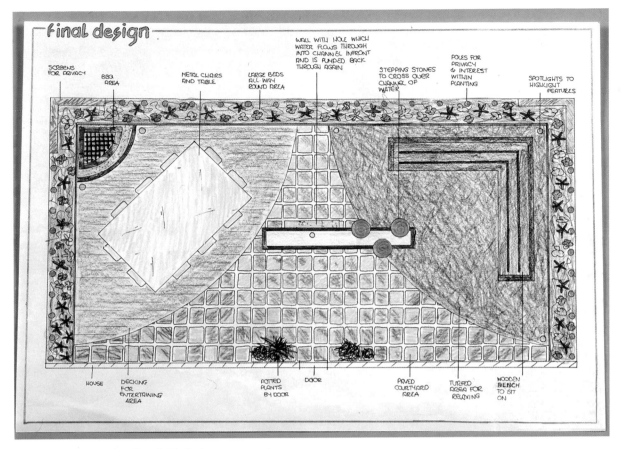

Figure 2.10 *Illustrating the final design proposal*

DOES IT MEET THE SPECIFICATION?

- Has a relaxing feel as the water flows into the seating area, soft surfaces add to the feel.
- Includes contemporary/modern materials and shapes.
- Does not make full use of the colour scheme. Only flowers and pots incorporate it.
- Has separate areas for relaxing and entertaining, both have seating.

- No storage, but does not include anything that would need it.
- No heating, but this could easily be included after, lighting to highlight certain features.
- Planters and metal struts provide a barrier and also privacy.
- Has running water to block out some city noise and add to relaxing feel.

Figure 2.11 *Evaluation of design proposals*

D Planning manufacture (10 marks)

1. Produce a clear production plan that details the manufacturing specification, quality and safety guidelines and realistic deadlines

You are expected to produce and use a production plan that explains how to manufacture both your 2D and 3D outcomes, within realistic deadlines. This involves providing details of how to produce both elements, taking into account quality and safety requirements. Realistic deadlines are those that are achievable; they should match the production of both outcomes to the time available.

The work that you have already done on modelling, prototyping and testing should help you to plan your manufacture. Prototyping enabled you to decide on the manufacturing processes you will use, plan materials requirements and work out the order of assembly of the individual component parts. Other planning decisions may involve:

- how easy the 2D/3D elements will be to produce in the time available
- if it is necessary to simplify anything
- how the 2D/3D elements could be manufactured if they were made by mass or continuous production
- if any special materials or tools are required and the safety procedures you need to observe.

Producing a production plan

Your production plan should be made before you manufacture your product. It is worth remembering that you will not be awarded marks for a production plan made after manufacturing the product. Your production plan should include a manufacturing specification and a production

schedule, which give clear and detailed instructions for making your product.

a) A manufacturing specification
 A manufacturing specification includes:

- a product description
- accurate exploded, orthographic, isometric or working drawings with clear assembly details, dimensions, sizes and tolerances
- quantities and estimated costs of materials and components
- finishing details.

b) A production schedule
 A production schedule can be produced in table form and should identify:

- the individual tasks within the manufacturing process
- the sequence of tasks in logical order
- the tools and equipment required for each task
- the estimated time each stage of manufacture/process will take
- critical control points in the product's manufacture where quality is to be

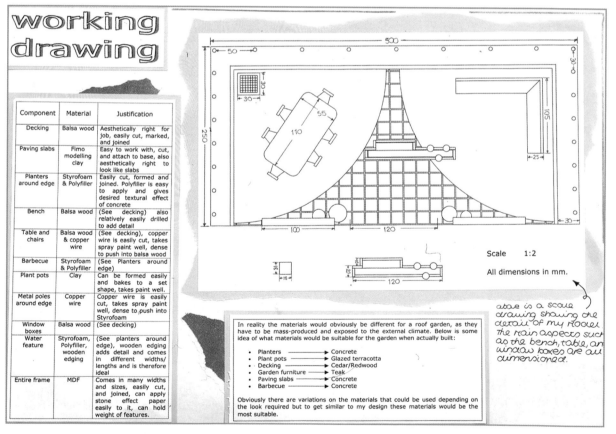

Figure 2.12 Part of a production plan

checked, e.g. during the final print run
- quality indicators, such as tolerances, to show how quality will be measured, e.g. using defined tolerance in a packaging net to ensure accurate fit
- safety requirements.

Part of a production schedule is shown in Figure 2.12.

The production plan is a key part of your quality assurance system, because it documents all the information required to manufacture your product. It identifies the key stages of production where you will check for quality to make sure that your product conforms to the specification. Using quality control ensures that if your product is made in high volume, each product is made to the same standard. Remember to record the critical control points where you will check for accuracy and the quality indicators you will use, such as the tolerances and dimensions you identified in your manufacturing specification. You should also record how you use testing, inspection and safety procedures during manufacture.

c) An estimate of production costs

Estimating product costs means producing an accurate price for the product, which would make it saleable *and* create a profit. In industry, cost levels depend on the method of production, which must be simple and fast so that labour costs are as low as possible. You can cost your product in set stages:

(i) Work out **direct costs**, like materials and labour costs (how long your product takes to manufacture at a set rate per hour).

(ii) Work out **overhead costs**, like rent, heat and electricity (these are often worked out as a set percentage of labour costs).

(iii) Add together the direct and overhead costs, to give the total manufacturing cost.

(iv) Work out your manufacturing profit (a set percentage of the total manufacturing cost).

(v) Add together the total manufacturing cost and the manufacturing profit, to give the selling price.

d) Setting realistic deadlines

Realistic deadlines are those that are achievable. Following a production plan will help you meet your production deadlines, but be prepared to modify your planning if you make any changes to your product or manufacturing processes during manufacture. Any modifications should be noted in your production plan, so make sure that you leave enough space to do this.

Think about this!

Planning manufacture
Use the following to help you plan manufacture:

- Ensure your materials are easy to work and handle.
- Check that the performance characteristics of your chosen materials meet the design specification.
- Forward order materials and components if necessary.
- Produce a manufacturing specification and a work schedule.
- Specify any safety requirements and procedures.
- Specify where and how you will check for quality.
- Make sure that you follow your production plan and record any changes you make.

To be successful you will:
- Produce a clear and detailed production plan with achievable deadlines.

2. Take account of time, resource management and scale of production when planning manufacture

Planning is an important part of any project and many of the activities that you undertake will overlap, because designing and manufacturing are complex and interrelated activities. In addition, the number of weeks that you have available will depend on your timetable. For example you may have a set number of weeks where you concentrate totally on your coursework project or you may have less time during the week but more weeks overall. In other words, your coursework project time may be short and fat or long and thin!

Planning materials

One of the most important reasons for planning is to make sure that materials, components, tools and equipment are available when required and in the appropriate quantity and quality. Check out your requirements well in advance. As far as quality of materials is concerned, you should ensure that they meet your specification requirements. For example, if you are planning the manufacture of a one-off product for a single user you may require the use of special materials or more time-consuming complex processes, so take this into account. However, if you are planning a manufacturing prototype you may need to consider using standard sizes or standard components to reduce costs, or to simplify manufacturing processes to ensure fast, cost-effective manufacture.

Remember to use your production plan as a working document in which you record any subsequent changes you have to make, if, for example, delays occur. Any realistic amendments to the production plan will provide you with useful data to refer to in your end of project evaluation.

Tasks

1 Planning

Use the following questions to help you plan your production:

- Will my materials, components, tools and equipment be available when I need them?
- How will my scale of production affect my manufacturing processes?
- Can I use standard components, parts or materials to simplify my task?
- Will the quality and quantity of my materials and components match my manufacturing specification?

2 Production planning

Using a Gantt chart is a good way of planning a project because it gives you a picture of the whole project at a glance. It can be used to plan:

- a whole project over the total number of weeks available
- the detailed manufacture of your product.

Produce a Gantt chart for your project:

a) Draw up a table with the dates across the top.
b) In the left-hand column, put in order a list of tasks to be done.
c) Note any tasks that can be done at the same time.
d) Plot the tasks against the time you have available.

To be successful you will:
- Demonstrate effective management of time and resources, appropriate to the scale of production.

Help! What if my production plan goes wrong?

The purpose of your production plan is to guide you through your product manufacture so that you use your time well.

- Sometimes you may come across delays, such as having to wait for materials to arrive, or for the use of a piece of equipment.
- The first thing you must do is not to panic, but to record any problems you have in your production plan. You can then adapt your manufacture to meet the changed circumstances.
- If you use a Gantt chart for planning you will easily see your progress, so you can monitor things as you go along.
- If you are held up for any reason, make a note of it in your production plan and get on with something else!

3. Use ICT appropriately for planning and data handling

The aim of using information and communications technology (ICT) for planning and data handling is to enhance your design and technology capability. You will not be penalised for non-use of ICT, although you should use it where appropriate and available.

When planning and data handling, good use of ICT might include using word processing, databases, spreadsheets or CAD software for a range of activities that may include:

- organising and managing data
- production planning using colour-coded Gantt charts, diagrams or flowcharts
- producing manufacturing specifications
- producing accurate **working drawings**
- working out quantities and costs of materials and components.

To be successful you will:
- Demonstrate good use of ICT.

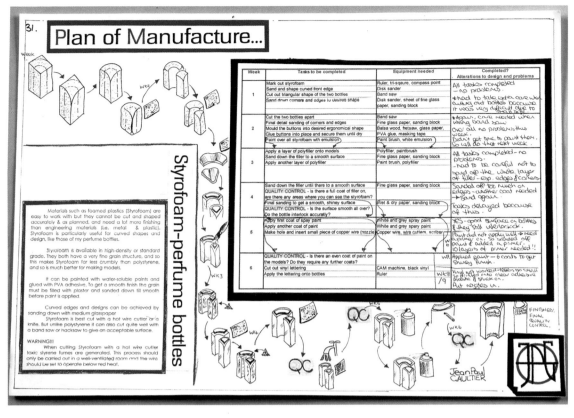

Figure 2.13 *Production planning*

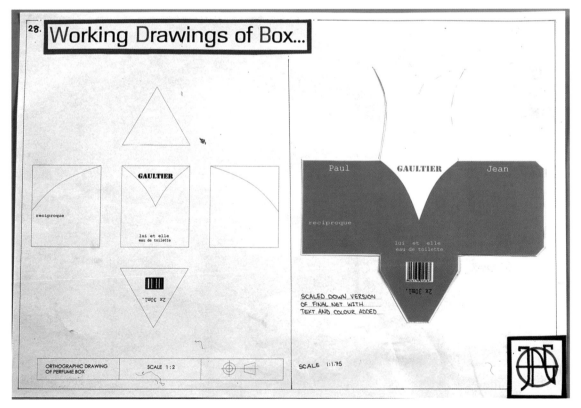

Figure 2.14 *Working drawing*

E Product manufacture (40 marks)

1. Demonstrate understanding of a range of materials, components and processes appropriate to the specification and scale of production

You are expected to demonstrate an understanding of the working characteristics of the materials and components appropriate to the manufacture of your product. The 2D outcome should be developed from traditional and modern graphics media. The 3D model or prototype product should be constructed from modelling materials *and* at least one resistant material. These could include some wood, and/or metal and/or plastic. The use of Styrofoam and foam board is not acceptable as the resistant material element and will not enable access to higher marks. Your 3D prototype product must be of a high quality for you to attain high marks.

To a certain extent you have already demonstrated an understanding of the materials, components and processes needed to manufacture your product as you:

- modelled, prototyped and tested materials, components and processes
- annotated your design proposals and explained the working characteristics of suitable materials, components and processes
- specified the materials, components and processes required to make your product.

You are now ready to further demonstrate your knowledge and understanding of materials and components through the actual manufacture of your product.

Scale of production

Your 2D and 3D outcomes may be a combination of one-off and high-volume levels of production. For example, if you made a 3D scale model of a robot for use in a film, you would be designing and making a one-off product to meet the needs of a client. The model may require working to a defined scale or the use of electronic components to create 'special effects'. Any printed material produced to support the marketing of the film may need to be designed for reproduction in a large batch.

Alternatively, you could design and make a 3D prototype robot toy that could be manufactured in high volume. You would still make this 3D prototype as a one-off, but it would require a different approach throughout its development. Its target market may be children (and adults who buy for children) – and its manufacture would require different materials or processes to those of the one-off film robot. Any point-of-sale advertising that you might produce to support the robot toy may also have to be printed in a large batch.

Although most of your practical work will involve making a one-off product, there may be times when two or more identical component parts will have to be made. Whatever scale of production you work to, your prototype will still need to be made and finished to the highest quality. There will also be other complications such as constraints related to the materials, tools and equipment available to you. If problems occur with the availability of resources, for example, you may have to change your original choice of materials or adapt the processes you use. If this happens, do not forget to record any changes in your production plan and justify any new choices of materials, components or processes.

Figure 2.15 *Remember to include photographs of 'work in progress' in your design folder*

Task

Approaches to manufacturing

A different approach is required when planning the manufacture of a product at different scales of production.

For one of the following products draw up a table to show the key stages of manufacturing the product as a one-off for an individual user, and as a product suitable for batch or mass production.

- A 3D architectural model and 2D presentation drawings.
- A 3D point of sale display and 2D advertising material.

For each scale of production, think about materials preparation, processing, assembly and finishing.

To be successful you will:
- Demonstrate clear understanding of a wide range of materials, components and processes.

---

2. Demonstrate imagination and flair in the use of materials, components and processes

Task

Practise the techniques and processes you aim to use during manufacture, so you can demonstrate your ability.

- Experiment with working, shaping, and joining materials.
- Use mock-ups to trial structures.
- Experiment with processes to improve aesthetic qualities such as materials finish.

Be prepared to modify your manufacturing processes if necessary or to adapt details of the design, e.g. changing the method of joining. Keep all your experimental work in a small box, so you can refer to it if necessary. You can evidence your experimental work by including a photograph in your folder.

One clear way that you can demonstrate imagination and flair is in the way that you handle materials and processes. An understanding of how materials behave and how processes work will enable you to show your skills and ability. This will result in the production of a quality product – one that:

- is attractive to the market
- is well made from suitable materials
- is enjoyable or fun to use
- would sell at an attractive price
- is manufactured for safe use and disposal, without harm to the environment.

This is quite a list of considerations, but if you check you should find that you have taken most of them into account in your design and manufacturing specifications. At this stage, prior to manufacture, you have the opportunity to hone your skills – to experiment with and trial materials, components, techniques and processes, so you can demonstrate your ability through the manufacture of your product.

To be successful you will:
- Demonstrate imagination and flair.

---

3. Demonstrate high-level making skills, precision and attention to detail in the manufacture of high-quality products

Demonstrating high-level skills involves making the best use of available materials and components appropriate to your design proposals. It also involves using tools and equipment with accuracy, confidence and skill. If you practise your existing skills before manufacturing your product, you should gain an understanding of your ability in relation to your expectations for your product. If your ability falls below your expectations, you have two options: either improve your skills or adapt the process. Improving your skills will result in improving the quality of your work, so you produce a high-quality product.

The making of high-quality products also depends on planning quality into your design and manufacturing process. Refer to your design and manufacturing specifications and to your production plan, where you should find references to quality. Your production plan should identify the key stages of manufacture, where you can monitor the accuracy of your work as it progresses, checking against the dimensions and tolerances you detailed in your working drawings.

4. Use ICT appropriately for communicating, modelling, control and manufacture

You can use ICT to help your product manufacture, where it is appropriate and available, but you will not be penalised for its non-use. You are not expected to know how to use specific equipment or programs, but you should understand the benefits of using ICT for manufacture. Different uses of ICT include:

- using software to model 'virtual' products on screen before manufacture, saving time and costs because it reduces the need to make expensive manufacturing prototypes
- communicating information between CAD software and computer numerically controlled (CNC) equipment
- using CNC machines for fast, accurate production processes
- communicating manufacturing information between the design office in one location and the manufacturing site in another.

If you do not have easy access to computer-aided manufacturing (CAM) equipment, you could use a printer or a plotter to print out technical drawings, or to cut out a component parts drawing and use it as a template for making identical parts.

If you do have access to specialised CAM equipment, you can use CAD to produce design ideas and then export the digital information to CNC equipment for producing accurate component parts for your product.

Factfile
Using ICT in manufacture
The increasing use of ICT through the use of CAD/CAM systems has had an enormous impact on manufacture. CAD/CAM enables the efficient design and manufacture of products and the control of manufacturing equipment. CAM automates production, repeats processes easily and precisely and enables the production of cost-effective products.

5. Demonstrate a high level of safety awareness in the working environment and beyond

Safety in manufacturing means the safe design, manufacture, use and disposal of products. Manufacturers must follow safety procedures and check standards, regulations and legislation related to product design.

This ensures that products are safe for the manufacturer, the consumer and the environment. Legal requirements, such as the Health and Safety at Work Act 1974 and Reporting of Accidents 1986, ensure that safe production processes are followed to prevent industrial accidents.

Safe production means identifying all possible risk and documenting safety procedures to manage and monitor the risk.

A question of safety
Ask the following questions at key stages of your design and manufacture:

- What could go wrong?
- What could cause things to go wrong?
- What effect would this have?
- How can I prevent things from going wrong?

You need to demonstrate an awareness of safety at all stages of design and manufacture, by making safety a priority in your work.

At the research and design stage you should take account of designing with safety in mind, both for you, the maker, for your intended user(s) and for the environment. This may involve researching safety regulations related to your product. Safety features should be identified in your design specification.

Your production plan should identify specific safety features related to manufacture, including safety guidelines for your chosen materials and the tools, equipment and processes you may use. Modelling and prototyping before production will enable you to test for safety against the criteria that you have identified.

During the manufacturing process you should follow safety guidelines related to safety with people, with materials, with equipment and machinery.

F Testing and evaluating (10 marks)

1. Monitor the effectiveness of the work plan in achieving a quality outcome

Your production plan is a key tool in monitoring the quality of your product. You should record any changes you make to your 2D or 3D outcomes, or to any processes you use during manufacture.

You may not need to make any changes, but if you do, however minor, they should still be recorded because they could have an impact on the quality of your product. Recording any changes will also enable you to make an identical product to the same standard.

Sometimes completely unforeseen problems can arise through, for example, using a process or technique that is new to you. Other reasons for making changes could be through not having the right materials, components, tools or equipment available when you want them, or because you are running out of time.

If you do have to make any changes, make sure that you explain what you have done and why – write it down straight away before you forget, or do a quick sketch to explain in your production plan any change in design or product manufacture. Recording any changes to your product will make it easier to evaluate its quality of design and manufacture (see Table 2.6).

Table 2.6 *Record any changes you make during manufacture*

Manufacturing process	Changes to process	Changes to product	Quality checks
			Check tolerances
			Check dimensions
			Check against specification
			Check finish

To be successful you will:
- Make effective use of your work plan to achieve a high-quality outcome.

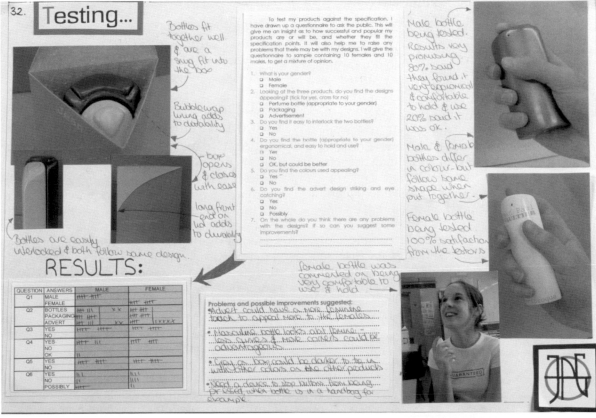

Figure 2.16 *Your production plan is a key tool in monitoring the quality of your product*

2. Devise quality assurance procedures to monitor development and production

Task

The meaning of quality

In industry quality means:

- conforming to the specification
- ensuring fitness-for-purpose
- making products with zero defects
- making products right first time, every time
- ensuring customer satisfaction
- exceeding customer expectations.

Use the following questions to help your quality planning:

- Do you aim for fault-free work?
- Do you know what standards are expected?
- Do you check the quality of your work against the specifications?
- Does your work meet the specification?
- Are you pleased with your work?
- Could you do it any better?

Quality planning is a key process during product development and manufacture and you should devise your own quality control procedures to monitor quality. Check that you can meet the quality requirements outlined in your design specification and check your product for quality at critical points in its manufacture, using quality indicators. These are categorised as 'variables' or 'attributes'.

- Variables are measurable characteristics that can vary within set limits, such as keeping within a tolerance of +/–0.5mm. Checking tolerances may involve the use of an accurate measuring device, such as a Vernier gauge or a micrometer. Variables can include length, width, height, diameter, position, angles and mass.
- Attributes are either acceptable or unacceptable characteristics. Make use of inspection using sensory tests of vision and touch to check characteristics such as colour or texture – for example, if the colour is meant to be blue, green won't do!

Remember to record how you use quality control checks and inspection during manufacture.

To be successful you will:
- Devise clear quality assurance procedures.

3. Use testing to ensure fitness-for-purpose

Testing to ensure fitness-for-purpose means testing that the product's performance meets the requirements of the specification and the user(s). Inspection and testing should be holistic in relation to your 2D and 3D outcomes – after all, the two elements are linked, with one supporting the other. For example:

- A prototype product, supported by point-of-sale advertising, would both be linked to the context for design, which may be corporate image. An evaluation of the one element without the other would be incomplete.
- Testing a board game and its packaging during manufacture may involve testing size tolerances, the clarity of reproduction techniques, the durability of the playing pieces, the quality of finish and the stability of the outcome.

You should record the results of any testing that you do during and after manufacture:

- Test the performance of the product against the design and manufacturing specifications.
- Test that the quality of the product is suitable for users – check against user requirements in the specifications.

To be successful you will:
- Make effective use of testing to ensure fitness-for-purpose.

4. Objectively evaluate the outcome against specifications and suggest appropriate improvements

You should be as objective as possible when justifying the success of your 2D and 3D outcomes against the design brief and specifications. Being objective means taking an unbiased view of the outcome. This can sometimes be difficult, as you will have been closely involved in the design, development and manufacture. It is easier to be objective if you use standards against which to judge your product – these standards are the design brief, the design and manufacturing specifications and the quality criteria identified in the production plan.

Make sure that you evaluate the success of your outcome against user needs, using feedback to evaluate how the product looks and how well it meets requirements. This should include:

- comments on the performance against

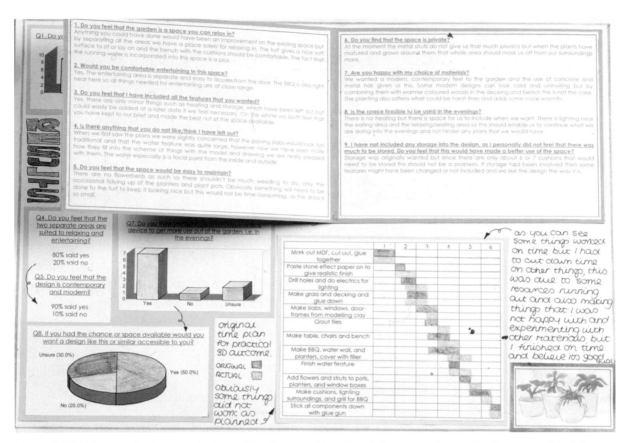

Figure 2.17 *Using evaluation against specifications and feedback from users is an effective way of judging the success of a product*

specifications, accuracy, design and manufacturing quality
• views of intended user(s) through questionnaires, surveys or user trials.

Objective evaluation should provide you with feedback on the success of your product, which will help you decide how and if it can be improved (see Figure 2.17). Your suggestions for improvement should be based around the product's aesthetic and functional success, its quality of design and manufacture and its fitness-for-purpose.

> **To be successful you will:**
> • Objectively evaluate the outcome and suggest appropriate improvements.

G Appropriate project (10 marks)

The G criterion has been included to reflect that your project meets all the coursework requirements (see 'What is a Graphics project?', page 39). It is very important to check the appropriateness of your project with your teacher or tutor at the start of your project to make sure that it will enable you to address all of the assessment criteria by which your project will be marked. After your teacher or tutor confirms that your project is appropriate, you will still need to keep an eye on the assessment criteria in order to achieve all the available marks. You will also need to take account of feedback from your teacher or tutor in order to improve your work as it progresses. Remember to include photographic evidence of your modelling and prototyping and the product manufacture, especially to highlight difficult techniques or hidden details.

Student checklist

1. Project management

- Take responsibility for planning, organising, managing and evaluating your project.
- Include photographic evidence to show hidden details or to demonstrate the processes you use at each stage of manufacture.
- Include only the work related to the assessment requirements.

2. A successful AS coursework project will

- Identify a realistic need and solve a problem for your specified user(s).
- Include relevant information that summarises your research.
- Show the influence of research on your design decisions.
- Demonstrate a variety of communication skills, including appropriate use of ICT.
- Include a 2D element developed using graphics media.
- Include a 3D model using at least one resistant material.
- Be a manageable size so you can finish it on time.
- Evidence understanding of industrial practices.
- Include clear photographs of modelling, prototyping, testing and manufacture.
- Detail the manufacture of one product and show how it could be manufactured in quantity.
- Allow time to evaluate your work as it progresses and modify it if necessary.
- Be well planned so you can meet your deadlines.

3. Evidencing industrial practices in coursework

- Use industrial-type terminology and technical terms.
- Use designing activities similar to those used in industry, i.e. develop a design brief and specification, use market research, modelling and prototyping, etc.
- Use manufacturing activities similar to those used in industry, i.e. use a production plan, quality control, test against specifications, etc.

4. Using ICT in coursework

- Develop the use of ICT for research, designing, modelling, communicating and testing.
- Develop the use of ICT for planning, data handling, control and manufacture.

5. Producing a bibliography

- Reference all secondary sources of information in a bibliography, e.g. from textbooks, newspapers, magazines, electronic media, CD-ROMs, the Internet, etc.
- Reference scanned, photocopied or digitised images. Do not expect to use clip art at this level.
- Do not expect marks for any work copied directly from textbooks, the Internet or from other students.

6. Submitting your coursework project folder

- Have your coursework ready for submission by mid-May in the year of your examination.
- Include a title page with the Specification name and number, module number, candidate name and number, centre name and number, title of project and date.
- Include a contents page and numbering system to help organise your coursework folder.
- Ensure that your work is clear and easy to understand, with titles for each section.

7. Using the Coursework Assessment Booklet (CAB)

- Complete the student summary in the CAB and *remember to sign it*! The summary should include your design brief and a short description of your coursework project.
- Ensure that the CAB contains a minimum of three clear photographs of the whole product, with alternate views and details.
- Write your candidate name and number, centre name and number and 6298/01 in the CAB by the product photographs and on the *back of each photograph*.

Help! What if my project goes wrong?

- If your Unit 2 coursework project doesn't meet your expectations, don't worry! You can retake the unit and the better result will count towards your final grade.
- If you find yourself in this situation your teacher or tutor will be able to advise you on the best way forward.

Materials, components and systems (G301)

Summary of expectations

1. What to expect

Unit 3 is divided into two sections:

- Section A Materials, components and systems
- Section B consists of two options, of which you will study only one.

Section A is compulsory and builds on the knowledge and understanding of materials, components and systems that you gained during your GCSE course.

2. How will it be assessed?

The work that you do in this unit will be externally assessed through Section A of the Unit 3 examination paper. You are advised to spend 45 minutes on this section of the paper.

3. What will be assessed?

The following list summarises the topics covered in Section A and what will be examined:

- Classification of materials and components
  - paper, card and boards
  - woods
  - plastics
  - metals and alloys
  - composites and laminates
  - components.
- Working properties of materials and components
  - hand and commercial processes
  - finishing processes
  - product manufacture.

- Testing materials
  - comparative testing of materials, components and processes
  - British and international standards.

You should apply your knowledge and understanding of materials, components and systems to your Unit 2 coursework.

4. How to be successful in this unit

To be successful in this unit you will need to:

- have a clear understanding of the topics covered in Unit 3A
- apply your knowledge and understanding to a given situation or context
- organise your answers clearly and coherently, using specialist technical terms where appropriate
- use clear sketches where appropriate to illustrate your answers
- write clear and logical answers to the examination questions, using correct spelling, grammar and punctuation.

5. How much is it worth?

This unit, together with the option, is worth 30 per cent of your AS qualification. If you go on to complete the whole course, then this unit accounts for 15 per cent of the full Advanced GCE.

Unit 3A + option	Weighting
AS level	30%
A2 level (full GCE)	15%

1. Classification of materials and components

Paper, card and boards

The production of woodpulp

Paper, card and boards are produced primarily from hard and softwoods, although other materials can be used such as cotton, straw and hemp, producing papers with different properties. Wood is made up of fibres that are bound together by a material called lignin. In order to produce paper these fibres must be separated from one another to form a mass of individual fibres called woodpulp. Softwood fibres are longer, offering greater strength. Hardwood fibres are shorter, offering a smoother, opaque finish. Woodpulp is produced by one of three basic methods – mechanical, chemical or waste.

Mechanical pulp

The logs of coniferous wood are saturated with water and debarked. They are then ground down, with the resulting pulp screened to accept 1–2 mm pieces with the larger pieces being recirculated. The resulting pulp can only be used for low-grade paper or packaging material, so the pulp is bleached with peroxide or sodium hydroxide. This method of producing woodpulp offers a high yield but contains greater impurities.

Chemical pulp

After debarking, the hardwood and softwood logs are cut into 2 cm chips along the grain. These are pounded into fragments and screened. The resulting pulp is stored and treated with acid or alkaline. The fibre yield is lower, but the fibres are longer, stronger and contain fewer impurities.

In semi-chemical pulp, the wood is processed as above, but the chips are treated with steam, which results in longer fibres. This offers a high yield and when the fibres are bleached they become shorter and similar to chemical pulp.

Waste pulp

Recycled paper and card used for waste pulp is often used for lower grades of paper, as its strength, durability and colour are not as good as virgin fibres. Waste pulp can be mixed with virgin fibres to produce better quality paper.

Manufacturers blend a variety of pulps and process them with bonding agents and pigments to produce paper with different qualities. These processes help to achieve a consistent colour and bind fibres to create a better surface. A sizing agent can also be added to improve water resistance and prevent ink from feathering on the surface.

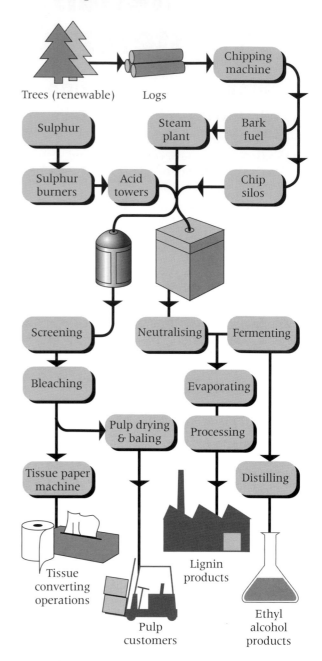

Figure 3.1.1 The mechanical and chemical production of wood pulp

The manufacture of paper

Paper is produced on a **Fourdrinier machine**. The pulp is diluted to 99.5 per cent water and held in a head box. A continuous stream of pulp is sent through an adjustable slit on to a moving, woven mesh belt which is vibrated to drain off the water and allow the fibres to interweave. The pulp is then pressed with a mesh on top to drain the water quickly and is passed under a dandy roll to smooth out the fibres, before it travels over suction boxes and rollers which draw out more water. At this stage the paper contains around 80 per cent water, allowing it to be unsupported by the mesh belt. The paper is then passed through a series of rollers and heated cylinders to produce a paper with 4–6 per cent moisture content. This is the desired content to allow a balance with the relative humidity of the atmosphere.

During the drying process, sizing agents such as starch, resin and alum can be added by spray or press. After this, highly polished rollers called calenders can be used to give a smooth, gloss finish.

The paper is wound on to a roll from which it can be placed into a precision cutting machine to produce the desired size, or directly loaded on to the printing press.

Weight and size

Paper is defined in weight as gsm (grams per square metre), with 80 gsm being the weight of average copier paper. Card and board are measured in micrometers (microns). A paper is usually classified as a board when it is greater than 220 gsm and more often than not is made from more than one ply. The thickness of card and board can be gauged by the number of plies it consists of (see Table 3.1.1).

Table 3.1.1 Common thicknesses of paper, card and board

Number of plies	Microns
2	200
3	230
4	280
6	360
8	500
10	580
12	750

Paper, card and board are most commonly available in metric 'A' sizes (i.e. A5 through to A0). However, 'B' sizes and old imperial measurements are also widely used.

Characteristics of paper, card and board

The choice of paper is essential to how printed materials are presented. Choosing the most appropriate paper for its intended application is a combination of personal preference and discussion with the client in order to determine the look of the end product. In general, the correct choice of paper must satisfy:

- the design requirements of the client brief, e.g. surface finish, colour, texture, size and weight

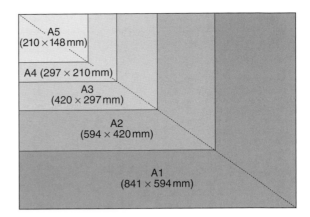

Figure 3.1.2 Common 'A' sizes of paper, card and board

- the demands of the printing process or surface decoration, e.g. will the printing inks used for offset lithography provide a quality finish on the paper?
- economic considerations, e.g. scale of production.

Paper, card and board can also be defined by their:

- durability
- colour and brightness
- texture
- opacity.

Cartonboard

Cartonboards are used extensively in the retail packaging industry, where specific properties are required. These boards must be suitable for high-quality, high-speed printing and for cutting, creasing and gluing using very high-speed automated packaging equipment.

Advantages of using cartonboard include:

- total graphic coverage and outstanding print quality
- excellent protection in structural packaging nets
- relatively inexpensive to process and produce
- can be recycled.

Task

Collect a variety of different types of paper, card and board. For each:

a) define what pulp may have been used to make it
b) compare its durability, texture and opacity to that of standard 80 gsm photocopier paper.

Table 3.1.2 *The properties and end uses of paper, card and board*

Type	Weight	Description	End uses	Properties	Cost
Layout paper	About 50 gsm	Thin, translucent paper with a smooth surface	Outline sketches of proposed page layouts; sketching and developing ideas; marker renderings	Translucent property allows tracing through onto another sheet; accepts most drawing media (except paints)	Relatively expensive
Tracing paper	60/90 gsm	Thin, transparent paper with a smooth surface; pale grey in appearance	Same as layout paper; heavier weight preferred by draughtsmen	Allows tracing through on to another sheet in order to develop design ideas	Heavier weight can be quite expensive
Copier paper	80 gsm	Lightweight grade of quality paper; good quality bleached surface	B/W photocopying and printing from inkjet and laser printers; general use for sketching and writing	Bright white and available in a range of colours; smooth finish for colour printing	Inexpensive when purchased in bulk
Cartridge paper	120–150 gsm	Creamy-white paper; smooth surface with a slight texture	Good general purpose drawing paper; heavier weights can be used with paints	Completely opaque; accepts most drawing media	More expensive than copier paper
Inkjet paper (coated)	80–150 gsm	Bright white, high density, ultra-smooth coated paper	Printing photo quality work with a matt finish, i.e. presentation materials, reports, colour reproductions, etc.	Suitable for 1200 dpi colour inkjet and laser printing; quick drying and recyclable	Expensive (usually sold in small packs)
Inkjet paper (photo glossy)	140–230 gsm	Bright white, professional quality, specially coated high-gloss paper	Vivid photo quality with maximum colour reproduction suitable for photo reproductions, graphic artwork and presentation materials	Quick drying; photo quality; heavyweight; two-sided photo gloss paper also available	Expensive (usually sold in small packs)
Card	230–750 microns	A thin variety of board, but thicker than paper	A range of uses from printing and drawing to 3D modelling and presentation work	Bright and fluorescent colours, duo-tones and metallics, and corrugated card types	More expensive than paper
Mounting board	1000–1500 microns	Extremely thick board with colour on one side only (white on back)	Mounting work for presentations and displays; work can be mounted flat or behind a frame mounting	Very high quality, strong and rigid board; available in a range of colours (wide range of pastel colours)	Expensive

Woods

Classification of woods

Woods can be divided into three main categories:

- hardwoods
- softwoods
- manufactured boards.

Both hardwoods and softwoods are produced from naturally growing trees, whereas manufactured boards are man-made using natural timber.

Hardwoods

Hardwoods are produced from broadleaved trees whose seeds are enclosed; examples include oak, mahogany, beech, ash and elm. Hardwood trees commonly grow in warmer climates such as Africa and South America and take about 100 years to reach maturity. They are usually tough and strong and because of their close grain, provide highly decorative surface finishes. Due to their age and location, many hardwoods are expensive to buy and may only be used for high-quality products. The exception is balsa wood, which has been used for modelling applications for many years as it is relatively inexpensive and easy to shape.

Softwoods

Softwoods are produced from cone-bearing

Table 3.1.3 *The properties and end uses of common cartonboards*

Board	Description	End uses	Properties	Cost
Folding boxboard	Usually consists of a bleached virgin pulp top surface, unbleached pulp middle layers and a bleached pulp inside layer	Widely used for the majority of food packaging and for all general carton applications	Excellent for scoring, bending and creasing without splitting; excellent printing surface	Inexpensive
Corrugated board	Consists of a fluted paper layer sandwiched between two paper liners	Protective packaging for fragile goods; the most commonly used box making material	Excellent impact resistance; has excellent strength for its weight; recyclable	Inexpensive
White-lined chipboard	Consists of a top layer of bleached wood pulp and a middle and back layer made from waste paper	Food packaging and general carton applications; white-back versions are known as Triplex board	Good for high-speed printing of automatically packed cartons	Inexpensive
Solid white board	Made entirely from pure bleached wood pulp	Packaging for frozen foods, ice-cream, pharmaceuticals and cosmetics	Very strong and rigid; excellent printing surface	Expensive
Cast-coated board	A heavier and smoother coating applied to white-lined chipboard and solid-white board	Luxury products requiring expensive looking decorative effects	Very strong and rigid; excellent printing surface; higher gloss finish after varnishing	Expensive
Foil-lined board	Consists of a laminated foil coating (can be used on all of the above boards); available in matt or gloss finish and in silver and gold	Cosmetic cartons, pre-packed food packages	Very strong visual impact; foil provides an excellent barrier against moisture	Expensive

conifers with needle-like leaves; examples include Scots pine (red deal), red cedar, parana pine and whitewood. As softwoods mature more quickly than hardwoods (about 30 years), they can be forested and replanted, which means they are in abundance and therefore cheaper to buy. Softwoods are also easier to work with and are lightweight, which makes them more suitable for modelling applications.

Manufactured boards

Manufactured boards can be manufactured either from laminating thin veneers of wood together, or from wood particles glued together and compressed. The advantages of using manufactured boards are that they are available in wide board sizes, which is not possible with natural timbers where the width depends on the width of the tree trunk. By running the grain of the veneers at 90 degrees to one another, some boards are given added strength. They are also very much cheaper to buy than natural timber and are of more use for model making, for example block modelling. There are several types of manufactured boards, including;

- plywood
- blockboard and laminboard
- particleboard (i.e. hardboard)
- fibreboard (i.e. medium density fibreboard or MDF).

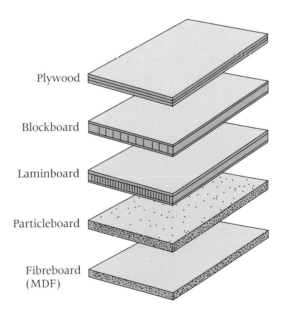

Plywood

Blockboard

Laminboard

Particleboard

Fibreboard (MDF)

Figure 3.1.3 *Manufactured boards*

Table 3.1.4 *The properties and end uses of hardwoods and softwoods*

Timber	Type	Origin	Properties/characteristics	End uses
Oak	Hardwood	Europe, USA, Japan	Very hard, tough, strong and durable; heavy; good finish; fairly easy to work; little shrinkage; contains an acid that corrodes steel	High-quality furniture, garden benches, boat building, veneers
Mahogany	Hardwood	Central and South America, Africa	Fairly easy to work; durable; good finish; prone to warping	Indoor furniture, interior woodwork, window frames, veneers
Beech	Hardwood	Europe	Hard, tough, very strong and straight; good finish; prone to warping; turns well	Workshop benches, children's toys, interior furniture
Ash	Hardwood	Europe	Tough; flexible (good elastic properties); works and finishes well	Sports equipment, ladders, laminated furniture, tool handles
Elm	Hardwood	Europe	Tough, durable and fairly strong; fairly easy to work; prone to warping; turns well	Garden furniture, construction work, turnery, furniture
Scots pine	Softwood	Northern Europe	Easy to work; knotty and prone to warping	Constructional woodwork (joists and roof trusses), floorboards, children's toys
Parana pine	Softwood	South America	Hard, straight grain (often knot free); fairly strong, durable and easy to work; smooth finish	Best quality internal softwood joinery
Whitewood	Softwood	Northern Europe, Canada, USA	Fairly strong but not durable; easy to work; very resistant to splitting	General interior work

Sources of different types of wood

The prime source of the world's supply of commercially grown softwood is the northern hemisphere, but particularly the colder regions of North America, Scandinavia, Siberia and parts of Europe. Conifers are relatively fast growing and produce straight trunks, which make for economic cultivation with little wastage. With careful management of forests it is possible to control the supply and demand of softwoods. Being relatively cheaper than hardwoods they are used extensively for building construction and joinery. What waste that is produced is used in the manufacture of fibreboards and paper.

The UK imports almost 90 per cent of its timber needs since it is one of the least wooded countries in Europe. The imported boards are usually supplied debarked or square edged ready for further processing at the sawmills.

There are thousands of species of hardwoods grown across the world and many are harvested for commercial use. Most broad-leafed trees grown in temperate climates such as Europe, Japan and New Zealand are deciduous and lose their leaves in winter with the exception of a few like holly and laurel. Those grown in tropical and sub-tropical regions like Central and South America, Africa and Asia are mainly evergreen which means they grow all year round and reach maturity quicker. Hardwoods generally are more durable than softwoods and offer much more variety in terms of colour, texture and figure. Since they take a relatively longer time to grow they tend to be more expensive than softwoods with the really exotic timbers being converted, into veneers, which allows for much greater use of a limited supply.

Tasks

1 Give two different examples of products made from the following raw materials:
   a) hardwood
   b) softwood
   c) manufactured board.
   For each product example, explain the suitability of the raw material.
2 Investigate the standard sizes for a range of boards using a timber catalogue.

The structure of woods

The characteristics and working properties of timbers can be broken down into categories. Knowledge in each of these areas will help you to identify timbers in your work. They are:

• weight
• texture

- durability
- colour
- odour
- ease of working.

A tree essentially consists of two major parts: the inner or 'heartwood', which gives rise to strength and rigidity, and the outer layers or 'sapwood', which is the region of growth where food is stored and transmitted. Growth is a seasonal process where layers are seen as concentric growth rings, known more commonly as annual rings. As the tree grows the wood tissue grows in the form of long tube-like cells, which vary in shape and size. These are known as fibres and are arranged roughly parallel along the length of the trunk and give rise to the general grain direction. This variation in cell size, shape and function leads to the botanical distinction of hardwoods and softwoods (see Figure 3.1.4):

- Softwood structure, e.g. Scots pine, red cedar, spruce:
  – Tracheids: these are elongated tubes which become spliced together in the direction of growth and make up the grain. Sap and food pass through smaller openings known as pits that harden as the tree grows older.
  – Parenchymas: smaller than tracheids, these make up the remaining cells.
  – Resin canels: evident in the majority of conifers. They carry away waste products in the form of resin and gum.
- Hardwood structure, e.g. oak, mahogany, beech:
  – Fibres: these constitute the bulk of hardwoods. They are not in any regular

pattern or formation and they are much smaller and more needle-like than tracheids in softwoods.
  – Vessels/pores: they form long tubes within a tree that carry food. They are used to distinguish one type of hardwood from another.
  – Parenchymas: these are more prominent in hardwoods and in oak they become quite thick, up to 30 cells thick, and are seen as the familiar 'silver' flashes within the grain.

Production of hardwoods and softwoods
First stage – conversion
Once a hardwood or softwood tree has been felled, it is transported to the sawmill where it is processed into planks or boards ready for use. The first process is called the conversion of timber – this is where the trunk is sawn up into usable sizes using large circular saws.

There are basically two methods of cutting used in the conversion process:

- slab, plain, or through and through
- quarter (radial) sawn.

Slab, plain, or through and through conversion is the simplest, quickest and cheapest of the two methods. The process makes a series of parallel cuts through the length of the log resulting in parallel slices or slabs. The thickness of the slabs can be varied as the log is cut and this type of cutting is frequently used on softwoods where the logs tend not to be that large in diameter.

Quarter (radial) sawn is a much more time-consuming process and involves much more manual handling. It is also a much more wasteful process. However, the timber produced tends to be better in quality and is much more stable in that it is less likely to move, warp, bow or twist.

Essentially, quarter sawing tries to make the annual rings as short as possible and at 90 degrees to the cut surface. This type of cutting results in the grain's figure being exposed and this is quite noticeable in oak where the silver grain is exposed (see Figure 3.1.5).

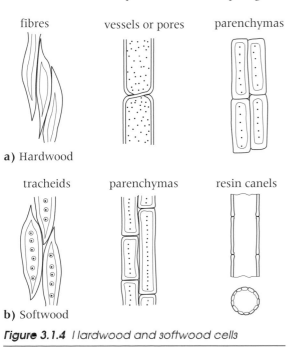

a) Hardwood

b) Softwood

Figure 3.1.4 Hardwood and softwood cells

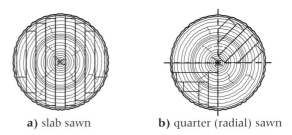

a) slab sawn   b) quarter (radial) sawn

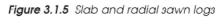

Figure 3.1.5 Slab and radial sawn logs

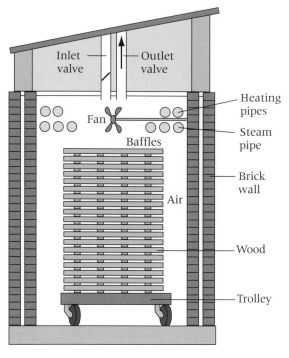

Figure 3.1.6 *A kiln allows the timber to be seasoned quickly*

Second stage – Seasoning

The second stage involves the seasoning of the timber by removing the excess water by drying it in either a kiln or in the open air. This is carried out to increase the strength and stability of the timber and its resistance to decay (see Figure 3.1.6).

Plastics

The sources of plastics

Thermoplastic and thermosetting plastics cover a wide and diverse range of substances that exist in both a natural and synthetic form. Natural resources such as cellulose from plants, latex from trees and shellac, a type of polish extracted from insects, play only a small part in the plastics industry. Synthetic resources, especially crude oil, supply the majority of the raw material for the production of plastics. This single resource of hydrogen and carbon accounts for the majority of plastics.

The refining of crude oil in a fractioning tower is the process that gives rise to the product hydrocarbon naphtha, which is subsequently cracked into fragments using heat and pressure to form ethylene and propylene. In naturally occurring compounds, the molecules, consisting of only a few atoms, are short and compact. In plastics, the molecules do not stay as single units but link up with other molecules to form large chains of giant molecules. This process is called **polymerisation**.

Classification of plastics

Plastics are subdivided into two main groups and one minor group, with the formation of the chains the key feature that separates them:

1. Thermoplastics. These plastics are made up of long chains of molecules with very few cross-linkages. The smaller cross-links are known as monomers and the polymer chains are held together by a mutual attraction known as Van der Waals forces (see Figure 3.1.7a). This physical attraction is weakened by the introduction of heat. As the molecules move, they become untangled and the material becomes pliable and easier to mould and form. When the heat is removed the chains reposition and the material becomes stiff once again. Thermoplastics have a plastic memory, which means they have the ability to return to their former state after heating provided that no damage or chemical decomposition has happened during the heating process. Acrylic, polystyrene and polypropylene are all examples of thermoplastics. Polythene is extensively used in the production of toys and carrier bags; polypropylene is used for containers with built-in hinges and chair shells where its good resistance to work fatigue is exploited.

2. Thermosetting plastics. Thermosets set with heat and thereafter they have little plasticity. During the polymerisation process the molecules link both side-to-side and end-to-end. This cross-linking process, known as covalent bonding, makes for a very rigid material, and once the structure has formed it cannot be reheated and changed (see Figure 3.1.7b). Polyester resin is used for paperweights. Urea

Figure 3.1.7a *Van der Waals bonding – low density polythene*

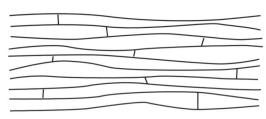

Figure 3.1.7b *Covalent bonding – polyester resin*

formaldehyde is a stiff, hard, strong plastic and it is used for electrical fittings.

3. Elastomers. This third group of plastics falls between the two basic groups. A limited number of cross-links allow some movement between chains. Rubber is a type of elastomer and it is used to make tyres for cars.

Acetate

Acetate, the collective name for cellulose triacetate and diacetate, was produced as a non-flammable material for the production of motion picture film. It is manufactured by reacting cellulose with acetic anhydride using a catalyst of sulphuric acid. The low flammability of acetate allows it to be used for a variety of purposes. It is still frequently used for motion picture films and overlay cells in animation. It is also commonly used for overhead transparencies and the preparation of images for printing.

Plastics in packaging

Plastics are widely used in packaging because they are:

- lightweight and versatile
- strong, tough, rigid, durable, impact and water resistant
- easily formed and moulded
- easily printed on
- low cost and recyclable.

Plastics can be identified by an internationally recognised coding system usually stamped on to the base of the package or on the label. This system enables plastics to be easily identified and sorted prior to recycling.

Each plastic has its own useful properties that make it suitable for use in different areas of the packaging industry (see Table 3.1.5).

SIGNPOST
'The Properties of Plastic' Unit 1 pages 18–19

Table 3.1.5 Common thermoplastics used in packaging

Thermoplastic	ID code	Properties	End uses
PET Polyethylene terephthalate	**1** PET	Excellent barrier against atmospheric gases and prevents gas from escaping package; does not flavour the contents; sparkling 'crystal clear' appearance; very tough; lightweight – low density	Carbonated (fizzy drinks) bottles; packaging for highly flavoured food; microwavable food trays
HDPE High density polyethylene	**2** HDPE	Highly resistant to chemicals; good barrier to water; tough and hard wearing; decorative when coloured; lightweight and floats on water; rigid	Unbreakable bottles (for washing-up liquid, detergents, cosmetics, toiletries, etc.); very thin packaging sheets
PVC Polyvinyl chloride	**3** PVC	Weather resistant – does not rot; chemical resistant – does not corrode; protects products from moisture and gases whilst holding in preserving gases; strong, good abrasion resistance and tough; can be manufactured to be either rigid or flexible	Packaging for toiletries, pharmaceutical products, food and confectionery, water and fruit juices
LDPE Low density polyethylene	**4** LDPE	Good resistance to chemicals; good barrier to water, but not gases; tough and hard wearing; decorative when coloured; very light and floats on water; very flexible.	Stretch wrapping (cling film), milk carton coatings
PP Polypropylene	**5** PP	Lightweight; rigid; excellent chemical resistance; versatile – can be stiffer than polyethylene or very flexible; low moisture absorption; good impact resistance	Food packaging – yoghurt and margarine pots, sweet and snack wrappers
PS Polystyrene	**6** PS	*Rigid polystyrene:* Transparent; rigid; lightweight; low water absorption *Expanded polystyrene:* Excellent impact resistance; very good heat insulation; durable; lightweight; low water absorption	*Rigid polystyrene:* Food packaging, e.g. yoghurt pots, CD jewel cases, audio cassette cases *Expanded polystyrene:* Egg cartons, fruit, vegetable and meat trays, cups, etc.; packing for electrical and fragile products

Task

State the impact on the environment that the production of PET bottles has – from production of plastic to disposal of bottles.

Metals and alloys

Classification of metals and alloys

The major proportion of all naturally occurring elements are metals and they form about one quarter of the Earth's crust by weight. Aluminium is the most common (8 per cent), followed closely by iron (5 per cent). With the exception of gold, all metals are found in the form of oxides and sulphates. The ores have no pattern of distribution around the world but some countries have larger deposits than others. Metals are divided into three basic categories:

1 Ferrous – the group which contains mainly ferrite or iron. It also includes those with small additions of other substances, e.g. mild steel and cast iron. Almost all are magnetic.
2 Non-ferrous – the group which contains no iron, e.g. copper, aluminium and lead.
3 Alloys – metals that are formed by mixing two or more metals and, on occasions, other elements to improve properties. They are grouped into ferrous and non-ferrous alloys.

The structure of metals

Metals usually have one or two loose electrons in their outer electron shell and therefore they are quite likely to become easily detached. The **metal crystals** have a regular arrangement held together by electrostatic attraction. It is this movement of electrons that accounts for metals' high electrical and thermal **conductivity**. This mobility also leads to a degree of plasticity in metals in the form of ductility and malleability. Once a bond is broken another is formed.

With the exception of mercury, all metals are solid at room temperature. In their molten form they are held together only by weak forces of attraction which mean they lack cohesion and will flow. As the metal solidifies, the energy is reduced within each atom, giving out heat, and the atoms arrange themselves according to a regular pattern or lattice structure. Their overall properties are affected by this lattice structure. Most metals crystallise into one of three basic types of lattice, as shown in Figure 3.1.8:

- close-packed hexagonal (CPH)
- face-centred cubic (FCC)
- body-centred cubic (BCC).

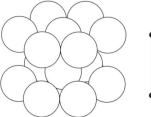

a) close-packed hexagonal (CPH)
zinc, magnesium
(weak, poor strength to weight ratio)

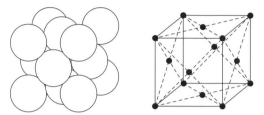

b) face-centred cubic (FCC)
aluminium, copper, gold, silver, lead
(very ductile, good electrical conductors)

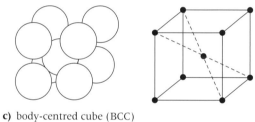

c) body-centred cube (BCC)
chromium, tungsten
(hard, tough)

Figure 3.1.8 Different metallic structures

Iron is a very important metal since it changes from BCC to FCC at 910°C. Above 1400°C it changes back to BCC again. In the FCC form it absorbs carbon, which is essential in the process of steel making. When cooling the changes occur in reverse.

A pure metal solidifies at a fixed known temperature with the formation of crystals, in either a cube or hexagonal structure. On further cooling, the crystals continue to grow as dendrites until each one touches its neighbour. At this point grains are formed and boundaries become visible when viewed under a microscope. Table 3.1.6 summarises the working properties and uses of common metals.

In order to obtain the metals in any useful form, they have to be extracted from the ore before processing can take place. Mining or

Table 3.1.6 *The working properties and uses of common metals and alloys*

Material	Melting point °C	Composition	Properties	Uses
Steels	1400	Alloys of carbon and iron	Dependent upon carbon content and other elements	
Low carbon steel		Less than 0.15% carbon	Soft, ductile, malleable	Wires, rivets and cold pressings
Mild steel		0.15–0.3% carbon	Ductile and tough; cannot be hardened and tempered	General construction steel, car bodies, nuts and bolts
High carbon steel		0.7–1.4% carbon	Hardness can be improved by heat treatment	Hammers, cutting tools and files
Alloy steels Stainless steel		Medium carbon steel + 12% chromium + 8% nickel	Corrosion resistant	Kitchen sinks, cutlery
High speed steel		Medium carbon steel + tungsten, chromium and vanadium	Retains hardness at high temperatures; brittle but can be hardened and tempered	Lathe tools, drills and milling cutters
Aluminium	660	Pure metal	Malleable and ductile; very conductive of heat and electricity	Aircraft, boats, window frames and castings
Duralumin		Aluminium + 4% copper + 1% magnesium	Work hardens, ductile and machines well	Aircraft parts
Copper	1083	Pure metal	Malleable and ductile; excellent conductor of heat and electricity;	Wire, central heating pipes and car radiators
Brass	927	65% copper + 35% zinc	Corrosion resistant; casts well; good conductor of heat and electricity	Casting, ornaments and marine fittings
Zinc	420	Pure metal	Ductile and easily worked; a layer of oxide prevents it from further corrosion	Coating for steel (galvanising), rust-proof paints, die casting

quarrying removes the ore from the ground, whereupon it is crushed to remove much of the unwanted earth, clay or rocks. The metal, now in a concentrated form, is roasted which causes the ore to change chemically into an oxide of the metal. The remaining stages of reduction break the chemical bond between the metal and the oxygen in the ore to leave a pure metal ready for further processing.

The production of ferrous metals – steel

Steel is produced from iron ore, which is widely found in the Earth's crust and mined. To produce steel, iron ore must firstly be processed into iron. The iron ore, limestone and coke are heated to very high temperatures in a blast furnace. Limestone is used to remove the impurities from the iron ore. The iron is then added to an oxygen furnace where it is converted into molten steel (see Figure 3.1.9). This molten steel is cast into ingots in preparation for further processing. In the case of the canning industry, a strip mill will

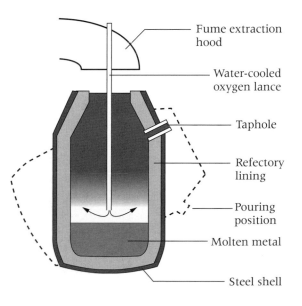

Figure 3.1.9 *The production of steel using a basic oxygen furnace*

- Fume extraction hood
- Water-cooled oxygen lance
- Taphole
- Refectory lining
- Pouring position
- Molten metal
- Steel shell

produce large coils of steel as a raw material to be used in the manufacture of food and drinks cans. One of the main problems with steel is that it can corrode. This is obviously a problem when making food containers, so to prevent it from corroding a layer of tinplate is added to its surface.

The production of non-ferrous metals – aluminium

Aluminium is a pure metal, which occurs naturally and is mined from beneath the land and sea. It is the most plentiful metal element in the earth's crust and occurs as the ore bauxite. Bauxite is very difficult to break down, so a process known as electrolysis is needed. The production of aluminium therefore requires large amounts of energy due to the expensive electrolytic process. There are two main stages:

1. The production of alumina from bauxite

Once the bauxite has been mined, crushed and dried it is refined into alumina. This is carried out in two stages. First, the bauxite is dissolved in hot caustic soda and then filtered to remove impurities, producing aluminium oxide. Then the aluminium oxide is roasted in a rotary kiln and a white powder, called alumina, is produced.

2. The production of aluminium using an electrolytic reduction cell

In the reduction cell the alumina is dissolved in molten cryolite using a steel furnace (see Figure 3.1.10). The furnace is lined with carbon (forming a cathode) and additional carbon rods (forming anodes) are suspended above the furnace. When a powerful electric current is passed through the heated mixture, aluminium is liberated from the alumina and deposited onto the carbon lining. This pure aluminium is periodically tapped off at the bottom of the furnace and is cast into ingots ready for further processing.

Task

Identify the materials that are used in the manufacture of a steel drinks can. Explain how the properties and composition of each material contribute to the function of the product.

Composites and laminates

Composites

When two or more different materials are combined by bonding, a composite material is formed. The resulting material has improved mechanical, functional and aesthetic properties and, as with most composites, it will have excellent strength to weight ratios. Composites consist of a reinforcing material that provides the strength and a bonding agent, termed the 'matrix', in the form of glues or resins.

Glass reinforced plastic (GRP)

Glass reinforced plastic (GRP), often referred to as fibreglass, consists of strands of glass that are set in a rigid polyester resin. The strands are woven into matting which is available commercially in different weights. The polyester resin, albeit a thermosetting plastic, exists in a liquid form that has a catalyst or hardener added to it along with a coloured pigment for decoration purposes. The glass fibre strands provide the basic strength while the resin with its additives bonds the fibres together and provides a very smooth surface finish.

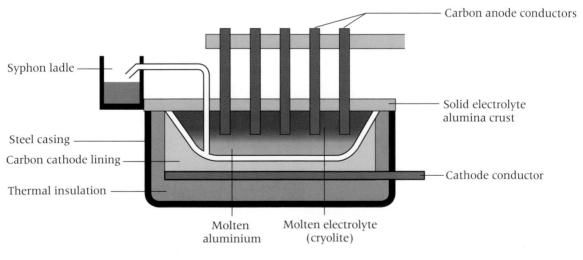

Figure 3.1.10 Electrolytic reduction cell

In order to achieve a high standard of finish from any GRP work, a high-quality mould must first be produced. The external or 'finished' side of the work must be finished to a very high standard since any defect or imperfection in the mould will be replicated in the finished piece.

The mould can be made from virtually any material but medium density fibreboard (MDF) and hardboard are often used with the additional use of wire and plaster of Paris for complicated shapes. If a porous surface has been used it is essential to seal the surface prior to use and a proprietary mould sealer should be used. It is common to make a full-size model of the finished piece which is then used to produce the GRP mould. In order to be able to remove the work from the mould it should be made with tapered sides and it should have no undercuts. A release agent is also essential and the mould should be coated several times before lay up proceeds. Good mould design should see no sharp corners and large flat areas should be avoided.

The stages in laminating follow a structured process. These are the basic stages:

- Polish mould with the releasing agent.
- Prepare matting into appropriate sizes.
- Mix gel coat with pigment, hardener and catalyst.
- Apply gel coat to an even thickness of 1 mm.
- Wait about 30 minutes before stippling matting over mould making sure it is wet through.
- Build up layers to the appropriate thickness.
- Full curing takes approximately 24 hours before separation can take place.

GRP is used in a vast range of products where its great strength to weight ratio can be fully utilised. This means that much stronger shapes and products can be built that weigh much less than when being produced by other means. It is also very resistant to corrosion. GRP is used to make sailing boats and canoes. Some high-speed train front nose cones are also made from GRP.

Carbon fibre

More recently, carbon fibre has been developed in a similar form to that of glass fibre. Carbon fibres are, however, much stronger and are used in structural components for aircraft, propellers, protective clothing, body armour and sports equipment such as golf clubs, skis and tennis and squash rackets. The extensive replacement of fibreglass with carbon fibre in the aircraft industry has led to major weight reductions of between 15 per cent and 30 per cent, which has resulted in better fuel economy. Carbon fibres are available in various forms but most frequently they are laid up using resin to produce strong lightweight structures.

Medium density fibreboard (MDF)

One of the most widespread and commonly used composite materials is medium density fibreboard (MDF). The fibres are made from wood waste that has been reduced to its basic fibrous element and reformed to produce a homogeneous material. Fibres are bonded together with a synthetic resin adhesive to produce the uniform structure and fine textured surface. There are various types of MDF board, which have a less dense central core but still retain the fine surface.

This type of fibreboard can be worked like wood and with a veneered surface it makes an excellent substitute. It finishes well with a variety of surface treatments and it is available from 3 mm to 32 mm thick and in sheets 2440 mm by 1220 mm wide.

As is the case with all composites there are some dangers involved in the use of them. As a result of the very fine fibres and synthetic resin adhesives great care must be taken when undertaking any form of cutting, drilling or sanding. Respiratory equipment should be used since the dust can cause irritation of the skin, throat and nasal passage and in school/college a dust extractor must also be used.

Laminates

Lamination is both a strengthening and decorative technique, which can be used to produce laminates in a range of materials and products.

Plywood

The development of plywood was an important process that improved the physical properties of timber. Plywood is made from thin layers of wood veneers, about 1.5 mm thick, called laminates. They are stuck together with an odd number of layers, but with the grain of each layer running at right angles to the last. This means that the two outside layers have their grains running in the same direction.

The interlocking structure gives plywood its high uniform strength, good dimensional stability and resistance to splitting. It is also available with a variety of external facing veneers and it can be made with waterproof adhesives, which means it can be used externally.

These veneers or laminates can also be stuck together over formers to produce curved shaped forms. This process is known as laminating and since different shapes can be formed, the strength of the material can be further enhanced by the shapes into which the material is formed. Once a former has been made it can be used over and over again to produce batches such as skateboard decks.

Thermoplastics

The lamination of thermoplastics, such as acrylic, can add strength (acrylic is extremely brittle) or more importantly a decorative edge. By gluing together several sheets of acrylic with Tensol cement, it is possible to create a multi-coloured 'liquorice allsorts' effect.

Paper and card

Laminating paper and card with polythene can enhance the appearance and add protection to printed materials. The laminated polythene surface can give paper a high gloss finish and enhance its strength and durability – especially when attempts are made to rip a piece of laminated paper. Laminates are also used in packaging to great effect where plastic films or coatings are added to cartons to give extra protection. A layer of polythene is added which protects the food or liquid contents from contamination by air or moisture. For an even longer life, a layer of aluminium foil is added.

Components

Pencils

Graphite pencils

Wood-encased graphite pencils and retractable graphite pencils (also known as clutch pencils) are available in a variety of grades, ranging from 8B, which is very soft, to 9H, which is very hard. The letter B denotes the blackness of the lead, while the letter H denotes the hardness. It is essential that the correct pencil is used for the task in hand, in conjunction with the correct paper.

Softer pencils, either used alone or in various combinations, are more suited to rough or grainy paper to create different effects when sketching or illustrating. Harder pencils are commonly used on smoother surfaces such as layout paper for technical drawing. HB or 2B pencils are the most suitable choice for rapid freehand sketching.

Coloured pencils

Coloured pencils are commonly made from a mixture of chemical pigment and kaolin, which is a type of clay. Professional quality, coloured pencils are available in an extensive range of colours, which vary in hue, tone and intensity. They are not usually graded in the same way as graphite pencils, but are often quite hard so that they can be sharpened to a fine point for use in detailed work. However, different types of coloured pencils can be used to create a variety of effects. Water-soluble pencils used in conjunction with a watercolour brush create colour blends and washes, while chalk-based, pastel pencils create the same effects as pastels and chalks but minimise the mess.

Pens

An extensive range of pens is available on the market, which are designed to fulfil a variety of functions. The majority of permanent or soluble ink pens are gravity fed. The design of the nib and the method of ink flow are the main features that distinguish these pens from one another.

Technical pens are predominantly used for precision and accuracy in technical drawing as they provide a consistent ink flow for detailed line work. The nib of the pen is made from steel tubing and ink is fed down it from a refillable cartridge. The size of the tube determines the width of the line and the interchangeable nibs range in size from 0.13 mm to 2.0 mm.

Ballpoint pens distribute ink via a steel or tungsten carbide ball that rolls within a plastic or metal skirt. The quality of the line will fluctuate depending on the type of ink-flow system used, for example some roller ball pens use hundreds of tiny beads behind the nib to maintain an even flow of ink

Plastic tip pens are used to produce accurate line widths, including very fine lines. A network of fine channels in the plastic nib is used to draw ink to the surface of the tip using a capillary action.

Fibre tip pens have a nylon or vinyl nib, which can vary from a firm tip for precise or finely detailed work to a suppler tip for sketching. The nib is usually constructed from synthetic fibres bonded in resin to make them more durable. The quality of fibre tip pens ranges from a standard fibre tip, colouring marker to examples similar to technical pens that use a tubular nib and are available in a similar range of nib sizes.

Fountain pens are traditional writing implements that have been used for many years; the design is derived from the quill. Fountain pens are predominantly used for writing, calligraphy and sketch work. The nib works by using a capillary action to draw the ink between a thin channel that looks like a hairline split in the middle of the nib. A reservoir of ink behind the nib maintains an even and constant flow. However, fountain pens are not always the most practical pen as they can release large deposits of ink if agitated.

Marker pens

The range of marker pens available is extensive. They can be divided into three categories:

1 Ink – the ink used in marker pens can be divided into two groups:

- water-based ink which is non-permanent and soluble
- spirit-based ink which is permanent and waterproof.

The drying times for these inks will vary but certain markers, usually spirit-based, will state if they are quick drying.

2 Composition of the nib – the most common materials used for the nib are:

- felt
- nylon
- fibre
- foam.

Felt nibs are easily damaged with heavy use whereas the other materials are designed to be more resilient.

3 Size and shape of the nib. These vary from fine to broad nibs and can create a number of effects and line thicknesses:

- round
- chisel
- square
- fine
- bullet
- brush.

In most marker pens the ink is soaked into a fibre core, which is in contact with the nib allowing the ink to soak up from the core to the nib. The most widely used marker pens for visual media and presentation work are known as studio or graphic markers. These may be referred to by their brand names such as Magic or Pantone markers.

Unlike alternative media such as watercolours or gouache, markers can be used instantly. They require no mixing or preparation and a wide array of colours is available. In addition to providing a comprehensive selection of hues and tones, another reason for this extensive variety of shades is to create matches with other materials like paper, overlays and printing inks. This allows consistency in colour at every design stage from initial sketches to the final product. Ranges of cold and warm greys are also produced to enable monotone visuals to be produced.

Studio markers are produced in spirit-based and water-based forms. Spirit-based markers are colourfast but tend to bleed and cockle paper if not used on marker paper. They give a flat, even colour allowing the colour to be applied in layers one on top of another. Water-based markers do not tend to bleed as much but in comparison to spirit-based markers they do not offer such flat and even coverage. In addition, water-based markers do not tolerate being overworked with layers of colour as this can result in a patchy appearance. As a result, spirit-based marker pens are the most commonly used.

Nib shapes for studio marker pens vary but the chisel tip is the most widely used as it offers both a broad line and a fine line when used on its tip. Many manufacturers offer fine tip markers for detailed work in the same colours as chisel tip markers, which has resulted in twin tip markers being produced with a fine tip at one end and a chisel nib at the other.

Task

Construct a series of pictorial drawings of a product of your choice, e.g. a mobile phone. Render using:
a) coloured pencils
b) marker pens.

Projected images using light

Projected images are used as a method of presentation and are extremely useful when displaying and communicating visual information to a large group of people. Two of the most commonly used projectors in schools and colleges are overhead and screen projectors.

Overhead projectors

Overhead projectors (OHPs) are used in conjunction with acetates or overhead transparencies (OHTs). The transparencies are placed on the horizontal glass surface of the projector and the light from the bulb underneath projects the image on to a screen using a mirror and a Fresnel lens to enlarge the image. Adjusting the height of the magnifying lens and the distance from the projector to the screen alters the size and sharpness of the image or text.

OHTs can be prepared using photocopying or printing methods. When using a black and white photocopier to prepare OHTs it is important that the correct acetate is used or it will melt in the process. Special inkjet or laser transparencies can be directly printed onto from a computer, enabling the use of colour. Permanent and water-based markers designed specifically for use with OHTs can be used to mark the acetates. This enables the user to add information to the OHT during a presentation to build up information, highlight or assist in explaining processes.

Screen projectors

Slide projectors use colour or black and white 35 mm photographic transparencies mounted in plastic or card frames. The projector passes light through the slide and through a lens that can be adjusted to focus the image on a screen. The distance between the screen and projector also determines the size and clarity of the projected image. Slides of artwork were usually projected because of their better clarity, as opposed to a colour OHT, but with the advent of digital technologies both types of projector are being used less extensively.

LCD projectors are now commonly used for presentation purposes, usually linked directly to a laptop computer for portability. The computer's LCD screen acts very much like a colour slide in a slide projector. The ability to create slides on computer, using software such as PowerPoint, has enabled people to quickly and easily produce high-quality presentations in full colour, incorporating text, images, animation, sound and even movie clips. LCD projectors use a bright light from a halogen lamp to illuminate the LCD panel. A lens in the projector connects the image from the image-forming element (in this case the computer screen), magnifies the image and focuses it onto a screen.

2. Working properties of materials and components related to preparing, processing, manipulating and combining

Aesthetic properties

The aesthetic value of a material relies upon how it is perceived in its appearance. An aesthetic judgement is made based on how pleasing the material is to the viewer. The materials used in the design and manufacture process may be evaluated on their own aesthetic appearance as well as the aesthetic contribution they will make to the product as a whole. The aesthetic properties of materials can be defined by their:

• colour
• style
• texture.

Colour

Most materials have an innate colour or colours. This can be enhanced or changed using various techniques (including heat, chemicals, pigments, dyes or natural finishes) to alter its appearance for functional or aesthetic purposes.

Style

The choice of materials used can also help the designer/maker to achieve a distinctive style for the product. A connection is made between the materials used and the aesthetic style that is being created; therefore, certain materials are more effective in conveying a style than others. For example, the manufacturer of traditional dining furniture would choose a material such as oak to achieve a traditional style rather than a contemporary or industrial material such as fabricated stainless steel. The use of stainless steel would communicate a very contemporary style to the viewer. However, a mixture of styles that would not normally be seen together or a traditional design that used contemporary materials would create a post-modern style for the product.

Texture

Like colour, most materials have an innate texture. The textual quality of a material can be achieved:

• visually – the look of a surface
• physically – the feel of the surface.

These textual qualities can be changed or enhanced for a desired effect.

Functional properties

Strength

The strength of a material is defined as the ability of the material to withstand forces without permanently bending or breaking. Strength can be broken down into five main areas:

• tensile strength – the ability to resist stretching or pulling forces
• compressive strength – the ability to withstand pushing forces
• bending strength – the ability to withstand the forces attempting to bend the material
• shear strength – the ability to resist sliding forces acting opposite to each other
• torsional strength – the ability to withstand twisting forces under torsion or torque.

Durability

The durability of a material relates to the type of object that is being produced. The choice of the material is dependent upon:

• the life span
• frequency of use
• demands placed upon the object (i.e. weather, corrosion).

A variety of materials can perform the same task. Therefore, when selecting materials, it is important to examine each material individually for performance and attributes.

Flammability

All materials can be changed and altered by the effects of heat. Materials that are susceptible to low heat or a naked flame are described as flammable. Paper, card and wood, due to their fibrous nature and low moisture content, are obvious candidates.

The shape and form of the material is also a determining factor in how quickly the material will burn. This is also accompanied by how well air can circulate in and around the material.

Regulations are stringent regarding flammable material in domestic and working environments. Materials that are flammable and release toxic gases when burnt need to be treated with fire retardant chemicals. Manufactured foams used in upholstery and soft furnishings are a good example of this type of material.

Mechanical properties

A mechanical property is associated with how a material reacts when a force is applied to it.

Plasticity – a material's ability to change shape permanently, when subjected to external force, without cracking or breaking.

Ductility – deformation through bending, twisting or stretching. Copper, aluminium and silver are all very ductile and can be drawn though a die into thin wires.

Hardness – a materials resistance to abrasive wear and indentation, which is important for cutting tools.

Malleability – the extent to which a material can undergo permanent deformation in all directions, under compression, without cracking or rupturing.

Please refer to 'Standard performance tests' on page 102 for a series of workshop tests for determining the mechanical properties of materials.

3. Hand and commercial processes

Drawing

Drawing can be broken down into pictorial, information and working drawings.

For all of these types of drawing the basic equipment used is a 90-degree set square with measurements and a drawing board with parallel motion or a T-square. These are used to ensure accuracy and consistency throughout the drawing process. Additional equipment may be needed such as a compass, protractor, etc.

Pictorial drawings

Pictorial drawing consists of a number of methods used to visualise objects in 3D form.

Perspective

Perspective allows an object to be drawn as it is viewed by the human eye. Parallel lines appear to converge at a vanishing point (VP) the further away they are. Lengths, heights and widths will appear to foreshorten as they recede into the distance. There are three types of perspective:

* **1-point perspective** – only one VP is used in relation to length or width
* **2-point perspective** – two VPs are used, one for length and one for width
* **3-point perspective** – three VPs are used for length, width and height (commonly used for drawing buildings).

The position of the VP will be placed on a horizon line or on eye-level. For length and width the VP must be on the same line. In the case of a third VP, it will be placed above or below the horizon line (depending on the view) and in relation to the centre of the object.

Where the object is drawn in relation to the horizon will determine what view is required (see Figure 3.1.11):

* Worm's eye view – the object is above the horizon and the underneath will be seen.
* Street level – the horizon line passes through the object; the top or underneath cannot be seen.
* Bird's eye view – the object is below the horizon line allowing the top to be viewed.

Isometric

Isometric drawing is an accurate method of showing the three faces of an object. All lines are

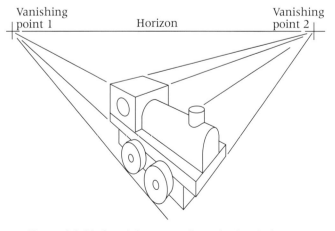

Figure 3.1.11 2 point perspective of a toy train

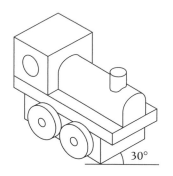

Figure 3.1.12 Isometric drawing of a toy train

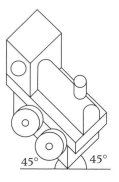

Figure 3.1.13 Planometric drawing of a toy train

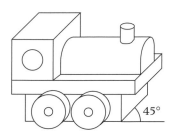

Figure 3.1.14 Oblique drawing of a toy train

true or scaled lengths; therefore foreshortening does not occur. All vertical lines are drawn 90 degrees to the parallel rule and horizontal lines on the object are drawn at 30 degrees to the vertical using a 60/30-degree set square. Isometric paper can be used to assist the process.

In addition, circles can be drawn using an isometric ellipse template, which offers a range of sizes. Compasses can be used for larger circles to strike arcs (using the correct method). Irregular curves can be achieved with a French curve or flexi-curves.

Planometric

Planometric, also known as axonometric, technique is often used for interiors and buildings as it allows a bird's eye view of the object.

The plan view is drawn 45/45 degrees or 30/60 degrees to the horizontal. All dimensions are true or in scale, thus circles and curves are drawn as they appear in view.

Vertical lines can also be drawn in the same way, but they can appear distorted and too high. Therefore, it is acceptable to reduce the vertical scale by three-quarters, two-thirds or even one half in relation to the plan.

Oblique

This method is not often used and has largely been superseded by the methods outlined above. Oblique projection is similar to planometric in that the face (front or side views) is true. The receding lines can be drawn at 45, 30, 60-degree set square angle from the face.

The face is drawn from a horizontal line with the vertical lines being 90 degrees to this line. The receding lines are drawn at half size to prevent the distortion of the drawing.

Enhancement

Enhancement techniques are used in both sketch work and formal pictorial drawings to produce a more realistic image and assist the viewer in

understanding the ideas being communicated. It can be divided into four key areas:
- line
- tone
- texture
- colour.

Line

Thick and thin line technique is a quick and effective way of enhancing a drawing. The outside edges or the lines that meet the background of the drawing are marked with a thick line. In contrast, the internal lines are thinner. The drawing is done in one line width (thin line), then a pen with a noticeable difference in nib width is used to trace over the outline of the drawing. This effect can also be achieved with a pencil if enough pressure is added.

Tone

All 3D objects are subject to a light source that causes the object to have a tonal range from light to dark. The surface closest to the light source will be the lightest, with its opposite side or the surface farthest away from it appearing the darkest. The 'tonal rendering' technique mirrors this, making the drawing appear more solid and realistic. When tonally rendering a drawing a decision must be made regarding where the light source is coming from. This is usually from over your left shoulder but it can be changed to vary the effect. Objects with flat planes (i.e. a cube) can be rendered with three separate tones: light, medium and dark. Surfaces that are curved (i.e. a cylinder) will show a more distinct tonal range that graduates from light to dark. The drawing can then be rendered in one of two ways:

- Graduation – a smooth and subtle transition from light to dark where the intensity of tone can be varied depending on the medium used (pencils, watercolour, acrylic, gouache, airbrush, marker pens, etc.). Pencils create a gradual blend through the varying amounts of pressure applied during application and paints can be altered through mixing and dilution. Marker pens can build tone in layers of application or by using a range of tones within the same colour.

- Solid – this method is used for media that do not provide subtle gradations in tone (pen, ink, dry transfer, etc.). It relies upon the use of application techniques to represent light and shade in alternative ways. For example, solid lines or dots can create a gradation depending on how closely they are spaced together or apart.

Texture

Incorporating the look or feel of a surface into a drawing can help to identify the material being portrayed. This can be achieved in monochrome, but the use of colour will add visual impact and along with tone will assist in building a realistic image. Various techniques and types of media are used in combination to render drawings in the likeness of materials such as wood, metal, plastic, glass, concrete and fabric. Some types of media are more effective in representing materials – for example, the reflective surface of chrome can be represented using an airbrush. Textures can also be applied using dry transfer sheets and computer-generated textures. These techniques are best suited to cover large areas.

Colour

Colour is used to enhance drawings aesthetically but it is also a useful tool to convey information. This can be done in a number of ways:

- Colour can be used to highlight sections of an illustration, drawing the viewer's eye to key areas of the visual.
- It can also be used to assist the viewer in understanding the information presented. This can range from the inclusion of simple directional arrows to the colour coding of a complex system, which distinguishes components or identifies working processes. Colour code systems are effective when used in conjunction with a key.

Information drawings

Information drawing is a process that is used to display data visually. The pictorial representation of data helps to simplify information, making it more immediate and comprehensible to the viewer. Information drawings can be displayed in a number of forms:

- charts
- tables
- diagrams.

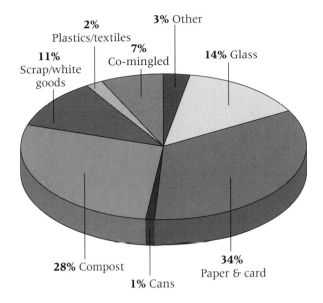

Figure 3.1.15 *Pie charts are used to dispay data visually*

The data has to be classified into groups or sections, and they need to be quantified by a number, scale or percentage. This allows the data to be displayed so that the sections are visibly comparable.

The use of icons, colour, texture and tone can assist in distinguishing sections, along with 2D and 3D forms. It should also be appropriate to the type of data being displayed.

The use of ICT software, such as spreadsheet packages, can produce a variety of pictorial information quickly and easily. Once the information has been entered, the user can cross-reference it with other criteria. This allows a wider analysis and a range of varying results. The outcome can be displayed as bar, pie and line charts.

Working drawings

Working drawings are produced to display all of the required information of a product in 2D form so that it can be produced or manufactured. The range of drawings includes:

- orthographic projection
- sectional views
- assembly drawings
- parts drawings.

A standardised method of producing working drawings has been defined by the British Standards Institution (BSI). All working drawings are laid out with a margin and title block. The title block should include:

- name
- date

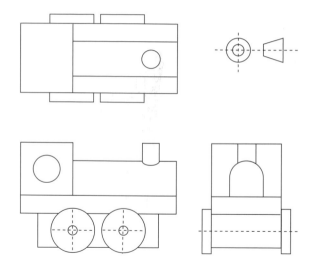

Figure 3.1.16 *3rd angle orthographic drawing of a toy train*

- title
- projection used
- scale
- drawing number.

Letters and numbers are written in block capitals to avoid confusion; letter stencils are used to ensure consistency. Drafting film is the best paper to use as it allows repeated corrections to be made in both pencil and ink. A hard pencil (2H) should be used to draw the object and construction lines. When the drawing is correct the object should be lined in with a technical ink pen. Photographic reproduction (blueprint/dye line) of the drawings can be made. This allows the original to be kept neat and untouched.

Task

Evidence a range of drawing techniques in your coursework project folder, including pictorial and working drawings of your final product. Assembly and exploded drawings can provide more information about the manufacture of your product.

Computer graphics

The use of computers for designing printed materials and products is now the industry standard, replacing the need for laborious and time consuming manual drawing and cut-and-paste techniques. The possibilities created by this new technology have led to rapid developments in the printed media, the Internet, animation and special effects for TV and film.

Desktop publishing (DTP)

Desktop publishing combines the features of word-processing, graphic design and printing all in one package. The modern print media all use DTP to produce the layout of their pages digitally, rather than the traditional process of typesetting. The simplified DTP process for producing a page layout consists of the following:

1 Text is either imported from a word-processing package or typed directly into the DTP package.
2 Illustrations or graphics are created within a specialised package, or scanned and manipulated as required.
3 The text and graphics are combined within the DTP package and inserted into a predetermined grid layout. Elements are cropped or manipulated to best fit the page layout and add visual impact.
4 A proof copy of the page layout is printed in colour for evaluation and/or modification.
5 The final page layout is sent digitally to the printers for a printing plate to be made.

Advantages of using DTP for modifying designs
- Text and images are easily manipulated.
- A range of different typefaces is widely available.
- Graphic tools are easy to use.
- Page layout grids and guides can be added on screen (they are invisible when printed).
- Cut and paste enables fast changes to page layouts.
- Can zoom in and out to facilitate detailed work.

Tasks

1 Using a page from a magazine, draw in the grid the designer may have used to construct the page.
2 Using a template on a DTP package, produce a business card promoting yourself as a graphic designer.

2-dimensional computer graphics

2D computer graphics and digital illustration are rapidly growing areas in the field of design. Many drawings that would have traditionally been hand drawn and rendered can now be easily achieved with the use of the appropriate software. These 2D software packages offer a range of tools to construct and manipulate images. The basic tools allow shapes and lines to be drawn, to which colour and texture can then be added. Effects and filters can be applied to

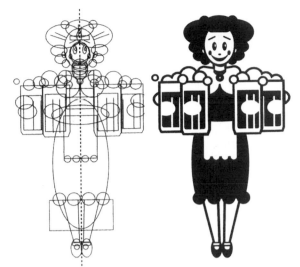

Figure 3.1.17 *Creating a 2-dimensional computer graphic from simple geometric shapes*

Figure 3.1.18 *3-dimensional computer-generated character for a computer game*

vary the look, such as blur or mosaic. Digitised images, such as scanned pictures and existing graphics files, can be combined and manipulated in the same way.

Many of the contemporary characters designed for T-shirts, album covers and illustrations, etc. are simply constructed using basic geometric shapes layered on top of each other with different colour fills.

> ## Task
> Have a go at producing simple cartoon-style characters using basic drawing tools and geometric shapes.

3-dimensional computer graphics
A 3D image can give a more realistic impression than a 2D image. Therefore, many designers will construct new products on screen, which, with skill, can easily be modified and manipulated. Product design teams can significantly decrease the time taken to design and develop a new product with 3D modelling, therefore saving developmental costs and time to market. These virtual products can be tested and evaluated without actually being manufactured, and design data can be directly outputted to a CAM system for modelling prototypes.

The rapid development in computer gaming has produced an impressive range of computer-generated characters interacting with life-like scenarios and landscapes. Rarely is a feature film released nowadays that does not contain computer-generated special effects. 3D animation starts with a wire frame model being created on screen, which can be viewed in all directions by the use of built-in camera angles in the computer software. The wire frame model can then be rendered, or given an appropriate surface finish or texture. This involves wrapping a 'skin' around the wire frame to give a photo-realistic image.

Typography
Typography is the way of designing communication by means of the printed word. This is done predominantly on the computer in desktop publishing applications such as Quark. Before this is done preliminary ideas will have to be produced to ensure an overall visual consistency. The four major elements are as follows.

Layout
The layout of the page will have to be consistent if several pages of text are produced. Therefore, the setting of the page size, margins and number of columns will have to be set as a template for the text to be applied. The position of the page number and footline (the chapter or title) will also have to be positioned and set.

Text
For the main body of the text an easy-to-read font should be chosen and maintained throughout. The size of the font is dependent upon what is being produced. For layouts where there is a bulk of text 10- or 11-point fonts should be used.

Headings and subheadings

The purpose of a heading is to draw the reader's attention to an article or text and give an impression of the theme or content. A subheading will draw the reader's attention further and perhaps give more information regarding the text. Think about the way headings and subheadings are used in newspapers and magazines. Headings and subheadings can also be used to break up the layout of a page or illustrate where a break has occurred in the text.

The typeface used for a heading is often the same as that used in the main body of the text. It can be different but the general rule is to restrict the use of typefaces to two. There are a number of effects or adjustments that can be made to create a contrast between the heading and the text:

- size
- italics
- lower and upper case
- colour
- character shape – serif/sans serif
- character width – expanded or condensed
- position on page
- density – solid/outline/positive/negative.

The subheading is secondary typography (heading is first, subheading is second, text is third) and should be smaller than the heading. There are two types:

- External – these subheadings appear directly beneath the heading to add further detail or explanation. They might also be placed in the margin alongside the text.
- Internal – these subheadings appear in the text at the beginning of a paragraph or section. They can be distinguished in the same manner as the heading. Lines or boxes can also be used to add emphasis.

Captions/breakouts

Captions assist in the explanation of photographs or images that appear within the text. The style of captions should be consistent throughout to give a distinctive and uniform look to the publication. The position of the caption is important as it should be near to the image but separated from the rest of the text to avoid confusion. The caption is usually placed directly beneath the picture and can be set apart with a line or box.

Typeface design

Typeface characters sit on an invisible baseline that does not move even when the font is enlarged or reduced (see Figure 3.1.19). Only the ascender and the descender lines shift when the font size is changed.

A huge range of fonts is available. A distinction can be made between fonts that are:

- serif – contain small counter strokes drawn perpendicular to the free ends of the character
- sans serif – characters without the stroke.

Several fonts are derived from a type family; these families consist of a number of typefaces that originate from the same style but have distinct variances like weight and proportion. For example, Latin bold, Latin condensed and Latin wide could be used for headings, subheadings and body text to achieve distinction but maintain visual grouping. Fonts can also be customised or created using software applications like Fontographer.

The size of font is measured in 'points', of which there are two types used:

- pica is equal to 0.35135 mm, which is divided into 12 picas
- computer (used in desktop publishing) is equal to is 0.3527785 mm (1/72 inch).

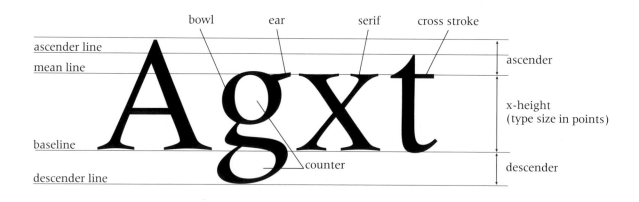

Figure 3.1.19 *Structure of a font*

The choice of font used will depend on its legibility as well as its aesthetic appearance. Legibility is the ease and speed by which the reader can decipher each character and word. For example, lower-case letters are easier to read than upper-case letters (capitals). The spacing between words should be the width of the letter 'i' and there should be more space between individual lines than words. The spacing can be adjusted between letters and lines by the following methods:

- Leading – the total distance from baseline to baseline. The term leading comes from traditional typesetting when strips of lead were inserted between the lines of type to fix the spacing. On computer applications leading is set at 120 per cent of the point size used for the font. In characters of higher points (such as those used in headings) 120 per cent may appear excessive and can be altered to 100 per cent or less.
- Tracking – controls the distance between each letter within a word. Tight tracking is used in large font sizes for headings or headlines, as they can appear too long or take up too much space on the page layout.
- Kerning – pairs of letters such as AV, ay, To, TT, wy, etc., appear to clash when they are placed together. Kerning is used to correct the visual disparity between these characters, particularly in large font sizes. Kerning enlarges or reduces spacing between these letters to improve their appearance.

Modelling and prototyping

Block modelling

Block modelling helps to determine shape, dimensions and surface details by constructing an accurate representation of the final product, usually out of styrofoam or a similar compliant material. Block models can be extremely useful in determining the ergonomic factors of many products. By constructing a number of block models of varying shapes and sizes it is possible for designers to literally get a 'feel' for the product. It will soon become apparent in 3D form the designs that are aesthetically pleasing or 'user friendly' and are worth developing – something that 2D images struggle to achieve.

Architectural models

Scale models of buildings and interiors are produced to allow the detail of architectural drawings to be visualised in a 3D form. Architectural models can vary from simple constructions made from white card to elaborate and highly detailed models that include surface

Figure 3.1.20 Block modelling at the early stage of the development of the Dyson vacuum cleaner

finishes or effects that give the viewer an idea of the materials to be used. Card and its derivatives are most commonly used as they can be shaped easily with basic equipment. With its plain surface, card also has the advantage of having no indication of scale; other materials such as wood have natural surface markings like grain and therefore would be out of scale in a 1:500 model.

Architectural models are often made with removable parts or sections so that the viewer can see the internal layout. Accessories and parts can be bought to assist in the construction of architectural models to save time while still including detail. These include scaled models of people, vehicles, shrubs, etc. or scaled surface effects such as paper printed with brick or tile designs.

Prototypes

Prototypes are the culmination of the design and modelling processes as they offer the most accurate representation of the final product.

Figure 3.1.21 A working prototype of the Dyson vacuum cleaner

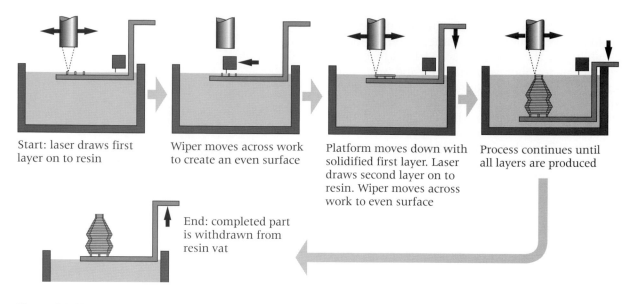

| Start: laser draws first layer on to resin | Wiper moves across work to create an even surface | Platform moves down with solidified first layer. Laser draws second layer on to resin. Wiper moves across work to even surface | Process continues until all layers are produced |

End: completed part is withdrawn from resin vat

Figure 3.1.22 *The process of stereolithography*

They are made from the same materials and components as the end product and include all external and internal details. Prototypes are usually made by hand to ensure that aspects of the design and production are correct before investment into tooling and machinery is undertaken.

Rapid prototyping using CAD/CAM

The need for manufacturing industries to reduce the time and costs involved in developing a new product has led to the use of rapid prototyping. This involves the creation of 3D objects using laser technology, which solidifies liquid polymers or resins in a process called stereolithography. In this process specialist software applications are downloaded on to a stereolithography machine, enabling 2D CAD drawings to be converted to 3D models. The process is based on the computer application slicing the 3D object into hundreds of very thin layers (typically 0.125–0.75 mm thick) and transferring the data from each layer to the laser.

The laser draws the first layer of the object on to the surface of the resin, causing it to solidify. The layer is supported on a platform, which moves down enabling the next layer to be drawn. This process of drawing, solidifying and moving down, quickly builds up one layer on top of another until the final 3D object is achieved. Most companies do not have this technology. Instead, they use rapid prototyping services from specialist companies. Stereolithography prototypes can typically be delivered within three to five days of the client's design data being received, therefore saving both time and development costs.

> ## Task
> Discuss how models contribute to the design process.

Photography and photographic processing

Photography

Cameras use three basic features to capture an image. These features may be adjusted manually or automatically depending upon the type of camera and the effect desired by the user.

Focusing

Adjusting the focus ring on the lens enhances the sharpness of the image. Various visual aids are incorporated into the lens design to assist the user to achieve sharp focus manually.

Shutter speed

The shutter speed and aperture both control the amount of light reaching the film. They are used in combination to affect the overall look of the photograph.

Shutter speed is adjusted by a control on the camera. Markings displayed are in fractions of a second (i.e. 60 is 1/60 of a second). This indicates the length of time the shutter will be open for and thus how long the film is exposed to the image. Other markings on the dial are:

- 'B' – allows shutter to remain open while button is pressed
- 'T' – allows shutter to open when button is pressed, and close when pressed once more

- 'Flash', which is the shutter speed to be used in conjunction with a flash.

The shutter controls the sharpness of the image. Slow speeds (60 and under) may result in blurring or camera shake unless the camera is in a fixed position. If the image is of a moving object, then quicker speed (500) will need to be used to capture the object as a still image.

Aperture

The aperture also controls the amount of light reaching the lens and affects the appearance of the picture. It is adjusted by a ring on the outside of the lens, marked with a series of numbers. The smaller the number, the wider the aperture, resulting in more light entering into the lens. The aperture also controls the depth of field (i.e. the distance between the subject and the camera). The further away the subject is, the greater the depth of field, and vice versa. The wider the aperture is, the less the depth of field and, conversely, the smaller the aperture, the greater the depth of field. This principle can be utilised to ensure that either the majority of the picture is in sharp focus or only a small section.

Together, the shutter speed and aperture control the amount of light and, therefore, allow the film to be correctly exposed. This is essential in producing a good negative. Most cameras will have a built-in light meter. The type of system used is dependent on the camera, as some automatically set both shutter and aperture. Others may just set one. Priority may have to be taken of one over another. For example, the shutter speed may have to be fixed to avoid camera shake; or the aperture closed for a greater depth of field. The speed of the film used will also determine adjustment.

Photographic processing

Black and white film

There are four stages in processing black and white film:

- development
- wash or stop bath
- fixation
- washing and drying.

First, the film is loaded on to a spiral. This has to be done in total darkness or in a changing bag. The spiral is then loaded into a development tank; this will allow liquid to be poured in without any exposure to light.

The correct amount of development solution for the kind of film and tank size is poured in at a temperature of 20 °C. This temperature should be maintained throughout the process. The length of time the film should remain in the tank is dependent upon the processing instructions for the film and the development solution. The development should be no shorter than five minutes. The tank will also have to be agitated for a duration of 15 seconds every minute. This is done to ensure an even development result.

Once the development process is complete, the solution is poured out of the tank. The stop bath is then poured in, agitated and left for 15 seconds. The solution is poured back into the container and then the fixer is added.

The time period for fixation will be outlined in the instructions. Agitation should take place every minute, as described in the development. The fixer can be poured back into the container for reuse. The number of times the solutions can be reused will also be mentioned in the instructions. The temperature of the stop bath and fixer should be within 4–5°C of the development solution.

At this stage, the spiral with film needs to be washed thoroughly with running water for 30–40 minutes. If the water is too cold it will cool the film too quickly resulting in marking. A gradual reduction in the film temperature to the level of the wash water is needed. After washing, the film can be removed from the spiral and hung in a dust-free environment to dry.

Colour film

Colour film development is a longer and more complicated process. It is usually done by machine. The film consists of three layers of black and white emulsions, which are blue, green and red sensitive. The film is produced as a black and white negative using silver halide. When this is removed a colour negative remains.

Production of nets

Constructing nets

A net, also known as a development, is a flat 2D shape than can be cut, scored and folded to produce a 3D shape. It is used when working with sheet material such as paper, card and board, metal and plastic.

The accuracy of the drawing is extremely important when constructing nets. When drawing nets, technical drawing equipment or CAD programmes should offer accuracy and consistency. To produce an accurate net the final 3D shape will need to be developed and drawn, either by hand or on computer. This will enable the shape, size and layout of the net to be drawn more easily. The net will need to show the following constructional information:

- Cut lines – a continuous line where the material is to be cut.

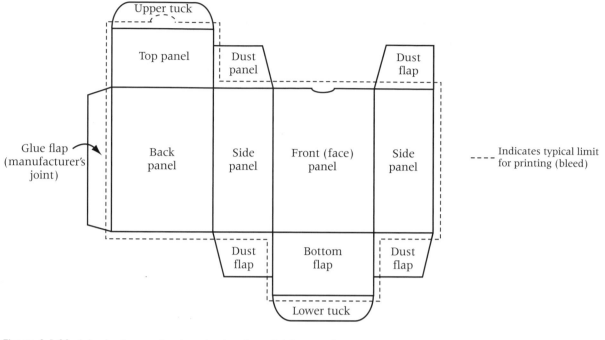

Figure 3.1.23 *A typical manufacturer's plan for a folding carton*

- Fold lines – a broken line where the material is to be scored, folded, bent or heated.
- Tabs – essential constructional information especially for paper, card and board, showing where glue is to be applied, or where dust flaps or tucks are required.
- Annotation – will assist in labelling edges, sides and features in relation to one another.

Structural packaging design

Nets are widespread in the production of packaging using cartonboards. The packaging of a product is extremely important as it carries all of the necessary information for the consumer to access and provides that vital first impression. Many companies produce standard packaging nets for designers to adopt and simply add individual graphic identity. This ultimately will save valuable time and costs in developing a product to market. There is also an internationally recognised system using diagrams of net constructions and symbols, which avoid the need for lengthy and complicated verbal descriptions especially useful in the global marketplace.

Commercial production of packaging nets

Making the die form

Packaging nets or cartons need to be cut to shape after printing, and before assembly or gluing. The cutters used are called dies and the stamping process is called die cutting. Making a die form is still a skilled hand-crafted process, which involves cutting and shaping hardened steel rulers and matrix strips that will either cut or crease the cartonboard on the die cutting machine. Ejector rubbers are added to the cutting rulers in order to push the cartonboard away from the rulers after a cut has been made. The die form is then mounted on the die cutting machine. This process will be the major cost in the setting up of the carton making job.

Die cutting

While the die is being made the cartonboard sheets for the job are selected and cut to size on a power guillotine. The cartonboard is then printed using the appropriate printing process, such as offset lithography.

After the die-form is mounted on the machine, a test sheet is cut and adjustments and corrections made. Once satisfied, each printed sheet is fed onto a platen, directly below the die form. The platen is then forced upwards, pushing the cartonboard hard against the rulers. The cutting rules cut right through the cartonboard, while the creasing rules push the card into the groove formed by the matrix strips. Once cut, the sheets are pulled to the next stage to be stripped of their waste. Ejector pins on the die form push the waste away onto a conveyor belt for disposal, whilst the stripped sheets leave the machine and are stacked ready for gluing.

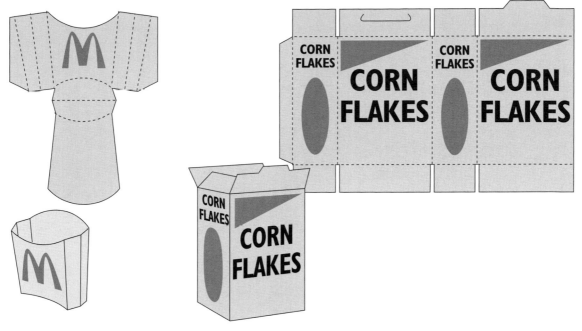

Figure 3.1.24 *Commercially available packaging nets and their products*

Folding and gluing

The carton blanks are then folded and glued on a highly automated gluing line. The first operation is to pre-fold one of the creases, and then the carton will travel to an automatic gluing station. The gluing module is 'timed' by a simple control system. When the front edge of the card blank breaks a light beam it triggers a device that squirts tiny drops of glue in quick succession. These form a row of adhesive that will soon be used to fasten the assembled carton. The final fold is made and the edges are held together under moving belt until they reach the final stage. The cartons are stacked up and compressed, using carefully controlled pressure, under another pair of slow-moving rubber belts. Here the adhesive has time to cure, before the cartons are removed and packed for transit. The cartons are then shipped to the customer's site for assembly and filling.

Task
Study a piece of packaging and, without taking it apart, try to sketch the net it is constructed from.

Working with materials

Resistant materials are an important aspect of the Graphics with Materials Technology course. Designers need to be able to determine the most suitable manufacturing techniques and processes in order to fully design and develop any product. Working with a range of resistant materials must be evident in product development Units 2 and 5; therefore a basic working knowledge of materials and processes is required.

Wasting processes are those that literally produce waste material. They include such processes as sawing, drilling, planing, chiselling, turning, filing, milling and screw cutting. Almost any shape can be cut out of a solid block of material using a wasting process. It is the role of the designer to adopt the most cost effective method of manufacturing their product by minimising waste and with a view to recycling materials.

Abrasive papers are used to produce a high-quality finish on a range of materials. As a general rule of thumb, it is best to use glasspaper on woods, emery cloth on metals and wet and dry paper on plastics. Abrasive papers can be attached to power tools such as disc, belt and orbital sanders when working on larger areas of materials.

Bending and forming processes can be used on most materials. Thermoplastics, for example, can be heated and bent using jigs and formed into interesting shapes using formers.

Casting

Casting is the process of pouring molten metal into a mould or pattern. In sand casting, damp

Figure 3.1.25 *Die casting is used to make products such as toy cars*

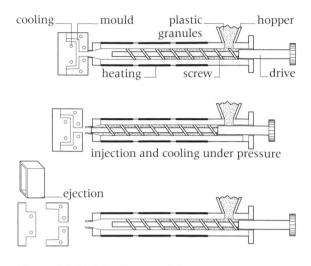

Figure 3.1.26 *Injection moulding*

sand is compacted around the mould or pattern. The mould is then removed to form a cavity into which molten metal can be poured. Most patterns are split along the centre line and are aligned using locating pins. Casting requires high temperatures to melt the metal and is often carried out within schools and colleges using aluminium, higher temperatures are needed to cast metals such as brass or bronze. Die casting uses low temperature alloys that are usually zinc based. There are two different methods of die casting: gravity die casting and pressure die casting. In gravity die casting, the molten metal is poured into the mould and gravity helps the metal to flow around the mould. In pressure die casting, the molten metal is forced into the die under pressure. This method is used where there are quite complicated products and components being cast.

Moulding
Injection moulding
Injection moulding is the most widespread and versatile process used for producing moulded plastic products, from bowls to television casings. Although the cost of a typical mould is high, the unit cost per component produced becomes very small for high volumes making it an ideal process for mass-produced components. The process is best suited to thermoplastics, but a few thermosetting plastics are used depending on the conditions the product is to be used in.

The process is simple. The material is heated to a plastic state and injected into an enclosed mould under pressure. The mould is opened and the product is removed with the use of ejector pins. Injection moulding is a highly automated process that produces high-quality products that require no further finishing other than to remove any sprue pins, gates and runners, which are chopped off and reused. The machine itself

consists of a hopper into which are fed the plastic granules, a heater and a rotating screw mechanism. The screw mechanism acts as a ram that injects the plasticised material into the mould before it is allowed to cool (see Figure 3.1.26).

Blow moulding
Blow moulding is the process used to form hollow products and components. A hollow length of plastic called a parison is formed by extrusion and is lowered down between an open split mould. The mould is then closed to seal up the free end of the parison and compressed air is blown into the mould forcing the plastic to the sides of the mould cavity where it is chilled and sets (see Figure 1.14 on page 29). Blow moulding is a highly automated process that produces little waste and requires only the flashing to be trimmed. An estimated 1.5 billion PET (polyethelene teraphthalate) plastic bottles are made and thrown away each year in the UK. Like injection moulding, the initial mould cost is high, as is the machinery, but with components being produced in the volume of PET bottles, it is easy to see how the unit cost is very low.

Vacuum forming
Vacuum forming is used to produce simple shapes from thermoplastic sheets. It is possible to vacuum form acrylic. However, the ideal degree of plasticity in acrylic is not reached until a temperature of 180°C. At 195°C acrylic starts to degenerate, therefore making it difficult to achieve a uniform heat across the whole sheet. However, 'Perspex TX' is an extruded form of

Table 3.1.7 Plastic moulding processes

Process	Advantages	Disadvantages	Plastics used	End uses
Injection moulding	Ideal for mass production – low unit cost for each moulding for high volumes Precision moulding – high quality surface finish/texture can be added to the mould	High initial set-up costs – mould expensive to develop and produce	Nylon, ABS, PS, HDPE, PP	Storage and packing containers; casings for electronic products
Blow moulding	Can form intricate shapes; can form hollow shapes with thin walls to reduce weight and cost of material; ideal for mass production – low unit cost for each moulding	High initial set-up costs – mould expensive to develop and produce	HDPE, LDPE, PET, PP, PS, PVC	Plastic bottles and containers of all sizes and shapes, e.g. soft drinks bottles and shampoo bottles
Vacuum forming	Ideal for batch production – inexpensive; comparatively easy to make moulds that can be modified	Mould must be accurate to prevent webbing occurring; large amounts of waste material produced	Acrylic, HIPS, PVC	Chocolate box trays; yoghurt pots, blister packs, etc.

acrylic and this becomes plastic at 150 °C making it more suitable for use in industry when products as large as baths can be formed. More commonly used materials include high-density polystyrene, ABS (acrylonitrile-butadiene-styrene) and a flexible grade of PVC (polyvinyl chloride).

Vacuum forming requires a mould of the finished component to be produced first and this must be to a high standard with the sides tapered slightly (to between five and ten degrees) to ease the removal of the formed component once completed. The process works by removing the air trapped between the mould and the sheet, thus reducing pressure below the trapped material. Atmospheric pressure pushes the heated plasticised sheet on to the mould. The plastic sheet to be vacuum formed should be clamped around its edges in an airtight plate with a rubber seal. Heat is applied by elements normally housed in a hood that is held above the material.

When a material has reached its plasticised state the heaters are removed and the pump is used to expel the air below. There is, however, one basic problem with vacuum forming, which is that on deforming the material can become quite thin. One technique used in industry to overcome this problem is that when the material reaches its plasticised state, it is blown to uniformly stretch the whole surface before the mould is raised on the table and the air expelled.

The whole process can be summarised into four basic stages:

1 The plastic is heated using radiant heaters.
2 The plasticised material is then blown to stretch it.
3 The table or platen is raised into the dome area.

4 The air is expelled causing atmospheric pressure to force the plastic material down on to the mould (see Figure 1.15 on page 29).

Vacuum-formed products range from acrylic baths to the plastic packaging found around Easter eggs.

Task

Critically assess the methods used in a variety of plastic moulding processes, using the questions below to guide you. If possible, watch a video or access a CD-ROM that shows industrial and commercial processes. Also try handling some products made by these processes.

- How is the force applied?
- How is the heat applied?
- How easily is the process controlled?
- How complex is the mould?
- Is any finishing necessary?

Joining

Joining processes can be categorised as follows:

- Permanent – once made they cannot be reversed without causing damage to the work piece.
- Temporary – although not always designed to be taken apart, they can be disassembled if needed without causing damage.
- Adhesives – these fall into two groups, natural and synthetic. The synthetic types tend to be toxic substances and therefore need to be handled with care. It is thought that most adhesive bonding can be classified as a chemical reaction.

Table 3.1.8 Methods and components for **permanent** joining of materials

Method/component	Diagram	Description	Materials
Nails		Nails are available in a wide range of shapes and sizes and are the quickest way of making a permanent joint in wood. The nail punches through the wood and the wood fibres grip the shank and resist attempts to remove it	Wood
Rivets		Snap, countersunk and flat head rivets are used to join flats of metal together. Pop rivets are used to join sheet metals together or to softer sheet materials, e.g. polypropylene (PP)	Metal, soft sheet materials
Joints		Joints are used in all aspects of the fabrication of products, e.g. traditional woodworking joints such as a dovetail joint	Wood, metal and plastics
Soldering and brazing	Brazing torch / Soldering iron	Soldering uses an alloy, which has a lower melting point than the metals being joined, to bond two pieces of metal together. Soldering includes soft soldering (for electrical components), silver soldering (for jewellery making) and brazing	Metal
Welding	Acetylene Oxygen	Welding is a process that works by melting the edges of the metal to be joined so that they fuse together. A filler rod of similar material is added to fill the weld. MIG, spot and electric arc are three common methods of welding for metals. Some plastics can be fusion welded, e.g. blow-up toys	Metal and plastics

Table 3.1.9 Components for **temporary** joining of materials

Component	Diagram	Description	Materials
Nuts and bolts		Bolts are commonly available with hexagonal heads, and, together with nuts of matching thread size and form, lock together to form strong mechanical joints	Wood, metal and plastics
Screws		Screws are available in a wide range of sizes and head shapes. The thread of the screw becomes enmeshed with the grain fibres to make a strong joint. Self-tapping screws are used for metal and plastics	Wood, metal and plastics
Knock-down fittings		Knock-down jointing methods allow strong joints to be made quickly and easily. The parts can also easily and quickly be taken apart, so that the whole construction can be flat-packed for easy transportation and storage	Wood

Table 3.1.10 *Adhesives for sheet and modelling materials*

Adhesive	Properties	Uses
Polyvinyl acetate (PVA)	Sold in a plastic container as a white, ready-mixed liquid. Easy to apply and sets in two hours, providing a strong joint. It is usually not waterproof	Woods
Epoxy resin	Sold in two tubes – one contains a resin and the other a hardener, which are mixed together in equal quantities. Chemical reaction hardens immediately, but full strength is not achieved until after two to three days	Expanded polystyrene to most materials
Contact adhesive	Usually used for gluing sheet materials. Sold in a metal tube for easy application	Metals and plastics. Dissimilar materials, e.g. plastic to wood. General purpose; fabric to most materials
Acrylic cement	Solvent-based adhesive for rapid bonding of acrylics. Sold in a screw-top metal can and needs a separate applicator. Always read manufacturer's instructions before use	Acrylic to acrylic
Polystyrene cement	Solvent-based adhesive that melts surface of pieces to be joined and causes them to weld together. Can use a brush to apply (water-like consistency) and it is absorbed into joint by capillary action	High-impact polystyrene (not expanded polystyrene)
Hot melt glue	Used in conjunction with a glue gun. Solid glue stick passes through a heating element and becomes a gel. Easy to apply from gun nozzle. Does not give a solid joint – more of a temporary fixing	Most materials for rapid joining (usually temporary)
Artwork spray	Adhesive sold in aerosol from. Excellent for presentations as work does not wrinkle and allows for repositioning	Paper, card and board

4. Finishing processes

The relationship between finishes, properties and quality is one that the designer should consider very early on as one of the major design considerations. Finishes in their various forms are used to improve the product's functional properties, aesthetic qualities and generally serve to improve quality overall. For example, tin cans are made from cheaper sheet steel but they are plated with tin to stop them rusting and contaminating the food, and restaurant menus may be encapsulated in plastic to provide a barrier against moisture and provide a wipe clean surface.

Surface coating

Anodising

Anodising is a surface treatment associated with aluminium and its alloys. The whole product is immersed in a solution of sulphuric acid, sodium sulphate and water. The product itself is used as the anode and lead plates are used as a cathode. When a direct current (DC) is passed through the solution a thin oxide film forms on the component. When finally washed in boiling water, coloured dyes can be added before the surface is finally lacquered. As a process it can be

used to finish components to a consistently high-quality finish.

Painting

Painting wood also involves the application of a number of coats. Prior to any painting, all knots should be sealed to prevent any resin from escaping. In between coats the surfaces should be rubbed down with fine glass paper. Topcoats are available in various forms:

- Oil based – commonly known as gloss paints, these are available in a wide range of colours. They are durable and waterproof, and are excellent for use on products that are outside such as window frames and doors. Certain paints used on boats never need rubbing down for repainting.
- Emulsion – available in vinyl or acrylic resin, these paints are water based but not waterproof.
- Polyurethane – tough and scratch resistant, these paints harden on exposure to air. They are used widely on children's toys.

Painting, other than spray painting, is very time consuming in its application. Great care also needs to be taken to ensure an even application over the entire surface. It does, however, allow you the flexibility to create original pieces by way of choice or mix of colours therefore creating a range of one-off finishes.

Varnishing

Varnishes are a plastic type of finish made from synthetic resins. They provide a tough waterproof and heatproof finish. Polyurethane varnish is available in a range of colours with different finishes; gloss, matt or satin. They are best applied in thin coats with a light rub down with wire wool in between. New varnishes based on acrylic dry more quickly, have less odour and brushes can be cleaned in water. As they do not use solvents, they are environmentally friendly.

Self finishing

Plastics require very little surface finishing because they already tend to be resistant to corrosion and general surface deterioration. The finish achieved on products such as fizzy drinks bottles and mobile phone casings are all due to the manufacturing processes involved, such as blow moulding and injection moulding. The high quality of finish is mainly due to the very high quality of the original mould. Surface details such as texture and low relief lettering can be added to the mould, which will appear on the final product; for example the brand name on the casing of an electrical product. Colour and tone are easily changed with the addition of chemicals and dyes into the raw plastic material. The die casting of metal is a similar manufacturing process in that it requires no surface finishing, other than sometimes painting. Molten metal is fed into a very high quality die to produce such products as toy cars and soldiers.

Surface decoration

The use of CAD/CAM is widespread in the production of signage for a variety of commercial applications. Vinyl stickers and engraving are two common methods of providing surface decoration where graphical images are required.

Vinyl stickers

Vinyl is the common term used to describe plasticised PVC. Vinyl stickers or graphics are ideal for one-off or batch production, from an individual sign for a shop or restaurant frontage, to a series of movie adverts covering the backs of buses. Many schools and colleges will have access to CAD/CAM facilities for cutting vinyl, which directly replicate that of the commercial process. The image is designed on computer and digitally sent to the vinyl cutter for contour cutting. Commercial sign-makers usually have

Figure 3.1.27 Vinyl graphics have been used on the roof of this mini

image banks and technically accurate dimensions for batch produced vinyl graphics such as logos or trademarks for large companies.

Once the cutting is complete, all of the background vinyl is removed using a process called 'weeding', leaving only the required graphic. The adhesive used on rolls of vinyl is contact or impact adhesive, which has been lightly coated on the back and covered with a treated backing paper to protect it and make the vinyl easy to peel off. Finally, a layer of application tape is applied over the graphic. The application tape will adhere to the vinyl so that it can be removed from the backing sheet and on to the required surface. The nature of vinyl gives the letters a soft and flexible quality that allows them to be adhered to a variety of contours. Once applied the application tape can be removed leaving only the vinyl graphic in place.

Tasks

1 Why are vinyl graphics used instead of the traditional hand-painted method?
2 If you have access to a vinyl cutter, design and produce a range of simple designs that could be applied to promotional items in your product development portfolio.

Engraving

Engraving is an extremely effective method of applying text or graphics to plaques, signs and name badges, etc. These signs have increased durability as the surface decoration is actually cut into the material, whereas printing inks or vinyl stickers will, over time, degrade. The CAD/CAM process for engraving is similar to that of vinyl stickers, but the output device in this case is an engraver, router or even laser cutter. Most engravers can engrave into a wide range of materials including plastics, stainless steel, brass, glass and even brick. Their services may include converting a client's file, scanning originals into a CAD programme, or creating new files that can be saved and used for repeat jobs.

Figure 3.1.28 *Engraving a sign*

5. Product manufacture

Computer printers

The two most widely used computer printers are inkjet and laser printers.

Inkjet printers deposit extremely small droplets of ink on to paper to form an image. The data to be printed are sent by the computer applications software to the printer driver. The printer driver translates the data and sends them to the printer via the connection interface. The printer receives the data and stores them in a buffer. The buffer can hold data so that the computer can be used for other tasks while the printing process takes place.

The control circuitry in the printer activates the motor to feed the paper, which is carried by rollers, into the printer. Once the paper is in position the motor activates a belt to move the print head across the page. The print head contains a number of nozzles that spray ink on to the page. The ink can be black, cyan (blue), magenta (red), yellow, or precise combinations of these to form any colour. The motor pauses between each spray of ink to form the shape of the characters and create spaces. However, this occurs at such high speed it appears to move in one steady continuous motion.

At the end of each printed line the motor moves the paper a fraction so that the next line can be printed. Finally, when a full page is printed the motor moves the paper through the printer on to an output tray.

Like the photocopier machine, the laser printer (see Figure 3.1.29) uses static electricity, an electrical charge that is built up on an insulated object. The laser system works on the principle that oppositely charged atoms attract one another and will cling together.

The laser printer receives the data from the computer in digital form. The printer's core component is a photoreceptor, which typically takes the form of a revolving cylinder or drum. The drum is constructed from photoconductive material that can be discharged by light photons.

The drum is positively charged by a roller or wire (corona wire) with an electrical current running through it. The drum revolves and as it does so a fine laser is beamed across the surface to discharge selected areas. By doing this the laser inscribes the image or text to be printed as an electrostatic image.

When the image is set on the drum, positively charged toner powder coats the surface. The toner having a positive charge is attracted to the discharged areas and clings to them. It is repelled

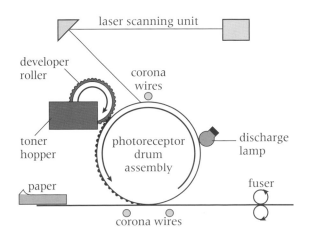

Figure 3.1.29 *The basic components of a laser printer*

from the positively charged background.

A roller or wire gives the paper a negative charge that is stronger than that of the electrostatic image. As the paper rolls under the drum its stronger charge pulls the toner away from the drum's surface. The drum is then discharged by exposure to light to remove the image allowing the process to begin again.

The paper, which now holds the toner powder, is also discharged and quickly passed through heated rollers that melt the powder, making it fuse with the fibres of the paper to fix the image.

To print a multicolour image on a laser printer a similar process takes place. However, it has to be done four times, once for each colour: cyan (blue), magenta (red), yellow and black.

Toner powder in the three primary colours and black is used in various combinations to create the full spectrum of colour. Depending on the printer model this can be achieved in various ways: the printer creates an electrostatic image for each colour and applies toner from separate units; all four colours are added to a plate; or in complex machines a separate printer unit is available for each colour and the paper moves from one to the next picking up each colour.

Commercial printing processes

Offset lithography

Offset lithography (litho) was developed from lithography for commercial use. Both processes are based on the principle that oil and water do not mix and will always separate.

A lithographic plate bearing an image is treated using a greasy medium, which will attract oil-based ink. The non-image area is treated with water, which repels the ink. In lithographic printing, paper is rolled across the plate and the inked image is transferred on to it.

Offset litho uses the same principle but the plate does not print directly on to the paper or the required surface. Instead, the lithographic plate is pressed on to dampened rubber rollers, which transfers a solution of water and gum arabic on to the plate (see Figure 1.10 on page 27). The water solution is accepted on the non-image area, while the image area repels it. Then the ink is transferred on to the plate; it is accepted by the image area and repelled by the non-image area by the water solution. The inked image is then 'offset' on to a rubber blanket cylinder that transfers the image to the paper.

To print a multicoloured image, separate blanket cylinders carrying each colour are required. In multicolour lithographic machines single units are joined together in a row to form multiple unit presses. The sheet is passed through the machine on a system of transfer cylinders, which print each colour separately.

There are a variety of printing plates used in offset litho. The most common of these are as follows:

- Pre-sensitised metal plate. A negative of the artwork/image is placed on to a thin sheet of metal coated with light sensitive material, before being exposed to ultra-violet light. The image is then fixed using chemical fixer and lacquer.
- Chemical transfer plate. A paper negative of the artwork is made and placed on to a paper, plastic or metal foil plate. This is passed through a processor containing developer which transfers the image on to the plate. The image is then fixed with chemicals before the plate is used.
- Electrostatic plate. The image of the artwork is reflected on to a special plate made of paper, plastic or metal foil. The zinc oxide surface layer acts as a semi-conductor and an electrostatic charge is created. Grease-receptive particles are drawn to the area that forms the image. The plate is then heated to fix the particles in place.

For small print runs, a basic hand-operated lithography machine can be used. Paper or plastic plates can be produced quickly and cheaply by drawing the image directly on to the surface. High-volume production processes require fully automated machinery, which is capable of producing vast quantities. High-quality electrostatic or pre-sensitised metal plates are used for this level of production.

Screen printing

Screen printing is a widely used and versatile printing process. It is used in graphics for signs, banners and posters and in manufacturing processes to print on fabrics, wall coverings, glass and ceramics. Despite being commonly referred to as silk screen printing, silk is rarely used to make screens. Modern hand and commercial printing screens are made from synthetic or metal gauzes that offer durability and stability.

Screen printing works on a relatively simple method of ink transfer. Ink is forced through holes in the gauze that form an image on the screen; essentially the screen functions as a stencil. Ink is pushed through the apertures in the fine mesh by pressure, applied using a plastic, rubber or steel blade often referred to as a squeegee. Whereas other printing processes (e.g. letterpress) rely on impact to print, screen printing is based upon a non-impact principle of ink transfer. This principle makes screen printing

Figure 3.1.30 Screen printing

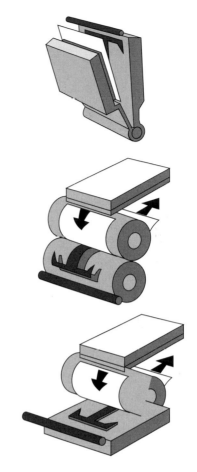

Figure 3.1.31 Letterpress

a versatile process because it can be used to:

- print on to a wide range of materials
- print on to surfaces that are curved or uneven
- print thick opaque ink deposits.

Fine gauze material is stretched at high tension over a metal or a wooden frame to make the screen. The three most common screen materials are polyester, polymide and stainless steel.

The image is formed on the screen using a photo-stencil. The screen is completely covered with a light sensitive polymer emulsion. When the screen has dried the opaque positive image is then placed on it and exposed to ultra-violet light. The areas of coating that have been exposed become insoluble in water, thus fixing the image. The screen is sprayed with water after exposure, dissolving the unexposed areas and leaving the stencil image.

Screen printing production methods vary according to the scale of manufacture required. They range from manual hand-bench methods for short runs of printing (for example one to 100 units) to fully automatic methods (rotary or cylinder) which are capable of producing high-volume run lengths (for example 7000 units and above).

To print a multicolour image a separate screen must be produced for each coloured section of the artwork. Precise multicolour images are best produced using manual, part or fully automated equipment such as carousel machines.

Letterpress

Letterpress is one of the oldest methods of printing. It is still in use but its applications are limited as it has been superseded by faster and more efficient processes such as offset litho.

A reverse image of the artwork/text is transferred on to a plate photographically. The unwanted areas surrounding the image are treated with acid and etched away. The non-

image area is now lower than the image area that stands in relief. Ink is applied to the raised surface area of the plate and this is then pressed on to the paper.

Letterpress was traditionally used as a large-scale manufacturing process for products such as newspapers and books. However, it is now used in high-quality, small-scale production for products such as limited edition books or stationery. Letterpress can be used to print monochrome and coloured images. To print multicoloured images, individual plates have to be produced for each section that is to be coloured.

Task

Try to organise a visit to a local printers to see first hand one or more of these printing processes in action. Alternatively, watch a video or access a CD-ROM detailing the processes.

Table 3.1.11 Printing processes

Process	Advantages	Disadvantages	Applications
Offset lithography	Good reproduction quality, especially photographs; can print on a wide range of papers; high printing speeds; cheap printing process; widely available	Colour variation due to water/ink mixture; paper can stretch due to dampening	Business stationery, brochures, leaflets, posters, magazines, newspapers
Screen printing	Economical for short print runs; stencils are easy to produce; able to print on virtually any surface	Difficult to achieve fine detail; low output; long drying times are required	T-shirts, posters, plastic and metal signage, point-of-sale displays, promotional items, e.g. pens
Letterpress	Dense ink gives good printing quality; less wastage of paper than in other printing processes	Expensive; slow process	Books with large amounts of text, letterheads and business cards

Black and white printing

Photocopying is a widely used form of electrostatic printing. Photocopying machines duplicate original images or text. The process is constantly evolving as technology develops and exact processes can vary depending on the machine being used.

- The original of the artwork/text to be reproduced is positioned face down on the glass surface of the machine and held in place with a cover.
- Inside the machine, a metal drum with a photoconductive surface receives an electrostatic charge.
- The original is illuminated and the reflected image is projected through a lens system on to the electrostatically charged drum.
- On the areas where light falls the electrostatic charge disappears. Charged particles remain only in the image area.
- As the photocopying process is dry, a resin-based toner is used instead of liquid ink. This is poured over the surface area of the drum. The powder has an opposite electrical charge to the image area on the drum surface and so settles on it, while the rest falls away.
- The copy paper is electrostatically charged; when it is positioned near the drum surface it draws the powder image on to its surface.
- Finally, the powder image is heated so that the resin melts, fixing it to the paper.

Finishing and binding ready for distribution

Finishing

Print finishing is a term that refers to a system of operations that prepare the printed product for distribution. There are numerous finishing techniques; the major ones are outlined below.

Imposition

A number of pages are arranged on to a large sheet so that when it is folded the pages will appear in the correct order. This is called a signature. A simple leaflet can be arranged so that the sheet is printed on one side, then turned over and printed on the other. Pages 2 and 3 are printed on the reverse of 1 and 4. This can then be folded to read as a four-page leaflet. Larger, more complex products like books require larger sheets that have to be folded and cut to make signatures.

Folding

Folding can be done by hand or using a machine depending on the amount of material that needs to be folded. For high-volume operations fully automated folding machines are required.

A sheet with a number of pages arranged on it is passed over a flat bed in which there is a long slot. A blunt blade designed to fold but not cut the paper pushes the sheet into the slot. A set of rollers folds the paper at the required point. This process is repeated until all the pages on the sheet have been folded to make a signature. The doubled page edges are cut or slit automatically by the machine.

Scoring

Heavy material such as board is difficult to fold and needs to be scored beforehand. The material is scored on the outside edge of the fold to break the top layer of fibres, making it more flexible. It is important that the scoring is not too heavy as it will crack or cut the material, or too light as this will affect the accuracy of the fold which may feather along the crease. Scoring can be done by hand using a ruler and blade, or on a machine if the material is too heavy.

Cutting

The instrument used for cutting is a guillotine. The size and degree of automation of the guillotine varies between models.

Hand-operated machines are used in classrooms and workshops for occasional and small batch work. A basic guillotine consists of a flat base board with a bar fixed on the edge to carry an encased blade that moves along the length of the edge. Some guillotines have an adjustable edge against which the material is positioned to ensure an accurate cut. Other models will feature a grid or measuring system to fulfil the same function.

Fully automated machines are used to cut large volumes of printed material in one operation. These machines can be programmed to precise specifications and can make a series of cuts using a number of blades. Automated machines use adjustable back and side gauges to position the material correctly for the cut.

Die cutting is a machine process that involves punching out a specified area of the sheet of paper or board for dramatic effect. The die is a sharp metal blade that can be manufactured to any shape in order to cut the appropriate shape or hole.

Gathering and collation

Gathering can also be referred to as collation. Gathering involves bringing together components of the product into the correct sequence to make up a complete product. This can be done by hand, for example piles of identical pages (i.e. page 1s, 2s, 3s can be laid out in a row and one sheet can be selected from each pile to compile a document). This process is suitable for smaller quantities; for larger quantities machines are used.

Fully automated machines work on a similar principle to the hand process. Items of the same printed material are placed in separate boxes or stations (also known as hoppers) along a belt in sequence. One item from each pile is fed on to the belt sequentially to compile one complete job. This can be done at high speed and can be incorporated into the stitching and binding process to streamline the task.

The term collation also refers to the process of checking the gathered sections to ensure they are in the correct order. This can be done by hand or by a machine using collation marks (also known as back or black step marks) which are placed in a set pattern so that when sections are compiled the marks appear in a sequence or progression on the reverse of the sheets.

Decorative finishing techniques

There are a number of finishing techniques that can be used to enhance the format of paper, card and boards and to provide visual impact:

- Laminating – applying a transparent plastic film to a surface to protect it and enhance its appearance.
- Encapsulating – similar to laminating with the addition of heat sealed seams, therefore fully covering the edges of paper and card, e.g. wipe clean menus.
- Varnishing – applying a gloss varnish to the entire surface or to elements of the surface to highlight areas of the design.
- Embossing – raising the surface of paper or cartonboard in the shape of an image or text so that it stands in relief.
- Hot foil blocking – applying a foil coating by means of a heated die to highlight areas of the design.

Task
Collect examples of each decorative finishing technique and discuss their individual visual impact.

Binding

Binding is a process used to fasten or hold together a number of printed sheets. Many products are bound such as magazines, books, reports and leaflets. Binding can range from the simplest forms, for example stapling, plastic or ring binding, to fully automated processes. There are various methods of binding. In addition to aesthetic considerations, the quantity of paper to be bound and the cost are determining factors as to which process is used. The three main methods are:

- Stitching and sewing. There are two types of wire stitching used in this process. Paper that is folded to make a signature requires saddle stitching down the spine while loose pages are attached at the spine by side stitching. The advantage of saddle stitching is that the book can be laid flat when read whereas the side stitched book cannot. Sewing is a more expensive option but it is neater and more durable – it is often used in hardback binding. Sections or signatures of the book are sewn together and a case of board is covered with material. The case is attached to the sewn sections with strips of adhesive tape that have been sewn into the spine. The strips are covered with endpaper, which are the leaves at the front and back of the book that cover the inner sides of the case.
- Mechanical binding. There are several types of mechanical binding. However, most involve

Table 3.1.12 Binding methods

Method	Description
Saddle-wire stitched	The simplest method of binding – pages are stapled through the fold
Side-wire stitched	Used when the document is too thick for saddle-wire stitching – staples are passed through the side of the document close to the spine
Perfect binding	Produces a higher quality presentation and the spine can also be printed on – pages are held together and fixed by the cover by means of a flexible adhesive, e.g. paperback books and magazines
Hard-bound or case-bound	Usually combines sewing and gluing to create the most durable method of commercial binding – stiff board is used on the cover to protect the pages, e.g. hardback books
Spiral or comb binding	Pages are punched through with a series of holes along the spine, then a spiralling steel or plastic band is inserted through the holes to hold the sheets together

drilling or punching holes through the paper and threading plastic or wire through to keep the pages together. Spiral and plastic comb binding are two of the most frequently used.
• Perfect binding. This is commonly used for

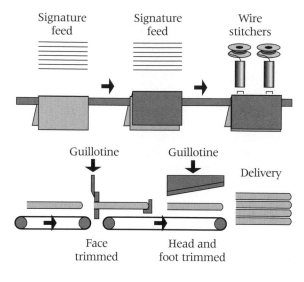

Figure 3.1.32 Folding, binding and trimming

paperback books. The sections of the book are cut with a guillotine and the spine is roughened so that it will accept the glue. After the sheets have been gummed along the spine the cover is creased and glued in to position.

Packing and distribution

Packing is an important area of the whole production process. Excellent printed materials can be produced on sophisticated machinery and yet arrive damaged at their destination due to inadequate packing. Printed materials can be shrink-wrapped on special machines, or work can be stacked on pallets and the whole pallet shrink-wrapped for protection. In all cases, detailed packing instructions must be given to the printer by the designer including labelling, type and size of cartons, type and size of pallets, and maximum height of printed materials on each pallet. Once palletised, orders can be delivered to the customer, retailer or wholesaler by commercial van or lorry.

6. Testing materials

The testing of materials and processes before manufacture ensures the production of high-quality products. Testing can include comparative testing using standard performance tests, quality control, the use of British and International standards and the use of ICT as a testing aid.

Standard performance tests

In industry, standard performance tests set out by the British Standards Institution (BSI), are regularly carried out in order to determine the

mechanical properties of materials and ensure their quality and fitness for purpose. There are two main methods of materials testing – by deformation and to destruction. Many industrial tests can be replicated within the school or college workshop.

Task
Carry out a series of standard performance tests in your school workshop using a range of different materials and cross-sections.

Table 3.1.13 Standard performance tests

Mechanical property	Industrial standard performance test	Reasons	Workshop performance test
Tensile strength A material's ability to resist stretching or pulling forces	British standard BS EN 10002 (2001) *Tensile testing of metals*	If a material is not strong enough for its purpose it would fail in use. The tensometer measures tensile strength of materials to determine what limits they can be put to	*Tensile strength experiment* 1 Clamp sample in vice and add a standard weight to the end of the bar 2 Measure the amount of deflection 3 Repeat experiment with other cross-sectional bars (bar length must be constant)
Hardness A material's resistance to abrasive wear and indentation	The Brinell BS 240 (1986), Vickers BS 427 (1990) and Rockwell BS 891 (1989) hardness tests	Materials must be capable of withstanding wear and indentation. If the incorrect material is chosen the product may fail. These three tests measure a material's ability to withstand scratching or resistance to indentation	*Indentation test* 1 Tighten a hard ball-bearing (tungsten-carbide) on to the sample using an engineer's vice 2 Measure the size or depth of the impression made
Toughness A material's ability to withstand sudden impact and shock loading without fracture	The Izod BS 131 (1961) and Charpy BS 131 (1972) notched bar tests	The measurement of resistance to shock loading. The material may break or fracture if not tough enough, leading to component/product failure	To replicate the *notched bar tests* 1 File a notch, or saw halfway through the sample to create a weak spot 2 Use constant hammer blows to 'feel' the amount of force necessary to bend or break samples
Ductility A material's deformation through bending, twisting or stretching	British standard BS 1639 (1964) *Methods for bend testing of metals*	If a material is too ductile it may stretch or deform, resulting in mechanical failure of the component or product	*1 Free bend test* a) Clamp sample in a vice and bend forwards and backwards past the neutral position b) Count the amount of bends before visible surface cracks appear or until it breaks *2 180-degree guided bend test* Bend sample around a former to measure the plastic deformation in one direction. Sample should be unbroken and free from visible surface cracks

Quality control

Quality control is carried out by industry before, during and after all manufacturing processes as part of their overall quality assurance procedures. Any manufacturer wishing to obtain an ISO 9000 International Standards award must monitor the quality of its products from its design and development, through to its end-use and degree of customer satisfaction. Therefore, the selection of appropriate quality materials and components is fundamental to the manufacture and performance of all products. Quality control systems must monitor and achieve consistently high standards by inspection and testing. Inspection will determine whether the materials, components or products are within specified tolerances – the degree to which it is acceptable in order to function in accordance with its design and manufacturing specifications.

Quality control in print runs

Paper is not an inert material – it reacts to changes in the environment, which may cause problems during a print run. Paper is affected dramatically by temperature and changes in humidity, which can cause it to curl and may affect the colour registration of the printed materials. The relative humidity of the print room has to be controlled to ensure curl stability. Usually, when paper stock arrives at the print shop, it is kept in storage for a while in order for it to adapt to the relative humidity of the print room. During the print run itself regular quality control checks are made to ensure the quality of the printed materials.

In order to aid the quality control in a print run, printer's marks are used. Colour bars, for example, provide vital information about the performance of both the printing press and the inks being used. They contain a whole range of tests, some are visual checks made by the operator of the printing press, but others can be performed electronically using a densitometer. This instrument monitors the thickness or density of the ink printed on the colour bar to ensure that it is of a consistent quality throughout the print run.

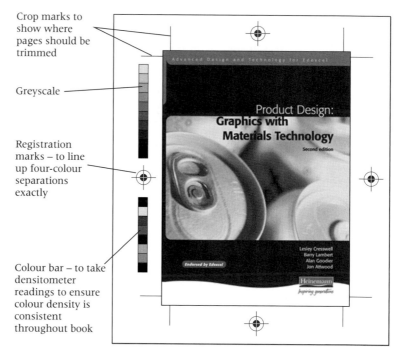

Crop marks to show where pages should be trimmed

Greyscale

Registration marks – to line up four-colour separations exactly

Colour bar – to take densitometer readings to ensure colour density is consistent throughout book

Figure 3.1.33 *Front cover of this book showing the printer's marks*

Table 3.1.14 *Quality control checks used during a print run*

Problem	Description	Quality control
Set off	Where the ink from one sheet smudges on to the underside of the following sheet	Use of sufficient anti-set off spray, use of sufficiently quick-drying inks, or use of better quality paper stock
Colour variation	This occurs where the printer does not maintain consistent colour throughout the run	Use of colour bars and regular densitometer readings
Hickies	These are small areas of unwanted solid colour surrounded by an unprinted 'halo' area, and are caused by specks of dirt, paper debris or ink skin on the printing plate or blanket cylinders	Regular washing of the blanket cylinder
Bad register	This is where colours protrude beyond the edge of the four-colour separations making the image look out of focus	Use and regular inspection of registration marks to line up the four colour separations exactly

> ## Task
> Discuss the reasons why so many quality control checks are used in a print run.

BSI and international standards

Products will be marked with the following testing standards:

- BS – British Standard
- BS EN – British adoptions of European Standards
- ISO – International Standards.

The Kitemark is a visual motif displayed on products and indicates stringent and regular independent testing, and quality testing to British, European and international standards.

The CE mark is a form of marking designating the manufacturer's claim that the product meets the requirements of all relevant European Directives. It is a legal requirement that a product covered by a European Directive must have a CE if it is to be sold in Europe.

The ISO 9000 family of International Standards are the goal of most manufacturers. These standards outline the duty of all manufacturers in designing, implementing and documenting rigorous quality management systems.

The purpose of these standards is to allow the customer to see if the product conforms to a level of safety, quality and fitness for the product's purpose.

Use of ICT

Computer-aided testing usually involves a machine with a probe, which makes contact with the surface of the material or component. The probe is moved across the component and its measurements are taken and fed into the computer for analysis. Other methods include the use of optical sensors as in the printing industry.

Advantages of using computer-aided testing

Provided ongoing quality control throughout manufacture.

- Different types of inspection processes can take place by using different probes or sensors, for example contact, non-contact and vision systems.
- Data from the probes or sensors is directly inputted into a computer system so reports can be generated.
- Errors can be identified, analysed and corrected early, reducing waste and costs.

Test programmes before manufacturing

ICT also allows manufacturers to run test programmes to assess the viability of making and manufacturing in the long term. They are able to assess tooling costs and times and how they might need to hold the work during machining.

Testing of products sometimes involves testing them to destruction and although this is appropriate for smaller products, it is not possible on aircraft or ships. In this case, computer models are used with results obtained from testing smaller scaled versions. Testing is carried out in wind tunnels or water tanks and the results gathered are used to build the computer models for the full-sized product.

Exam preparation

You will need to revise all the topics in this unit, so that you can apply your knowledge and understanding to the exam question. In preparation for your exam it is a good idea to make brief notes about different topics, such as 'Product manufacture'. Use sub-headings or bullet point lists and diagrams where appropriate. A single side of A4 should be used for each heading from the Specification.

It is very important to learn exam skills. You should also have weekly practice in learning technical terms and in answering exam-style questions. When you answer any question you should:

- read the question carefully and pick out the key points that need answering
- match your answer to the marks available, e.g. for four marks, you should give four good points that address the question
- always give examples and justify statements with reasons, saying how or why the statement is true.

Practice exam questions

1 a) Using notes and sketches, describe two functions of vessels (pores) in hardwood trees. (2)
   b) Explain the advantages of using pine for vacuum forming moulds. (3)

2 a) State the meaning of **two** of the following terms:
   - alloy
   - anodising
   - engraving. (2)
   b) Using one product example, briefly describe the term 'self-finishing'. (3)

3 A plastic bottle has a line running down each side and across the bottom. The bottle top has a small circular indent at its centre. Using annotated sketches, describe a process used to make:
   a) the bottle (3)
   b) the bottle top. (3)

4 The modern print media makes use of desktop publishing (DTP) rather than using the traditional process of typesetting to produce printed pages. Describe the advantages of using DTP for modifying designs. (5)

5 a) Explain what you understand by **one** of the following terms:
   - pictorial drawing
   - block modelling
   - offset lithography
   - binding. (2)
   b) Describe one practical application where screen printing could be used. (3)

6 a) Explain why testing before manufacture is an important process. (3)
   b) Describe the purpose of British Standards. (2)

7 Explain why computer modelling is increasingly used as a testing aid. (5)

Total: 36 marks

Design and technology in society (G302)

Summary of expectations

1. What to expect

Design and Technology in Society is one of the two options offered in Section B of Unit 3, of which you will study only one. This option builds on the knowledge and understanding of how design and technology affects society, which you gained during your GCSE course.

2. How will it be assessed?

The work that you do in this option will be externally assessed through Section B of the Unit 3 examination paper. The questions will be common across all three materials areas in product design. You can therefore answer questions with reference to **either**:

- a Graphics with Materials Technology product such as an Art Nouveau perfume bottle
- a Resistant Materials Technology product such as a Bauhaus table lamp
- **or** a Textiles Technology product such as a cushion using a William Morris fabric.

You are advised to spend **45 minutes** on this section of the paper.

3. What will be assessed?

The following list summarises the topics covered in this option and what will be examined:

- The physical and social consequences of design and technology for society
  - the effects of design and technological changes on society
  - influences on the development of products.
- Professional designers at work
  - the relationship between designers and clients, manufacturers, users and society

- professional practice relating to the design management, technology, marketing, business, ICT
  - the work of professional designers and professional bodies.
- Anthropometrics and ergonomics
  - the basic principles and applications of anthropometrics and ergonomics
  - British and International Standards.

4. How to be successful in this unit

To be successful in this unit you will need to:

- have a clear understanding of the topics covered in this option
- apply your knowledge and understanding to a given situation or context
- organise your answers clearly and coherently, using specialist technical terms where appropriate
- use clear sketches where appropriate to illustrate your answers
- write clear and logical answers to the examination questions, using correct spelling, punctuation and grammar.

There may be an opportunity to demonstrate your knowledge and understanding of Design and Technology in Society in your Unit 2 coursework.

5. How much is it worth?

This option, together with Section A, is worth 30 per cent of your AS qualification. If you go on to complete the whole course, then this unit accounts for 15 per cent of the full Advanced GCE.

Unit 3A + option	Weighting
AS level	30%
A2 level (full GCE)	15%

1. The physical and social consequences of design and technology for society

The effects of design and technological changes on society

Design and technology has improved the lives of millions of people worldwide, but the changes brought about by developments in technology have resulted in far reaching physical and social consequences.

The benefits of technology are countless – imagine your life without electricity, television or computers! Many of the products that we take for granted today are available as a direct result of developments in mass production and technology (see Figure 3.2.1). Mass production may have improved our lives, but it has also brought unforeseen consequences – one simple example being a concern about the adverse influence of computer games on children. On a larger scale, there is growing awareness of the need for sustainable technology. More people are now aware of the problems arising from the over-use of the world's natural resources, the consequences of pollution and the impact that global manufacturing has on the lives of many people worldwide.

Mass production and the consumer society

The history of design as we know it began with the industrial revolution and the invention of the steam engine in the mid 1700s. As a result of this invention, coal mining, iron and steel and machine production took on a new importance and set the scene for the development of industrial mass-production. The steam engine led to the development of machinery, transport and trade. This fostered the need for the design of new products, but design was no longer to be in the hands of individual craftspeople. Expensive and time-consuming hand work could now be replaced by machine work. Products, once exclusively made for the rich, could now be made at an affordable price for ordinary working people.

The steam engine led to the development of machinery, transport and trade. This fostered the need for the design of new products, but design was no longer to be in the hands of individual craftspeople. Expensive and time-consuming hand work could now be replaced by machine work. Products, once exclusively for the rich, could now be made at an affordable price for ordinary working people.

Mass production

The main characteristic of mass production was that the production process was simplified into a limited number of tasks. The craftsperson became 'redundant', to be replaced by low-skilled factory workers who performed simple repetitive tasks.

What started out as a wonderful opportunity for ordinary people to find work and gain access to inexpensive consumer products ended in misery for many. The reduced requirement for skills meant lower wages and the employment of women and children in 'sweatshop' factories. The resulting poverty led to worker uprisings and the development of trade unions, aimed at combating poor living conditions, poverty and the increasing pollution brought about by industrialisation.

Assembly lines

The story of mass production would not be complete without reference to the introduction of the moving assembly line. In the United States of America, industrial production developed at a much faster pace than anywhere else and the first automatic assembly lines were in use by the mid nineteenth century. The first assembly lines were used in slaughterhouses, then in the sewing machine industry and later in automobile manufacturing. Assembly lines made mass production easier and faster, enabling products such as cars to be made more cheaply and be available to a much wider market.

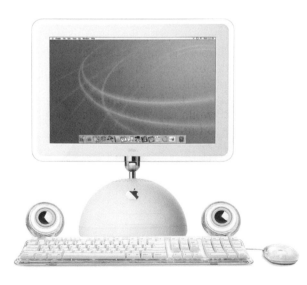

Figure 3.2.1 *Computer technology has revolutionised the workplace*

Figure 3.2.2 *Heavy, overdecorated Victorian furniture*

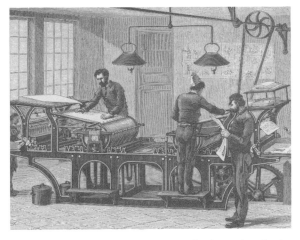

Figure 3.2.3 *The first steam-operated mechanical printing presses were used early in the nineteenth century*

Design tradition

Although the technical advances of the nineteenth century brought about new production methods and new products, there was no tradition of design for mass-produced products. Instead, poor quality utensils and furniture were decorated to imitate the traditional styles of handmade products. You may have seen examples of the heavy, over decorated furniture that was typical of the Victorian era (see Figure 3.2.2).

Design for mass production

During the second half of the nineteenth century trade unions demanded changes to the poor living conditions of factory workers and the design of simple, inexpensive consumer goods that were appropriate for their lives.

The poor quality, over-decorated, over-sized furniture that was currently being mass produced was totally inappropriate for the cramped housing conditions of the majority of workers. As a result, furniture for factory workers became a new area of design. These changes in the concept of design for mass-production were long overdue. They mark the beginning of modern design history. The connections between form, function and products began to be recognised. They are the same ones that we use today in design and technology.

Design in the USA

In contrast to Europe, the concept of design in late nineteenth century USA was more technical and functional. The increasing mechanisation included not only production methods, but the products themselves. Products such as the sewing machine, the microphone, the telephone and the mechanical typewriter were mass produced. Many of these products were totally functional and lacked any kind of decoration. The concept of 'styling' didn't exist at that time, mainly because there was no competition from similar products and therefore no competing market to sell into.

The development of the consumer society

In both the USA and Europe, the early years of the twentieth century were decades of uncertainty. War, revolution and economic crises resulted in massive unemployment, poverty and housing shortages. For the well-off, however the lifestyle was much the same and culture, entertainment, sport and social life continued. In the USA, jazz, swing and the Charleston dominated the dance halls. Popular culture and the age of Hollywood would become an enormous influence on lifestyle, fashion, design and even morality.

As international commerce and transportation systems developed, new opportunities for product design came about, such as luxury ships, aeroplanes, hotels, theatres and department stores. The gradual spread of electrification brought about the design of innovative new products using new technology and new materials. In the 1930s, most middle-class American homes owned consumer goods such as radios, refrigerators, toasters and washing machines. As the standard of living improved, the demand for new products increased. Advertising became an important new industry, using market research, packaging and product

109

styling to sell the new products. Design became an important marketing tool. This was used to great effect after the Wall Street crash of 1929, when product 'styling' was used to motivate people to buy. Designing aesthetically appealing products was one strategy used to promote consumerism. Another was called 'streamlining', which developed from research into aerodynamics for cars and planes. After the 1930s, the availability of new materials such as plywood, plastics and sheet metal enabled rounded 'streamlined' shapes to be applied to many products and it came to represent 'modernity', 'speed' and 'improved efficiency'.

Another more controversial strategy was that of 'planned obsolescence', in which advertising played a major role. New updated product models were continually brought onto the market to increase demand for the 'latest', technically up-to-date, model. Some products were even designed to wear out after a time!

Figure 3.2.4 *The Psion Series 3 palm-top computer provides an address book, diary and data transfer using infer-red technology to another computer or printer*

Task

These days legislation and standards ensure that good-quality products are manufactured. This was not the case in the past when planned obsolescence was designed into some products. Consumers were encouraged to throw away the old and buy the new, better product. Explain your views on consumerism and the 'throw away' society.

The 'new' industrial age of high-technology production

In the twentieth century, developments in materials technology, together with changes in lifestyle, revolutionised product design. New materials such as aluminium, stainless steel, heat-resistant glass, polyester, polypropylene and silicon enabled new ways of designing and manufacturing. In particular, the development of digital computers in the 1940s and the silicon chip in the 1960s enabled cheap portable computer technology, which transformed modern industrial society.

Miniaturisation

The most important technological development in recent years has been in the field of microelectronics. Not only have products have become smaller through advances in microchip technology, but previously unimaginable multi-functional products have been developed.

These include the first transistor radios and televisions in the 1950s and colour video recorders in the 1960s. By far the most influential product in the 1980s was the Walkman, which had an unprecedented impact on people's lifestyles. Current developments include portable CD players, multi-functional fax, e-mail and Internet browser telephones and palm-top computers.

In the past it was easy to recognise the function of a product, such as a typewriter or a watch. Nowadays, the impact of miniaturisation may mean that products effectively 'disappear', in that the product no longer visually represents its function, or the function is not clear. The designer then has to convey this function by other means. If a calculator is built into a wristwatch, is the product designed as a calculator or a watch? How is the purpose of such a product made clear?

Case study: Swatch of many functions

Swatch watches were first marketed in the early 1980s when the Swiss watch industry was fighting for survival against Japanese market expansion. Swatch decided to target the lower end of the market in order to produce affordable watches that were designed around the concept of style and fashion. The technological aspect took a back seat.

Now things have changed and the marketing climate is different. Technology is now a major driving force for design, having become seriously 'trendy'. The most recent developments by Swatch are interactive watches with so-called 'access technology'. This type of watch can act as a ski pass and metro ticket. It has already been in use in European ski resorts.

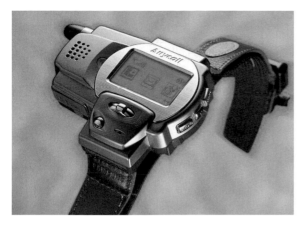

Figure 3.2.5 *Interactive watches, similar in style to this one, will allow the wearer to access the Internet and 'talk' to a PC*

Figure 3.2.6 *Globalisation includes having to design products for someone in another country. These 'adaptive spectacles' were designed for the developing world. The lenses are easily adjusted by the wearer to correct their vision – no sight test is required*

The next development for Swatch is the 'Internet Swatch', which has the potential for you to e-mail, access data from a PC, book tickets and browse the web from your watch! Swatch is also working on 'Swatch Talk', a watch that also acts as a mobile phone. The aim is to make people who wear the Swatch brand feel 'up to the minute', with the opportunity to be permanently connected to the net, allowing mobility and interactivity.

Task

Investigate other high-tech products that are interactive. How many functions does the product have? Why is it necessary for one product to have more than one function? Explain how the styling of the product enables the customer to understand the purpose of the product.

The global market place

The need to be competitive means that many companies sell their products all over the world. It can sometimes be a problem to design for unfamiliar markets or design products that will sell across different countries. Many companies have design teams situated throughout the world so they can design for a particular local market. Other companies use focused market research to discover the needs of specific niche markets.

Products that sell across the global market sometimes have to be remodelled to include different design features depending on where they are to be sold. Remodelling may involve many factors including:

- increasing the amount of recyclable materials used in the product
- whether a plug is fitted or not
- the fitting of different visual displays
- using devices to suppress noise levels.

Increasingly products are sold under the same name in different countries. For example, a kitchen cleaner recently changed its name to Cif to harmonise with the rest of Europe.

Global manufacturing

Global manufacturing is closely linked to the growth of multinational companies, which operate in more than one country. In the past, multinationals were mainly associated with mineral exploitation or with plantations, such as for cotton or food. Since the 1950s, many multinationals have been involved in global manufacturing, especially of cars and electrical goods. Today, global manufacturing is growing at an increasing rate, mainly due to international competition and developments in Information and Communications Technology (ICT). Global manufacturing covers a wide range of activities, such as:

- petroleum (BP, Exon)
- motor vehicles (Ford, General Motors)
- electrical goods/electronics (Philips, Sony, Hitachi)
- financial services (Barclays)
- food and hotels (Coca Cola, McDonalds, Trust House Forte)
- textile and garment manufacturing (DuPont, Marks and Spencer).

High-speed revolution

The high-speed information revolution that has come about through developments in ICT will continue to increase international competition. For many companies, competition means reducing labour and material costs. Global manufacturing is a means of doing just this, since moving manufacturing to another country can make use of lower labour costs, thus reducing one of the highest costs of manufacturing. The trend towards global manufacturing often includes designing products such as books or computer game covers in one country and printing them in another. Design studios, equipped with state of the art computer technology and broadband can receive briefs in any format, speeding up the turnaround from concept to the finished design. Systems are on-line 24 hours a day to allow continuous contact between clients, the design team and the film planning, plate making, printing and finishing departments.

Issues related to global/local production

Issues related to local/global production are concerned with the effects of the global economy and of multinationals on quality of life, employment and the environment. While the head offices of many multinationals are often located in developed countries (such as Western Europe), some multinationals are based in newly industrialised countries (NICs) such as Singapore or Taiwan. Developing countries, such as those in Africa or Asia, have generally welcomed multinationals and the benefits that locating manufacturing there brings. However, there are disadvantages for both developed and developing countries in global manufacturing.

Advantages of global manufacturing for NICs and developing countries:

• It provides employment and higher living standards.
• It may improve the level of expertise of the local workforce.
• Foreign currency is brought into the country to improve their balance of payments.
• It widens the country's economic base.
• It enables the transfer of technology.

Disadvantages of global manufacturing for NICs and developing countries:

• It can cause environmental damage.
• The jobs provided may only require low-level skills.
• Managerial roles may be filled by employees from developed countries.
• Most or all of the company profits may be exported back to developed countries.
• Multinationals may cut corners on health and safety or pollution (which they could not do in their home country).
• Multinationals can exert political pressure.
• Raw materials are often exported or not processed locally.
• Manufactured goods are for export and not for the local market (where many could not afford them anyway).
• Decisions are made in a foreign country and on a global basis, so the multinational may pull out at any time.
• With increased mechanisation, there is a reduced need for the local workforce.

Influences on the development of products

The design and manufacture of products is a complex affair. Why some products are successful, why they are made as they are and how they are used and disposed of are issues that affect every one of us. How do we choose which products to buy? Do we really need all of them? How do we recognise a well-designed product? Why is design important?

The Design Council recently conducted a survey of 800 UK manufacturers about the contribution of design to the UK economy. The results were very clear. Ninety two per cent of businesses

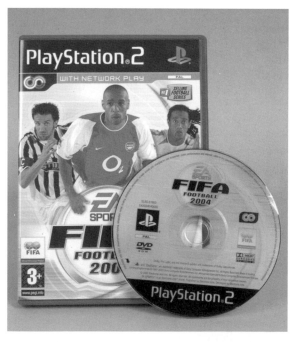

Figure 3.2.7 *Computer game covers can be designed in one country and printed in another*

agreed that design helps to produce a competitive advantage and 87 per cent believed it increases profits and aids diversification into new markets.

Design is clearly important. It is interesting to note that in sixteenth-century England the word 'design' meant a 'plan from which something is to be made'. Today, we normally use the term 'design' to mean the drafting and planning of industrial products. One outcome of industrialisation has been the requirement for the profession of the 'designer'.

Task

Global manufacturing highlights the many moral and ethical questions, the so-called values issues, that are inherent in product design and manufacture. Investigate further one of the following issues related to global economy manufacturing, to find out who are the winners and losers in terms of jobs, quality of life, and the environment:

a) What are the effects of the global economy on lifestyles in developing countries? Is it ethical to advertise products that many people can't afford? Should developed countries impose their values on traditional cultures?

b) Do multinationals have a responsibility to society and the environment? What are the effects of building new factories and transporting raw materials and products? What are the effects on energy demands in NICs and developing countries? Should multinationals follow sustainable manufacturing practices?

c) The effects of deforestation and over-use of the world's natural resources are plain to see. How can we avoid the effects of global warming or the loss of bio-diversity in many countries?

d) Moving manufacturing away from developed countries causes unemployment there. As competition increases, many NICs are also affected by further relocation of manufacturing, such as the move from Hong Kong to China. How can the threat of unemployment be overcome in a global economy?

e) Developing countries have to pay off debts to world financial institutions rather than spend money on food and the development of their own countries. This may mean a country exporting food to pay debt during a famine. Is this ethical?

Influences on design

For a company to make a profit, product design must be market led and market driven. It must take into account the needs of the target market group and the various influences of market trends, including colour and style. For many consumers these days, design has become an important means of self-expression. Consumers choose products not just for what they do, but for what they tell the world about them. Products are no longer simple, functional artefacts. We buy a perfume because it is a 'unique brand' and its packaging can express a sense of style and luxury. We buy a chair not just as a comfortable product to sit on, but also because it can express a sense of tradition or modernity.

Now that products are mass produced and sold in millions, the primary purpose of the designer is to inject a sense of personality in a mass-produced object. For example, the first telephone or early typewriters were the result of mechanical solutions and nobody had any preconceived idea of what they should look like. These days, when products are mass produced and sold in millions, it is the primary purpose of many designers to give the product a sense of personality.

Task

Investigate the changing brand identity of a packaged food item, such as breakfast cereals. What did the earliest packaging design look like in comparison with the current brand identity? Explain the differences in terms of the materials used, the shape, typography, styling and colour.

Product reliability is no longer an issue. Most products carry guarantees and there is consumer legislation to support product quality and safety. Most brands within a given product category perform equally well. The main reason for buying a product is how it looks – its aesthetic qualities. The product's function relating to its performance is often taken for granted. The job of the designer, then, is increasingly to give a product and therefore the consumer an image. Should the product look traditional, retro, or high-tech? At the same time, the consumer should understand how the product works, so that it is obvious how to operate it.

Form and function

The connection between form and function has been one of the most controversial issues in the history of design. When products were first mass-produced in Victorian times they were highly decorated to look like handmade products, whether the decoration was appropriate or not. The development of 'reform' groups such as the Arts and Crafts Movement gradually brought about a change in the concept of design. The form of products was to be simplified and the products well made from suitable materials. At the turn of the nineteenth century, developments in materials and technology enabled the production of innovative new products such as the telephone. Many of these products were so innovative that no one knew what they should look like! The development of mass production required that products be standardised, simple and easy to produce. The supporters of functionalism therefore suggested that the form of a product must suit its function and not include any excessive decoration.

From the functionalism of the early twentieth century, as far back as even the 1970s, the functional requirements of mass production were used as a benchmark for the form of an industrial product. In other words, for a product to be mass produced at a profit, it needed to be simple and easy to produce.

The argument about form and function still continues. As a student designer, you are asked to simplify your designs for ease of manufacture. One of the many reasons for doing this is to give you time to manufacture your product. The concept of form and function is complex. Think about the design of modern products and of the development of styling to market them. Think about developments in the technical design of products. Think about marketing and image. These days a product's packaging is thought to have three basic functions:

- its practical and technical function. Does the packaging contain, protect and preserve the product?
- its aesthetic function. Does the packaging advertise and describe the product?
- its symbolic function or the image it gives the user. Does the packaging project a desirable brand identity?

It is recognised today that design fulfils not only technical and aesthetic functions, but it is also a means of communication. Design tells the onlooker something about the user.

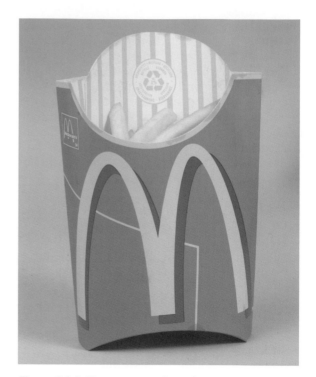

Figure 3.2.8 *The success of modern products is often related to brand image*

Task

Well-known brands use logos, symbols and typography to create the product image. Choose three branded products and their packaging. For each product explain:

- the target market
- the lifestyle of the people who buy it
- how the brand identity persuades the consumer to buy the product.

- Aesthetic properties relate to how a product looks, matching its styling with its end-use. As a student designer you need to develop an 'eye' for aesthetic characteristics, such as balance, line, shape and form, texture and surface pattern. Scale too is an important factor; some designers develop products by using change of scale as a starting point.
- Colour is a most essential characteristic of many products, as it is often the first thing we notice about the product. Colour is a powerful marketing tool, which can encourage consumers to buy. Colour in homeware products changes less frequently than fashion products.
- Mechanical properties relate to product performance, e.g. using ductile and tough mild steel for car body parts.

Task

Explain why image and style are increasingly important to the design of products. Choose two products that illustrate your answer. Explain why both are attractive to the consumer.

Design and culture

Throughout history, designers have been influenced by what they see and the products they use. All designers need starting points for design, whether they specialise in industrial, graphic, textile, or fashion design. The work of other artists and designers, both today and through history, have often provided such starting points.

The design movements that are included in the next section are some of the key ones in relation to the development of design. In a book of this type, it is impossible to include every possible influence on design – added to which all designers are influenced by different things. You will therefore need to read around the content of this section. You may also already have favourite designers or design influences that you will be able to draw upon when you answer the exam questions for this unit.

The Arts and Crafts movement

William Morris (1834–96) believed that it was essential to understand the production process in order to be a designer. This belief became one of the founding principles of the Arts and Crafts Movement with which Morris was closely associated. In 1875 he set up Morris & Co, which produced furniture, stained glass, wallpaper, fabrics, carpets and pottery. Morris was opposed to mass production because he believed that manufacturers were more concerned with quantity than quality. Morris saw the effects of industrialisation as polluting the environment, giving poor working conditions and producing low quality mass-produced products. He and others in the Arts and Crafts Movement believed in a return to workshop production using traditional techniques and the importance of the craftsperson creating the product. Morris believed in artistic design based on simple natural forms.

The Arts and Crafts Movement introduced a new direction in design, reviving an interest in simple, practical design. Comfortable wooden chairs, like the ladderback or Windsor, which had been used to furnish simple country cottages, now found their way into the homes of the wealthy. This type of simple undecorated

Figure 3.2.9 In 1864 William Morris began designing printed wallpapers which were hand printed from wooden blocks. His designs are still available today, surface printed with the patterns run slowly out of the machines, to emulate the soft edges and depth of colour of the hand-blocked originals

furniture created a demand for simplicity of design and fitness for purpose, representing the values of 'truth to materials and form'. The Arts and Crafts Movement had an important influence on Art Nouveau and the Bauhaus Movements. In this sense, it can be said to form the basis of modern design and, through its influence on Scandinavian design, to have influenced much of the design in Europe (see Figure 3.2.9).

Art Nouveau

Art Nouveau was an important design movement in France around 1900 and soon developed into an international movement. The main characteristic of Art Nouveau styling was its use of flowing lines, stylised climbing plants and water lilies. In Scotland, Charles Rennie Mackintosh was also inspired by geometric forms based on Japanese art.

Art Nouveau was an influential decorative style used in architecture, glass, jewellery, fabrics and wallpaper. The entrances to some metro stations in Paris are typical of the Art Nouveau style (see Figure 3.2.10). Although Art Nouveau designers were much more interested in using modern materials and mass production, some considered themselves to be artists and rejected industrial production methods. Their designs were only available to the wealthy, such as Rene Lalique jewellery made from semi-precious stones, glass and gold. Art Nouveau could be described as fitting in somewhere between art

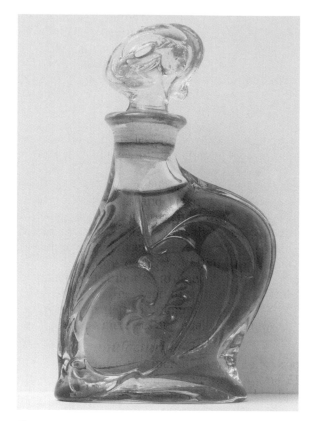

Figure 3.2.10 *The design of this perfume bottle was inspired by the Art Nouveau decorative style*

and industry. Although it encouraged a return to craft work and used some mass production techniques, some say that it delayed the development of modern industrial design. Around 1910 the Art Nouveau style virtually disappeared.

The Bauhaus (1919–33)

In Germany, the Bauhaus school became the centre of modernism and functionalism. It laid down design principles that still influence the teaching of design and industrial design. Many of its products still look modern today. The principal aims were to use modern materials and to combine the concepts of form and function.

The Bauhaus was founded and run by Walter Gropius between 1919 and 1928. The basis of Bauhaus education was a preliminary apprenticeship centred on free experimentation with colour, form and material. The goal was to offer an equal education in artistic and handicraft skills that were linked to industry. After the preliminary course, students chose one of the commercial workshops for carpentry, pottery, metalwork, textiles, stage design, photography or commercial art. Bauhaus training was given by important artists such as

Johannes Itten, Lyonel Feininger, Paul Klee, Georg Muche and Oscar Schlemmer. This was the first time that so many artists were involved to this extent in the teaching of design.

In 1922, Laslo Maholy-Nagy was appointed director of the Bauhaus metal workshop and instructor of typography. He designed all the Bauhaus books and opened a graphics studio in Berlin in 1928. He experimented with light, film and plexiglass, developing the dimmer switch in 1930.

As the Bauhaus developed its influence, the first industrially useful designs were produced in the metal workshop. Students were encouraged to use new materials such as steel tubing, plywood and industrial glass. The aim was to produce economical mass-produced products, such as table lamps, metal teapots and chairs.

The Bauhaus was closed in 1933 by the Nazis in an effort to stop modernism in design. Gropius and Maholy-Nagy moved to the USA in 1937 and the New Bauhaus was founded. There the Bauhaus style became known as the 'International Style', which was mainly concerned with architecture. However, architects such as Le Corbusier and Alvar Alto also designed furniture that followed the principles of the Bauhaus.

Figure 3.2.11 *This Bauhaus table lamp, first made in 1923–24, was one of the most successful to come from the Bauhaus metal workshop. It was made from industrially produced metal and glass. Although originally hand made, it is mass produced today*

Figure 3.2.12 *In 1920s Paris, many artists were inspired by developments in art and technology. Fernand Leger was also influenced by the teachings of the Bauhaus and by Cubism, as can be seen in his 1924 painting* Elements Mecaniques

Figure 3.2.13 *This 1925 enamelled Art Deco buckle combines geometric blocks of solid colour, influenced by Cubist art*

Task

The work of other designers and artists has always been used as a starting point for design work. It is a part of design development to build on what has gone before. Choose an artist whose work you find attractive or inspirational. Try to decide what it is that you like about it. Make sketches of the aspects of the work that you like and try to integrate these aspects into design work for your next project.

Art Deco

Art Deco originated in France and its design style was thought to be highly modern and elegant. It was influenced by the Cubist painting of Picasso and Braque, and African and Egyptian art. Art Deco focused on exclusive designer-made products, using geometric shapes, zig-zag patterns, chevrons and sunbursts, and expensive ivory, bronze, crystal, and ebony (Figure 3.2.13). The movement was at its height in the 1920s and 30s and although it was initially opposed to mass production, its extravagant style gradually made use of more modern materials. It influenced the design of many mass-produced products made from new materials such as aluminium, chrome, coloured glass and Bakelite. Art Deco was a popular style for interior design and architecture.

New Art Deco materials

Aluminium, plywood and Bakelite were used extensively in mass-produced Art Deco products,

Figure 3.2.14 *The Vespa scooter was famous all over the world*

such as jewellery, cigarette cases, perfume bottles and radios. Bakelite, invented in 1907, was first used as a substitute for wood and was carved out of solid blocks. It was, however, perfect for moulding into the smooth, streamlined shapes of 1920s and 30s electrical products because it was malleable, durable and inexpensive.

Figure 3.2.15 *In this 1967 record cover for Sgt Pepper's Lonely Hearts Club Band, the famous stars were made from life-sized cardboard cut-outs and wax models*

Modern design and styling

In the 1950s the development of the supermarket meant that food packaging became a marketing tool. Packaging had to be instantly recognisable to the consumer. 'Italian style' products became all the rage. The Vespa scooter, the Fiat 500, the Lancia, Alpha Romeo and Ferrari cars, furniture, espresso coffee machines and fashion products were set to become sought after design classics. In the 1950s and 60s new synthetic materials were used to create colourful furniture which was durable, easy to clean and relatively inexpensive. In the 1960s, design was

Figure 3.2.16 *Many Memphis products, such as this teapot, were inspired by children's toys*

seen as being an essential part of Italian culture and designers differentiated products by giving them names. For example, the Sottsass 1969 typewriter was called 'Valentine', giving the product a sense of personality.

Youth culture

In the 1960s youth culture had an enormous impact on fashion and design. For the first time young people were seen as an emerging group of consumers, different from their parents. Powerful television advertising and the use of new materials, shapes and bright colours led to the development of mass consumerism. Pop music and the 'hippie' movement influenced graphic design, fashion and interior design (Figure 3.2.15). Images from daily life such as soup cans (Andy Warhol) became a part of art. In packaging design, bold colours and strong visual images were used to communicate the message to buy. Cellophane, aluminium and plastics enabled the development of new types of packaging such as 'throw away' cans with ring pulls.

By the early 1970s other influences had an impact on design. The combination of space exploration and science fiction films suggested infinite technical opportunities for product design. For example, the film *2001: A Space Odyssey* and television programmes like *Star Trek* inspired designers to create futuristic products. Improvements in packaging technology included the Tetrapak and moulded plastic containers that were lighter and cheaper to transport than breakable glass.

Memphis

Memphis was an Italian design group led by Ettore Sottsass. Originally an architect, Sottsass became a consulting designer for the typewriter manufacturer Olivetti. Memphis was the most important design group of the 1980s and designed a variety of furniture, glass and ceramic products specifically for mass production (Figure 3.2.16). Memphis designers loved the fast changes brought about by fashion and their witty, stylistic design was influenced by comic strips, films and punk music. They combined materials such as colourful plastic laminates (such as melamine and formica), glass, steel, industrial sheet metal and aluminium. Many products were inspired by children's toys. In the 1980s the Memphis group introduced a new importance to design, so that the status of design itself grew and took over a key role in the development of individual lifestyles.

Design after Memphis

A number of designers, such as Ron Arad, Jasper Morrison and Tom Dixon, moved away from functionalism and focused on one-off design, using concrete, sheet metal, plywood and rubber. Other designers, concerned about built-in obsolescence, started to use recycled materials and to design more energy-efficient products. For example, the French 'super designer' Philippe Starck designed some furniture that can be recycled through the use of screw fixings rather than gluing the component parts together. Some of his other products use ecologically sound materials, like his television set made from MDF, which gives the product a more human face than using hard edge black plastic.

Task

Investigate the work of one design group. Produce annotated sketches to show the type of products that were made by designers in that group. For each product describe its style and function. Explain how the materials used in the product influenced its design.

New materials, processes and technology

In the twentieth century, developments in materials technology and changes in lifestyle revolutionised product design. New materials such as aluminium, stainless steel, heat-resistant glass, polyester, polypropylene and the silicon chip enabled new ways of designing and manufacturing. This led to the development of multi-functional products that look 'technical'. These high-tech products are more durable, easier to clean, and more exciting to use. Designers like Michael Graves and Philippe Starck have created many multi-functional products that have 'street cred'; products that provide function with image.

The development of digital computers in the 1940s and the silicon chip in the 1960s, has enabled cheap portable computer technology, which has transformed modern industrial society. The most important technological development is in the field of microelectronics, enabling products to become smaller through advances in microchip technology.

Smart materials

Smart materials can be plastics, metals, textiles, ceramics or liquids. Their physical properties can be altered in response to an input or changes in their surroundings. Examples of smart materials include the following:

Figure 3.2.17 *Piezo-electric actuators are used in greetings cards to produce a sound when the card is opened*

- Piezo-electric actuators, used in greetings cards, to produce a sound from an electrical signal when the card is opened.
- Polymorph is a modern material that becomes soft and pliable in hot water (62 °C) and hardens when cool. It can be used for rapid prototyping graphic products.
- Light emitting plastics (LEPs) – flat, flexible, plastic displays, made by sandwiching a thin layer of polymer between two electrodes. When a low voltage is applied, polymers with different properties can emit red, blue and green light. LEP displays are more robust, thinner and more energy efficient than current LCD displays and are likely to be used in hoarding advertisements, safety signage, mobile phones, CD players, TV and computer monitors.
- Smart ceramic material, which absorbs light energy and re-emits it. Used for glow-in-the-dark products such as watches, emergency signs or torches.
- Thermochromatic materials, containing billions of microscopic liquid crystal capsules that change colour at specified temperatures. Used in kettles, feeding spoons, thermometers and battery test strips.

119

Modern production techniques

Modern materials require the development of new manufacturing processes and components, such as the following:

- The combination of powder metallurgy and injection moulding to produce very small accurate components with superior mechanical properties.
- A new technique for manufacturing self-chilling aluminium cans. The can has a layer of water gel on its inner surface, sealed inside a vacuum. Underneath the can is a second part containing a clay drying agent. Twisting the two parts allows vapour to flow from the water gel to the clay drying agent, where it is absorbed. This takes the heat out of the drink and deposits it in the clay absorber – so the drink becomes cold and the absorber becomes hot.

Developments in plastics

In the 1950s the use of plastics such as acrylic, PVC or polypropylene revolutionised the furniture-making and packaging industries. Plastic could be made in any colour and was malleable, light, inexpensive and modern. Plastic products were durable and could be produced in any shape or size using a moulding machine.

The popularity of plastic had a downturn after the 1973 oil crisis. It was no longer called modern or high tech, but cheap, tacky, tasteless and unecological. Plastic's popularity may improve in the future with the development of new eco-friendly biopolymers:

- Enpol – a fully biodegradable plastic with comparable strength and cost to polythene, uses 2.5 times fewer materials to achieve the same performance as conventional plastics.
- d2w™ – a degradable polythene packaging already used by some supermarkets. Although it degrades totally by heat or sunlight in landfill, it can be recycled in the same way as non-degradable plastics.
- Eco Foam – made from chips of foamed starch polymer, is water soluble, reusable and free from static. It can replace polystyrene packaging materials.

Computers and design

The development of powerful user-friendly home computers and Information Communications Technology (ICT) has enabled changes to the function of many consumer products. For example a multi-functional wrist computer, designed in 1988 included a compass, watch, telephone and city map. Digital technology features in many modern products like refrigerators and cameras, but its biggest impact is in the use of CAD/CAM in design and manufacture. Modern software programs enable product simulation on screen and the modification of technical data, without the need to produce physical prototypes. The use of CAD graphics has led to new approaches to typography, layout and image making. The 'Mac-to-plate' process has revolutionised the digital press, and digital printing is increasingly used in fabric printing.

Miniaturisation

The invention of the silicon chip has transformed modern industrial society. Without microchips a computer would be the size of a living room and a pocket calculator the size of small car! By the 1970s, thousands of electronic components could be printed on top of a single silicon chip measuring only 5 mm square. This requires a tiny amount of sand and very little energy in use.

Micro and Nano technology are being used in exciting developments, such as the world's smallest silicon gyroscope. This is tough and rugged because it is solid state with no mechanical moving parts. In medicine Nano machines can travel round the body and clear blood clots. The interface between humans and machines is set to become one of the most challenging aspects of the future.

Eco-design and environmentally friendly processes

Consumers have grown accustomed to the benefits of mass production and the supply of inexpensive, well designed, easily available products. Unfortunately mass production is not sustainable. Environmental legislation, such as the banning of lead from electronics by 2008, is encouraging the use of low-tech sustainable technology, recycling and the reuse of materials.

For a product to be eco-friendly, it must be durable and recyclable and have an aesthetic appeal. For example:

- with a few minutes winding, the Freeplay flashlight produces a steady, reliable light source without the need to consume toxic batteries (see Figure 3.2.18)
- the recycled pencil is made from recycled polystyrene cups
- electric and solar cars have been developed to reduce emissions, and reduce the weight of the car through an increased use of plastics and alloys. The well-designed car uses little fuel, produces few emissions, lasts a long time, is easily repaired and at the end of its life can easily be broken down and recycled.

Figure 3.2.18 *The Freeplay wind-up flashlight*

2. Professional designers at work

The relationship between designers and clients, manufacturers, users and society

The role of the designer

These days product designers need to be creative and have an understanding of technology, so they can develop products that match the quality, price and availability requirements of the consumer. Product design incorporates the identification of needs and opportunities, generating design ideas, using an understanding of materials and process technology and satisfying consumer demand.

In large companies designers usually work in a design and production team. A small company may employ only one designer, but very few work totally alone. It is the responsibility of everyone in a company to get the product to market on time and to budget.

Good designers take account of their company design policy, resources, target market profile and marketing objectives. For many industry sectors effective marketing is about branding, image and developing a competitive edge.

The role of design and production teams

Design and production teams include creative and open-minded thinkers, problem solvers, technicians and financial planners. A designer fulfils some of these roles but it is the responsibility of the whole team for ensuring that the product can be produced efficiently to time and budget. The team also has to make sure that consumers are assured of safe, high quality products.

In order to develop a product that matches the quality and price requirement of the target market, designers and production teams need to undertake some or all of the following activities:

1 Identify needs and opportunities:
  - use market research to identify target market needs
  - research existing and new materials, processes and technology
  - develop a design brief and specification, including aesthetics, performance requirements, production constraints and quality standards relevant to the product.
2 Design, including the use of CAD:
  - generate and develop ideas
  - test the feasibility of ideas, models and prototypes against specifications and market needs.
3 Production planning:
  - produce working drawings and specification sheets
  - plan the timing of key production stages
  - produce a production schedule
  - plan resource requirements and production costs
  - plan quality checks during production.

The aesthetic role

Many consumers are concerned with how a product looks and the image it gives them. With this in mind, the role of the designer is increasingly to give a product a 'personality'. The aesthetic role is therefore one of creating and developing innovative, attractive products that meet market needs. Designers also need to be aware of future market needs, moral, cultural, social and environmental issues and the competition from other products.

Function and technical role

All products must be designed with function and performance in mind. Some products, such as seat belts, are produced specifically for their functional performance rather than for aesthetic qualities. The functional role of the designer therefore is to keep up to date with technical information about materials, processes and finish and to understand the competition from other functional products.

Economic and marketing role

Consumers want to buy innovative, attractive products at a price they can afford. The economic and marketing role of a designer is therefore to design marketable products to a price point. In other words, designers must work to target production costs, which are established at the start of a project. Target production costs are based on a study of the design, development and manufacturing costs of the product. They are also checked against the cost of existing similar products. In this role the designer needs to be aware of the market into which the product is to be sold and should have a clear understanding of production processes and costs.

Organisational and management role

In the past, the design development process was consecutive, with each department contributing to the overall process before handing over the product to the next department on its way to production. In this scenario, the organisational and management role of the designer is limited. Many companies these days use the concept of concurrent manufacturing. This brings together all the different departments (marketing, design,

production, quality assurance, etc.) to work concurrently on product development. Designers are increasingly involved in the whole organisation and management of the product and generally work as part of a design and production team. This team shares all the information about the product, using software such as Product Data Management (PDM). This enables fast and easy communication between design, production, suppliers and clients and results in a faster time to market of products that meet customer needs.

Case study

The airline Go was initially set up to find a gap in the market between the national carriers and the 'cheap and cheerful' airlines. HHCL, one of the country's most respected advertising agencies, worked closely with Go to develop and launch it as the 'low cost airline from British Airways'.

Identifying the need and opportunity

The Go brand launch and its advertising campaign was achieved through close collaboration between the client, Go, and the advertising agency. The client/agency project team was briefed to identify the meaning of the brand to the consumer and to develop a complete corporate image. The team identified two very distinct types of airlines; the quality national ones and the low-cost carriers. They found that modern travellers wanted something in between, so decided that the Go brand should be a quality, low-cost airline. The team then focused on creating an appropriate brand image.

Design development

HHCL worked with the design agency Wolff Ollins to develop a strong, simple, clever corporate image that would appeal to the modern

Figure 3.2.19 *The Go brand image*

traveller. The resulting brand image of coloured circles made the Go brand instantly recognisable to the target market. The design team developed a complete corporate package, from the headed stationery to the plane's livery. The advertising aimed to be simple and easy to manage, in order to keep up with expected customer demand.

The success of the product

The collaboration between the client and agency teams resulted in fast generation of ideas and a hectic, but smooth, launch. Every media opportunity was taken, from television to sandwich bags, to ensure that advertising kept up in a highly competitive business environment. This resulted in the target market recognising and understanding the brand and advertising. The product was so successful that it was bought out by its rival Easyjet.

Professional practice relating to design management, technology, marketing, business, ICT

Design and marketing

Although market research is often done by the marketing department, all designers need to be aware of market trends. Sometimes individual designers establish a product need, such as James Dyson who developed the Dual Cyclone vacuum cleaner.

One of the key features of successful marketing is the development of a marketing plan aimed at the needs of the target market group. It involves developing a competitive edge through providing:

- well-designed, reliable, high-quality products at a price consumers can afford
- products with a desirable image that will appeal to the market.

Target market groups

Successful product development needs a clear target market group. This is often decided by market segmentation, which divides the target market into different groups. There are many ways of dividing up consumers into market segments, including:

- age – using demographic information to target a specific age group
- level of disposable income – targeting the available spending power of consumers
- lifestyle – targeting the lifestyle and brand loyalty of specific consumers
- product end use – targeting market-specific products such as multi-purpose tools or furniture.

Marketing plan

A marketing plan can be used to promote products and brands using retailers, newspapers, TV, radio and the Internet. A successful marketing plan uses market research to find out:

- consumer needs and demand
- the age, income, size and location of the target market group
- the type of product customers want and how much they will pay for it
- economic trends affecting the market, such as spending power in relation to home mortgage interest rates
- the competition from existing products
- the time required to develop and market the product, such as launching a product just before Christmas.

Efficient manufacture and profit

On average, 80 per cent of a product's cost results from its design. Design management is therefore an essential element of efficient manufacture. It provides the profit that is essential for funding the research and development of new or improved products. Efficient manufacture involves the efficient use of resources, using the most suitable materials and manufacturing processes.

Capacity and efficiency

Organisations generally look at the forecast demand for a product and use this to find the production capacity needed. If capacity is less than demand, orders are not met and potential customers are lost. If capacity is greater than demand, there is spare capacity and under-used resources. The basic measure of manufacturing performance is therefore capacity. This is defined as the maximum numbers of products that can be made in a specified time. All processes have limits on their capacity, such as factory output per week or machine output per hour.

If, for example, a manufacturing process has a capacity of 200 units per week, this is the maximum number of units that can be made. If the manufacturing process is idle for half the time and actually makes 100 units, the utilisation of its capacity is 50 per cent. If it uses 25 hours of machine time to make these 100 units, the productivity is four units per machine.

When measuring the efficiency of a manufacturing process a comparison needs to be made between the actual output of products with the possible output. If, for example, an office worker can process five documents per hour, but actually processes four documents, their efficiency is $\frac{4}{5} = 0.8$ or 80 per cent.

Modifying design and manufacture to achieve efficiency

Many companies analyse the manufacture of existing products to find out if the product design can be modified to improve manufacturing efficiency. The aim is to reduce production costs by creating designs that use less material or energy during manufacture and to reduce the production of waste.

Changing the product design and manufacture to increase efficiency may involve using:

- a simpler design with fewer components to reduce assembly time
- different materials to reduce their weight or the quantity used
- materials that result in less waste
- materials that use less energy during manufacture
- different shaped components to make them easier to machine or mould
- different machining of components to reduce the production of waste
- a simplified or different speed production process so it is more efficient.

Figure 3.2.20 *The new '202' Coca-Cola can is a good example of 'lightweighting'*

Factfile

'De-materialisation' is the process of reducing a product's material content and achieving the product function in another way. 'Lightweighting' is making a product lighter while maintaining its function.

Case Study

Coca-Cola Enterprises Ltd (CCE Ltd) produces and distributes around two billion soft drinks cans per year. In the early 1990s, many people switched to plastic bottles, putting pressure on can manufacturers to reduce costs. As further lightweighting was not thought possible with the existing '206' can design, a new '202' can was designed (Figure 3.2.20). It had a reduced end diameter, but the same can body diameter. The change of design had to take account of the can manufacture, the filling process, the can strength and stackability for distribution. CCE Ltd secured the agreement of European can manufacturers to a common specification for the '202' can end and body.

Before the redesign of the '206' can there were no common specifications for cans and can ends in Europe. This meant that seaming machines had to be reset for different suppliers' cans and different end profiles. The achievement of a common specification for the '202' can has eliminated this problem.

The new design reduced raw materials costs and simplified the production processes.

- Changing from the '206' to the smaller '202' can gave cost savings of over £1 per thousand cans and reduced metal use worldwide in the canned drinks industry.
- CCE Ltd saved around £2.3 million/year from 1995 onwards with a payback period of less than two years.
- The use of lightweighted cans enables more products to fit on a single truck, reducing the number of truck movements.

Task

Choose a product that is easy to take apart, such as a torch. Carefully disassemble the product and keep a record of how it fits together, so that you can put the product back together again! Analyse the materials, components and processes used to manufacture the product. Redesign the product to make it more efficient to manufacture or use.

Aesthetics, quality and value for money

Aesthetics

Industrial products, such as tyres or oil filters, are made solely to meet performance requirements.

However, the impact of aesthetics on the design of industrial and technical products is increasing. In the car industry, for example, colour and style forecasting is used to predict the future aesthetic needs and wants of consumers.

Quality and value for money

In a competitive market no product will sell if it is of poor quality. Consumers expect to be able to buy reliable, high-quality products at a price they can afford. Good quality for the consumer is often described as 'fitness-for-purpose'. This can be evaluated through a product's performance, price and aesthetic appeal. All of these criteria must be met if a product is seen to be 'value for money'.

Quality for a manufacturer means meeting the product specification and finding a balance between profitable manufacture and the needs and expectations of the consumer. This often involves juggling the competing needs of function, appearance, materials and cost. Quality is therefore an important issue because there often has to be a compromise between quality and cost. There will always be a need for products at different cost levels, to meet the different spending needs of consumers.

Values issues related to design

Values issues are inherent in designing and making and in many ways are a 'driving force behind design'. One example is the increasing interest in the environment and the fact that one of the many roles of the designer is to design with recycling in mind.

As with other aspects of designing, there is often a conflict between values held by the designer, the client and the user.

- Cultural and social values are related to the way in which aspects such as fashion and lifestyle affect design. Trends in colours used for clothes, for example, influence colours in cars, interior and packaging design. Social influences from the media such as film, television and music also influence design. Exhibitions also stimulate a revival of interest in design influences such as a recent exhibition of Art Nouveau.
- Economic issues mean that there is a need to reduce costs and maximise profit if companies are to survive. This, and developments in ICT, have led to a shift in manufacturing away from traditional areas to new developing countries where labour costs are cheaper.
- Values issues are often very sensitive and companies need to approach these issues with great care. The clockwork radio, designed by Trevor Davis, was designed to help those living in poorer countries of the world by providing a communications link at low cost. When production started, it was based in South Africa and was largely done by people with physical disabilities. In December 1999, the company involved announced that production was being moved to China to save on production costs. This one decision covered a wide range of values issues.
- When operating in the global market place companies have to be careful to use product names which do not cause offence to religious or cultural views. Consideration also needs to be given to the ethics of imposing values from the traditional industrialised nations on those countries which do not have the same tradition. For example, computers that are advertised on television worldwide are no use to people who do not have electricity, but who require the basic essentials of life. Is it moral to advertise such products and to create a demand for them?

The work of professional designers and professional bodies

Professional designers

It is impossible in a book of this type to include information about every influential product designer. The following designers are therefore intended as a starting point. It is not necessary for you to study every designer listed, you could study the work of two or three designers whose work you admire. Keep a sketch book with a collection of images and notes about their work, including the kind of materials they use, the types of product they design and the market they design for.

Ron Arad (1951–)

Arad originally studied architecture in Israel and later studied in London. His view of designing is that it is essentially a form of expression for himself and for the client. Arad's early work was intended to represent the decay of the post-industrial scene and this accounted for his choice of materials – industrial materials and recycled parts, making extensive use of concrete and steel sheet metal. Arad came into the public eye with the design of his 'Rover' chair in 1981 – a chair which used recycled leather seats from Rover cars (see Figure 3.2.21). In 1981, Arad founded One-Off Ltd, producing individual pieces of furniture, and he is now one of the most creative designer-makers (see Figure 3.2.22).

Figure 3.2.21 *The 'Rover' chair designed by Ron Arad in 1981*

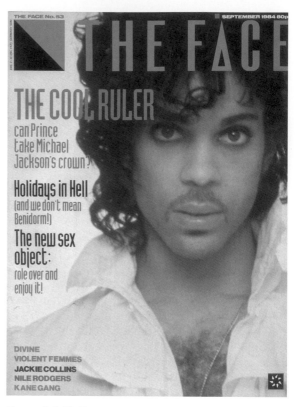

Figure 3.2.23 The Face *magazine had a cult following in the 1980s*

Figure 3.2.22 *The 'bookworm' bookcase, designed by Ron Arad in 1992. As with many other successful ideas, this is a very simple idea of a continuous strip of sheet metal formed in a one-piece curve*

Neville Brody (1957–)

Neville Brody studied graphic design at the London College of printing and started his career designing record covers. In 1981 he was appointed art director of the ground-breaking music and style magazine *The Face*. The introduction of the Apple Mac in 1984 enabled Brody to experiment with design elements such as letters, symbols, logos and surface texture. He manipulated existing typefaces and used them as dramatic graphic images. His 'Brody font' was *the* graphic style of the 1980s and his distinctive page layout and type design attracted a cult following. His work influenced the design of magazines, record covers and packaging. In 1990 Brody and the writer Jon Wozencroft started a new digital magazine called *Fuse*, in which designers were invited to develop new fonts. Fuse was in the form of a computer disk that included copyright-free fonts so that the user could also experiment with typefaces.

Peter Saville (1955–)

The London-based graphic designer Peter Saville was the founding partner and art director of Factory Records where he designed corporate ID for music. He has created the visual identity of bands such as New Order, Joy Division, Ultravox,

Figure 3.2.24 *Suede's 'Coming Up' album sleeve*

Figure 3.2.25 *Tom Dixon's 'S' chair*

OMD, Wham!, George Michael, Pulp and Suede. Saville's style uses images from art, fashion and photography and then manipulates them to abstraction. Saville bases his work on his knowledge of the client and their music, and on what he feels the group's audience want to see. For the Suede album, Saville and Suede's lead singer, Brett Anderson, came up with an image that would inspire the 'Coming Up' sleeve. Saville invited the fashion photographer Nick Knight to shoot the image, which he then manipulated using Paintbox software to encapsulate Suede's world of 'Filmstars' and the 'Beautiful Ones'.

Tom Dixon (1959–)

Tom Dixon was one of a group of designers in the 1980s, including Ron Arad and Jasper Morrison, who were inspired by Punk and worked with recycled materials and welded metal. These designs combined aspects of art, craft and sculpture and were known as 'creative salvage'. Dixon was self-taught and developed more commercial designs in the late 1980s. The organic curve of the 'S' chair is based on a sketch of a chicken and he worked on over fifty prototypes (see Figure 3.2.25). These were made using rush, wicker, old tyre rubber, paper and copper.

Throughout his career, Dixon was not approached by any British manufacturer, but in 1987 the Italian company Capellini bought the 'S' chair design and mass produced it. In 1994,

Dixon opened his SPACE shop and in 1996 launched a new product range called Eurolounge. In 1998, Dixon was appointed head of design at Habitat. He is re-instating a Habitat workshop for the design team to work on new ideas and prototypes. At Habitat, Dixon aims to reproduce some design classics and to put into production designers' prototype products.

James Dyson (1947–)

One of Dyson's early products was the Rotork Sea Truck, a marine vehicle that could carry a three-ton load at 50 mph. Other products include the Wateroller, a lawn roller which used water to give it weight and the Ballbarrow, a redesign of the traditional wheelbarrow.

Dyson spent five years and made more than 5000 prototypes to develop the highly successful Dual Cyclone vacuum cleaner, produced for the mass market in 1993. The cyclonic system uses the principal of centrifugal force to suck up air and revolves it at eight hundred miles an hour through two cyclone chambers, until the dust drops to the bottom of a transparent cylinder.

Wolff Olins

In 1965 graphic designers Wally Olins and Michael Wolff set up the design agency Wolff Olins. Over a number of years, the design agency developed the corporate image for a wide range of companies such as P&O, Q8, British Telecom and the airline Go. Wolff Ollins were able to develop strong, simple brand images that

Figure 3.2.26 *The design agency Wolff Olins designed the corporate image for Q8*

appealed to the client and the consumer. They developed complete design packages, from the company logo to the stationery and company uniforms. In the late 1980s Wolff Olins designed a new and somewhat controversial corporate image for British Telecom, which used the initial letters BT together with the outline figure of a piper.

Michael Graves (1934–)

Graves is another architect who has also worked on the design of products, mainly furniture and ceramics. His early work was inspired by ideas from art history, particularly classicism and cubism. Graves is one of the leading exponents of post-modernism and has designed in the Memphis style for Alessi.

Jasper Morrison (1959–)

Morrison trained as designer at the Kingston School of Art and Design in London. During the early 1980s, he produced a range of designs including his 'Handlebar' table, constructed of wood, two bicycle handlebars and a circular glass top. Morrison designs for companies such as Vitra, Cappellini and Artemide. Some of his most important pieces were his storage units, known as the 'Universal' range, designed for the Italian company Cappellini.

Morrison is a designer who is committed to practical, simple ideas, which he has successfully developed, paying close attention to detail and high-quality making (see Figure 3.2.28). He is known for his minimalist plywood furniture with untreated surfaces.

Philippe Starck (1949–)

The French designer Philippe Starck was described as the 'super-designer' of the twentieth century. He happily borrows design styles from the past, such as 'streamlining' or Art Nouveau. Starck uses unusual combinations of materials such as plastic with aluminium, glass with stone or fabric with chrome. In the early 1980s he designed the interior of the Café

Figure 3.2.27 *The Graves kettle, produced in 1985 for Alessi, is made from steel with a polyamide handle. The plastic bird mounted on the spout sings when the kettle boils*

Figure 3.2.28 *Wine rack by Jasper Morrison*

Figure 3.2.29 *This MDF and plastic television designed by Phillipe Stark gives the product a more human face than using hard edge black plastic*

Costes in Paris, including a three-legged chair, which has come to signify modern restaurant design. He designs saleable, relatively inexpensive products for mass production. These include many well-known products such as his Juicy Salif Lemon Press, lighting (the Miss Sissi Table Lamp) and everyday products such as furniture, televisions, water taps, clocks and toothbrushes. Starck also designs some furniture that can be recycled and often uses ecologically sound materials such as in the MDF television set shown in Figure 3.2.29.

Ettore Sottsass (1917–)

Sottsass is one of Italy's best-known designers, whose work spans over 40 years. In each decade, he has produced innovative work that seems to express the period. In 1981, Sottsass launched the design group Memphis, producing exciting and witty new designs that sometimes look like funny children's toys. They mixed plastic laminates with expensive wood veneers in bright primary colours. Sottsass's influences came from 1960s pop and his fascination with the ancient structures of Indian and Aztec art. Sottsass has worked on a wide range of products such as typewriters, furniture, computers, glass products, lighting and many more, for international companies including Knoll, Cleto Munari and Artemide. Many younger designers have worked with Sottsass, who has been described as one of the most influential designers of the late twentieth century.

Professional bodies associated with designing

The Crafts Council

The Crafts Council is a national organisation for the promotion of crafts. It provides exhibitions, a shop selling crafts and books, a Reference Library, the National Register of Makers with contact information for over 4000 UK craftspeople and Photostore, a visual database containing over 40,000 images of selected high-quality craft work. Major craft exhibitions often go on tour around the country.

The Crafts Council can help designer-makers by providing practical help and advice on business matters and as a source of information and reference materials. Potential clients can obtain information about craftspeople and see illustrations of their work. Students can also use the same information to research contemporary design and craft work as a source of products for evaluation and inspiration.

The Design Council

The Design Council aims 'to inspire the best use of design by the UK, in the world context, to improve prosperity and well being'. The Council does this through the use of events, publications, research, educational resources and case studies which highlight examples of design and innovation in action. Annual events include Design in Business Week and Designers into Schools Week, which brings the real world of design into the classroom by enabling Design & Technology students to work with professional designers. This initiative gives students a unique opportunity to build their creativity and problem solving skills and helps them see Design & Technology as a creative force shaping everyday experiences.

The Engineering Council

The Engineering Council (EC) was created in 2002 as the main professional body for engineers, technologists and technicians. The Council supports designing and making in schools and colleges through the Neighbourhood Engineer Scheme and the Women into Science and Technology programme (WISE).

3. Anthropometrics and ergonomics

The basic principles and applications of anthropometrics and ergonomics

Ergonomics

Ergonomics is the science of designing products for human use, and matching the product to the user. It has a role in designing everything from toothpaste to cars and its application makes products usable. Ergonomics is therefore an essential part of the design process. Sometimes products are matched to a single user, where the product is customised to suit one person. The main objective for ergonomists is to improve people's lives by increasing their comfort and satisfaction when using products. When ergonomics is applied to the workplace it can help improve productivity and reduce errors and accidents.

Anthropometrics

Anthropometrics is a branch of ergonomics that deals with measurements of human beings, in particular their shapes and sizes. For many products, complex data is required about any number of critical dimensions relating to the human form, such as height, width or length of reach when standing or sitting. In some products a single critical dimension is all that is needed. In general, products are matched to a target population of users who come in a variety of shapes and sizes.

Using anthropometric data

When applying anthropometric data to a design problem, the designer's aim is to provide an acceptable match for the greatest possible number of users. This is achieved by the use of anthropometric data. Many companies use anthropometric data charts from the British Standards Institute (BSI), which are available in simplified form from the *Compendium of Essential Design and Technology Standards for Schools and Colleges* (see your department copy or visit www.bsi.org.uk/education). Simple data charts relating to measurements for men, women and children can also be found in the clothing sections of mail order catalogues. In your own designing you may need to collect your own data if you are designing for a specific individual or for a small target market group.

Statistical data available from the BSI is associated with average heights, in which only one person in twenty is shorter than average (the fifth percentile) and only 5 per cent of people are taller than average (the ninety-fifth percentile).

The BSI anthropometric data therefore covers the 90 per cent of the population who fall between the fifth and the ninety-fifth percentile. For example, if you design a chair which has to be comfortable for 100 people with a range of heights from 1.5 metres to 1.9 metres, designing to fit the whole range of sizes would be difficult. According to the principles applied by the BSI, you would need to ignore the smallest five people and the tallest five people and design the chair to fit the remaining 90 people in the group. This principle is applied to the design of most products.

Ergonomic considerations

Ergonomic methods are used to improve product design by understanding and/or predicting how humans interact with products. The methods used focus on different aspects of human performance, taking either a quantitative or qualitative approach.

- The quantitative approach predicts the physical fit of a product to the body, encompassing workload, speed of performance and errors.
- The qualitative approach predicts user comfort and their satisfaction with the product, so it has the optimum interaction with the user.

When designing a product it is very important to decide on the kind of people you are designing for. For example, what is the target market and market segment? Are you designing for men, women or children, are the users able-bodied or disabled, are they young or old, are they normally sighted or do they have impaired vision, are they expert users or first-time buyers?

The profile of the end-user population can often help when designing a product. For example, The Royal National Institute for the Blind (RNIB) developed a battery-operated or mains-driven Talking Scientific Calculator, designed to meet the needs of visually impaired students. The idea for the calculator came from the RNIB's work with visually impaired people and from requests from mathematicians, who needed a device that would help visually impaired students when taking exams. The easy-to-use Talking Scientific Calculator helps students to study and take exams without any help. The calculator is designed with no visible display, but includes 50 functions on a custom-made, 40-key tactile keypad. The PIC-based technology allows the calculator to be programmed to 'speak' different languages. The custom-designed membrane keypad uses many visual and tactile keys for ease of use. The calculator can be connected to a computer or printer to provide a record of an exam.

The interaction between users, products, equipment and environments

All types of products that are designed so their dimensions match those of their users will suit their end-user better. Equipment needs to be designed so it can be operated easily and safely by 90 per cent of the population. Safety considerations require easy operation but must also ensure that equipment is not operated by mistake.

Figure 3.2.30 shows some of the measurements that need to be taken into account when designing products. It is very important when designing office furniture, for example, to make sure that the correct ratio is used between the height of a chair and table and the position of a computer monitor and keyboard, to enable users to work comfortably. Many people who are outside the fifth to the ninety-fifth percentile have to learn to be adaptable when using products. Conditions such as back pain or Repetitive Strain Injury (RSI) are commonly caused by using equipment that is not the correct size for the user.

In a well-designed table the headroom, knee room and elbow room must accommodate the dimensions of the largest person. The height of the table from the floor should therefore be not less than the knee height of the ninety-fifth percentile user. Similarly, in a well-designed chair, the height of the seat from the floor should not be greater than the knee height of the fifth percentile user. This will enable a short person to reach the floor with their feet to enable good blood circulation.

- The height of a table from the floor that is suitable for a ninety-fifth percentile user will also suit 95 per cent of the population.
- The height of a seat from the floor that is suitable for a fifth percentile user will also suit 95 per cent of the population.
- However, even with these dimensions in place, problems will still occur for very tall and very short people.

Human dimensions are also important in the design of environments from bedrooms to kitchens to primary school to sports venues. When designing vehicles, designers need to take account of the different height and reach of male and female drivers. The height and position of seats, the steering wheel and mirrors must be adjustable to allow the driver to be

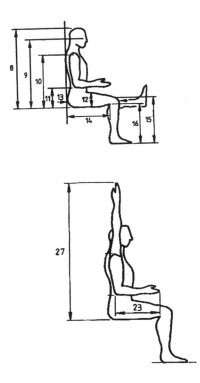

Figure 3.2.30 *The numbered dimensions on these human figures can be found in BSI anthropometric charts in the* Compendium of Essential Design and Technology Standards for Schools and Colleges

comfortable for long periods of time and to easily see the road in front and behind. All the controls, switches and foot pedals must be suitable for a range of sizes of hands and feet (see Figure 3.2.31).

Figure 3.2.31 *One of the main features of the Ford Ka, apart from innovative design, is its exceptional ergonomics which make the car extremely practical*

British and International Standards

The British Standards Institution (BSI) is the world's leading independent standards and quality services organisation. It works globally with manufacturers to develop British, European and International standards. The BSI belongs to the International Organisation for Standards (ISO). CEN is the European Committee for Standardisation, which harmonises technical standards in Europe, in conjunction with the BSI.

Setting standards

Most standards are set at the request of industry or to implement legislation. The setting of standards, testing procedures and quality assurance techniques enable companies to meet the needs of their customers. Tests are carried out against set standards for a wide range of products manufactured in the UK and overseas. Some British Standards are also agreed European and/or International Standards. Any product that meets a British Standard is awarded a Kitemark, as long as the manufacturer has a quality system in place to ensure that every product is made to the same standard.

The relationship between standard measurements and the design of products

Whatever project you are working on, standard measurements probably exist in relation to the product you are making. Sometimes a single critical dimension is all that you need, whereas other designs may need the lengthy and complex calculation of data. When designing a one-off or custom-made product, it may be necessary to use specific dimensions. For example, fitted furniture for an awkward alcove would need to be made to critical dimensions for it to fit in the space available. When designing mass produced products, however, it is usually necessary to match the product to a range of users who come in various shapes and sizes. Information about standard measurements may be found in the *Compendium of Essential Design and Technology Standards for Schools and Colleges* and through the BSI website on www.bsi.org.uk/education.

Ergonomic considerations for designs and models

There are a number of considerations to be taken into account when designing.

- It is a fallacy to think that just because your product is the correct size for you it will be right for everyone.
- Designing for the 'average' user doesn't mean that a product will be suitable for everyone. Since average dimensions only take account of 50 per cent of the population, the other half may find the product unsuitable or difficult to use.
- Although people are adaptable, it should not be used as an excuse for bad design. Problems like back pain, for example, are often the result of using furniture that requires unsatisfactory working positions.
- Many products are sold for their aesthetic properties and may be designed without reference to anthropometric data, because it is expensive to buy.

UNIT 3 B2 CAD/CAM (G303)

Summary of expectations

1. What to expect

CAD/CAM is one of the two options offered in Section B of Unit 3, of which you will study only one. This option builds on the knowledge and understanding of the use of ICT and CAD/CAM that you gained during your GCSE course, depending on which focus area you studied. If, however, these materials are new to you, don't worry. This unit will take you through from first principles.

2. How will it be assessed?

The work that you do in this option will be externally assessed through Section B of the Unit 3 examination paper. The questions will be common across all three materials areas in Product Design. You can therefore answer questions with reference to **either**:

- a Graphics with Materials Technology product such as digitally printed signage
- a Resistant Materials Technology product such as a kettle component produced by rapid protoyping
- **or** a Textiles Technology product such as circular weft knitted fabric.

You are advised to spend **45 minutes** on this section of the paper.

3. What will be assessed?

The following list summarises the topics covered in this option and what will be examined:

- The impact of CAD/CAM on industry
  - changes in production methods
  - global manufacturing
  - employment issues
  - trends in manufacturing using ICT.
- Computer-aided design
  - CAD techniques
  - common input devices
  - common output devices.
- Computer-aided manufacture
  - CNC machines
  - the use of CAM when producing products
  - advantages/disadvantages of CAM.

4. How to be successful in this unit

To be successful in this unit you will need to:

- have a clear understanding of the topics covered in this option
- apply your knowledge and understanding to a given situation or context
- organise your answers clearly and coherently, using specialist technical terms where appropriate
- use clear sketches where appropriate to illustrate your answers
- write clear and logical answers to the examination questions, using correct spelling, grammar and punctuation.

There may be an opportunity to demonstrate your knowledge and understanding of CAD/CAM in your Unit 2 coursework. However, simply because you are studying this option, you do not have to integrate this type of technology into your coursework project.

5. How much is it worth?

This option, together with Section A, is worth 30 per cent of your AS qualification. If you go on to complete the whole course, then this unit accounts for 15 per cent of the full Advanced GCE.

Unit 3A + option	Weighting
AS level	30%
A2 level (full GCE)	15%

1. The impact of CAD/CAM on industry

The need for companies to develop competitive products or services is vital for their own economic survival, for the prosperity of their workforce and for the other businesses in their community or supply chain. Eventually, most products can be designed better or updated because of advances in technology or produced more economically by improvements in production methods. The purpose of any production system, including those in the graphics environment, is to ensure that the correct personnel, software, processes and systems are employed to ensure the best possible outcome for the client or customer and designers.

In all product sectors, designers are able to use the computer-based tools and features within a computer-aided design (CAD) system to create, develop, communicate and record product design information. Computer-aided manufacture (CAM) is a rapidly evolving set of technologies that translate design information into manufacturing information. **Computer integrated management (CIM)** systems are used increasingly to plan and control automated manufacturing processes at all levels from batch to mass or continuous production. The system of process planning combines a range of sub-systems including computerised sensing and control systems, robotics and computer-driven equipment; CIM is how the two systems CAD and CAM are integrated. The type and degree of integration will depend on the scale of production and other operational or commercial considerations.

Increasingly important in the printing industry is the use of computers and microprocessors to ensure that the printing stream works smoothly for the customer as well as the print room operator. Computer-based document management systems have been developed to allow automatic handling of the printing and finishing processes. It is possible to make easy and rapid interventions in order to prioritise printing jobs, choose printers and other output devices or take control of specialist printing jobs. Related developments include automated accounting functions that generate the data needed for tasks like cost allocation, reporting and resource planning. The easily available administrative and management data that are generated by these systems allows for the constant monitoring of manufacturing capacity and costs to generate management level reports and maximise efficiency.

Electronic document management

Once everything in a production process can be carried out electronically the strategy of electronic document management can be introduced. This is quite simply the scanning of physical documents and the importing of computer-generated documents so they can be searched and retrieved at a workstation anywhere, perhaps alongside an existing database system. This allows graphics and printing companies to offer their clients an immediate response to their queries or interest taking account of all existing correspondence. Search criteria can be met in a number of ways, either through document titles, built-in key words, or through **optical character recognition (OCR)** technology on words held within the documents. This procedure allows a company to research its documents and archives in many ways, allowing them quickly to draw together the information they require. In comparison with more labour intensive and traditional paper-based methods an electronic document management system is quick, flexible and cost efficient in time and resources, both physical and human.

The impact of CAD/CAM on design companies

In graphics-related operations, whatever their scale, the introduction of CAD/CAM and related digital technologies means that they are able to offer a comprehensive service to their clients.

In large companies creative teams are increasingly multidisciplinary in order to reflect this convergence between the manufacturing technologies and the creative industries. Larger companies might need to have graphic and product designers, digital media specialists and other specialists such as interior designers and architects on their staff. Smaller sized companies are unable to employ such an extensive range of specialists so they tend to focus on particular market segments. It may be that a company employs both product and graphic designers on their staff or interior designers and architects could come together in a creative enterprise. Freelance graphic designers have to work closely with their clients and with their printing companies. If there is a print production problem, the designer can turn to these printers for advice based on their specialist experience. The key factor that unites these operations is the use of digital processes and technology to provide faster, more efficient customer service characterised by smooth workflow, short job turnaround times and the ability to meet special requests at any time.

Most business owners and marketing department heads do not have the time to stay on top of the technicalities of printing and website creation, let alone deal with coordinating the work of photographers, writers and illustrators. Whatever the organisational arrangement, these new computer-based operations all need the services of digital media specialists. These could be employed directly, but for many companies the most cost-effective way of enlisting specialist technical support is to make use of the skilled people employed in companies providing bureau services. These bureaux also have a vast range of specialist equipment that is too expensive for most companies who would not need it on a day-to-day basis. Typically, they can provide digital scanning or large format plotting and printing of electronic files that have been created from a variety of CAD applications. These files can usually be plotted or printed in monochrome or full colour on various media including paper, drafting film, vinyl for banners and signs, and textiles.

CAD/CAM – creating or responding to change?

CIM is an increasingly effective way of integrating CAD/CAM technologies into the production of a range of products including those that incorporate graphics such as product cartons or point-of-sale packaging. **Die cutting** is the manufacturing operation for the cutting and creasing of flat sheet materials so that that they can be formed into 3D cartons and packages. The **die-cutting tools** can be designed and tested on a CAD system. In high-volume manufacturing operations, computers and microprocessors provide the control systems that integrate the different production stages to ensure continuous production. Across all industrial sectors the development of computer-based design and production technologies is providing a driving force for change. The same technologies also provide an effective response mechanism by allowing companies to react quickly to changes in external factors such as a fluctuating market demand (see Figure 3.3.1).

The pressures for change

The twenty-first century shopping experience is fast, competitive and demanding. Producers and manufacturers of mass-produced items

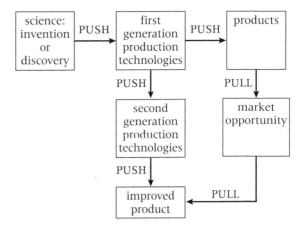

Figure 3.3.1 *Changes in production methods are either pushed by the development of new technologies or pulled by the demands of the market*

market and sell their products across the world. This creates pressure on the design processes, the manufacturing processes and the organisation of the production and logistics facilities. A product may have to be remodelled to include different design features depending on where it is to be sold. For example, a point-of-purchase merchandising display for photographic film may retain the same shape and form but the language and styling of the text may have to vary from one international market to another. In some countries local legislation also imposes restrictions on manufacturers. Multinational manufacturers often face a challenge from local manufacturers of similar products. These product differences are classed as design variants. They range from the amount of recyclable materials used in the product, to whether a plug is fitted or not, to the fitting of different visual and graphic displays. Another pressure for change is that product lifespans are becoming shorter, especially in areas such as telecommunications and information systems.

In a fast moving sector like graphic design and its allied disciplines in manufacturing, the deadlines set by clients are often incredibly tight, creating further pressures on the design process. The application of computers has changed design practice from the traditional 'design for print' approach to interactive design. This new field of design encompasses many areas ranging from web design through to digital television. Web design involving designers in interactive communication design has emerged in a relatively short period of time. This means that

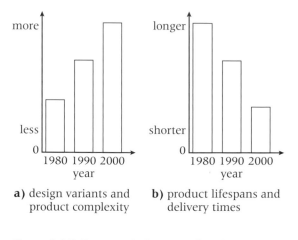

a) design variants and product complexity

b) product lifespans and delivery times

Figure 3.3.2 *Pressures to improve the design and production process*

designers are now faced with the prospect that they need to keep learning about new software and how to operate in new design arenas. Keeping up with changes and upgrades to existing software also takes up an incredibly large amount of time and creates further pressures for change especially for those working in the freelance graphic design market. New technologies and the introduction of smart materials are also increasing the technical complexity of designs and how they can be manufactured. Figure 3.3.2 shows the effect of pressures to improve the design and production process.

Task

Select a common graphics-based product such as a computer game.

a) List and explain the design factors or the design constraints, such as the specific technical or legal requirements, that have to be considered by the company.

b) Consider and compare the different requirements that the product buyer might have and write down the constraints that they might also impose on the design of a successful product.

Changes in design and production methods

A major design innovation has been the development of desktop publishing (DTP). This is a process whereby a designer sitting at a desk can create camera-ready copy and final film/artwork on a personal computer. Another significant change has been the continuing evolution of more powerful CAD systems for use on personal computers. For graphic designers in particular, a fully featured CAD system provides the means to develop design concepts and model or test ideas from the simplest to the most complicated product. DTP applications support presentation graphics and multimedia in the preparation of design for reproduction. These applications allow for the accurate positioning of pictures and blocks of text. Typography can be accurately controlled such as when editing and rotating text. Colour separation features allow the opportunity to specify percentage tints either as spot or process colours, which can then be output into separated film for printing processes.

Using software to design for manufacture

Design for manufacture software such as PowerSHAPE™ produced by Delcam allows users to take concept designs or designs prepared in other CAD systems and add specific and complex manufacturing features such as fillets or split and draft surfaces that are needed for the trouble-free manufacture of plastics-based products.

These fully featured CAD systems also allow the production of photorealistic images and 'virtual products' that can be used to 'sell' the products to potential clients or customers through different media such as magazines and the worldwide web.

Photorealistic images

The growth in the use of the Internet and the development of websites to advertise and sell products has increased the pressure on software designers to improve the 'real life' quality of product images. These digital images are described as photorealistic. Modern CAD systems allow greater control over the quality of these images, typically through the use of an object or image browser. One of the key features for graphic designers is the ability of desktop publishing software to manipulate images and represent different material qualities such as metallic, non-metallic and textured surfaces.

Page make-up and layout applications such as Adobe Photoshop™, Illustrator™, Aldus PageMaker™ and QuarkXpress provide interactive 'virtual studios' in which the designer can control the direction and intensity of light falling on a product image as well as the type of camera lens in order to create a photorealistic image. Designers can choose daylight conditions or flood lighting. Camera lenses available include fisheye, wide-angle and telephoto. These determine how much of an image can be seen and from what distance. Images can be manipulated electronically; they can be retouched, masked, textured and resized to fit the space available.

Virtual products

The comparatively recent introduction of 3D CAD systems allows the creation of interactive virtual reality environments and products. A virtual product can be viewed from any angle at any distance. Products with moving parts can be seen in operation 'on screen' as animations. These animations are generally kinematic motions that do not take account of the mass of the object or the other physical forces acting on it. They do not represent a complete model of how the product will behave in reality.

We will return to the subject of photorealistic images and virtual products later in this unit.

Design management

In the more sophisticated CAD applications that require larger amounts of computer memory the software keeps track of what are called design dependencies. This means that when the one value or element on a drawing of a part or an assembly is changed, all the other values that depend on it are also automatically and accurately redrawn. This is known as **parametric designing** and is particularly useful in applications such as the packaging of consumables for special promotions such as '10% extra free'. This means that although the relationships of the geometric shapes making up the package stay the same, all the dimensions have to be changed to allow the package to hold the increased volume of product. One simple change to a key measurement such as the height of the package causes all the other dimensions to be altered to the correct sizes. The graphics can then be added by cutting and pasting from the original drawing and resizing as required to allow room for the promotional text to be added.

The benefits of CAD/CAM product modelling

Combining 3D CAD systems with computer-aided modelling techniques such as **Rapid Prototyping (RPT)** allows the creation of physical models as soon as the 3D digital model is designed. This reduces potential communication problems in the product development team. A further benefit is that any potential errors or technical and tooling

problems are found out more quickly. Changes to a product design in the later changes of its development are potentially very costly in terms of time and reworking costs.

Architectural model making is one area of product modelling to benefit from the application of CAD/CAM. Designs developed on screen can be quickly redeveloped to accommodate design changes arising from discussions with clients. Computers are now used to produce anything from a simple design study model costing relatively little to an illuminated, mechanised marketing model costing thousands of pounds. Model makers, such as Pipers of London, now use CAD/CAM in a variety of ways. The company is able to accept digital drawings direct from an architect and then use a range of computer-controlled cutting techniques to increase the accuracy, efficiency and quality of the final model.

Task

Construct a simple organisational chart indicating the main stages in the design process, as you understand it. Identify how you think computers could be used at each stage in the process, indicating what features might make them useful, or how they might have a limited application.

Computer numerically controlled (CNC) machines

CNC is in widespread use across industry and especially in automated production systems. Computer numerically controlled (CNC) machines are controlled using number values written into a computer program. Each number or code is assigned to a particular operation or process. In manufacturing equipment such as CNC milling and press machines, improvements in the machine-operator interface have made them easier and more intuitive to use. Computer programs, sometimes referred to as 'wizards', have largely eliminated the need to learn elaborate CNC machining or programming codes. Most CAD/CAM software is now capable of generating the required NC machining codes known as 'G' and 'M' codes from the digital data created from a drawing. Examples of 'G' and 'M' codes used by manufacturing equipment are shown in Table 3.3.1.

These codes can be shown on a screen as the product is either 'virtually manufactured' as a

Table 3.3.1 *Examples of 'G' and 'M' codes*

Code no.	Type	Description
G00	Rapid traverse	The tool moves from point 1 to point 2, along the shortest path available The feed rate (speed of movement of the tool) is usually set to run as fast as possible
G01	Point-to-point positioning	The tool moves from point 1 to point 2, in a straight line, with a controlled feed rate
M00	Stop program at this point	
M02	End of program	
M03	Spindle on clockwise	Code is followed by a number prefixed by the letter S, denoting the spindle speed
M05	Stop spindle	
M06	Change tool	Code is followed by a number prefixed by the letter T, denoting the number of the tool you wish to change to
M30	Program end and rewind to start of program	

simulation or in reality on the CNC machine. The numbers or codes are easily changed from within the CAD software when required. Most CNC machines also have the capability to be programmed manually from an adjacent keypad.

The benefits of CNC machines are that:

- graphic images and products are produced accurately, quickly, with consistent quality
- they provide increased operational flexibility as they can be employed in batch and mass-production systems such as high- and low-volume print runs
- they can be used reliably in processes requiring continuous operation such as carton manufacturing
- they can be used in conditions that are hazardous to human operators such as the guillotine cutting and trimming of large printed documents
- they are economic to operate over time even though they have a high initial cost.

Computer integrated manufacturing system (CIM)

In both small- and large-scale production processes, traditional approaches to designing and manufacturing involve a linear or sequential approach in which the image or product passes through a series of predefined stages.

In this system, often referred to as 'over the wall', process planning and the handling of production data are relatively straightforward activities, but design errors or manufacturing problems can occur at many points along the line depending on the number of people involved. This extends the time taken to design and manufacture a product and bring it to market.

As we have seen earlier, product life cycles are reducing significantly and with the globalisation of manufacturing there are severe competition pressures. These are forcing companies to look at more efficient ways of operating such as concurrent engineering.

In this manufacturing system, a product team is organised so that all the specialisms within a large company are represented. These multidisciplinary teams share their expertise right from the start of a product's life. They work together at the development stage to reduce errors and draw on the skills and expertise of others in the team. A print production engineer will immediately be able to tell a graphic designer whether what has been designed can be produced rather than waiting until the 'completed' drawing is received.

Desktop manufacturing systems
The role of computers in flexible manufacturing systems (FMS)

Software applications in a flexible manufacturing system (FMS) allow a central computer to process production data in order to sequence and control a network of machines and materials handling systems in order to meet the order book more efficiently. For instance, the printing industry is in flux; new market requirements are calling for new solutions to meet shorter deadlines, shorter production runs with reduced set-up times, last-minute changes, improvements in quality and reductions in waste. At the most basic level the trend towards smaller, better-quality print runs means that it is very important to get the press inked up and achieve even inking conditions quickly. Several companies have changed to printing presses that use microprocessor-based systems to measure print quality, evaluate colours and use the data to automatically control the inking process.

Task

A cosmetics company is developing a new range of point-of-sale merchandising displays involving more modern graphic images and fonts designed to attract a younger customer to a well-established skincare product. If you were putting together a product development team to produce this new range, what areas of specialist expertise might need to be represented?

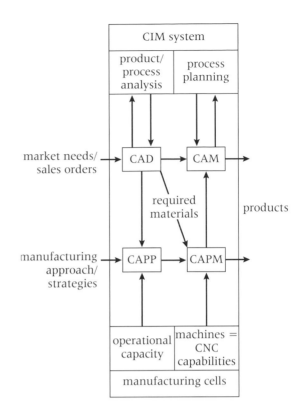

Figure 3.3.3 *Software applications within CIM*

The production data generated by computer-based production systems can also be used in many other ways. They can be used to judge progress against 'world-class manufacturing' quality criteria such as 'right first time every time'. The move to total quality management (TQM) means that product quality is no longer the responsibility of one department or one person. This new organisational culture of continuous improvement (CI) through concurrent engineering cannot exist without effective computer-based communications.

CAD and CAM systems generate vast amounts of digital data that can be used at different stages in an integrated manufacturing system. Relational databases and other data storage methods enable data to be tracked, stored and retrieved as required. Data can be used to plan CNC and other manufacturing operations – computer-aided process control (CAPP). These 'new' data combined with other data generated from the CAD model can be used to manage the production processes, a technique known as computer-aided production management (CAPM). Figure 3.3.3 shows a systems model of the use of software applications within CIM.

Electronic Data Management

Speed, reduced cost of design creation and revision, accuracy, flexibility and the increased scope of drawing production have ensured that CAD has been rapidly integrated into design studios and drawing offices worldwide. Many graphic design companies, architects and manufacturers have large and difficult to store archives of past and sometimes present-day projects on media such as paper. Drawings were once produced on paper or other media and stored in plan chests and filing cabinets but they are now created on computers and are stored as electronic files. Many companies find it cost effective to use the technical facilities provided by specialists firms called bureau services who

can convert design information stored on paper copies into digital files that are easier and more convenient to store. These files can then be made available online in an information retrieval intranet or electronic data management (EDM) system.

Global manufacturing

We are currently living through a high-speed information revolution that is not only making the world a smaller place, but it is also increasing international competition and creating pressures on manufacturing companies to respond or die.

The trend towards cooperative and integrated working is increasingly more cost effective because of improved communication and data exchange systems that are not limited by geographical boundaries.

There are now many examples of products that are designed and engineered by teams working across the globe and assembled using component parts that, because of increased specialisation, are sourced from supply chains in different countries.

The London Eye is a feat of modern engineering and is an example of this method of

Figure 3.3.4 The London Eye – an engineering project that drew on design disciplines from around the world

Task

The London Eye was conceived and designed by Marks Barfield Architects. Visit *The World of the Eye* on the Internet at www.londoneye.com to find out how it was designed.

Design and produce a graphic storyboard to illustrate one aspect of the design or the construction of the Eye. You might want to analyse the major components that make up the London Eye. Find out where they were designed. Where were the parts manufactured? How was the manufacturing organised? You could collect examples of the promotional literature used to raise and inform public awareness of this product and the service or experience that it provides. What are the key messages that are being given? How are the messages of the corporate sponsors conveyed?

design and manufacturing (see Figure 3.3.4). The project management team drew on specialist expertise and suppliers based in many different countries.

Employment issues

CAD/CAM technologies are directly affecting patterns of employment in all manufacturing sectors right across the globe. The numbers of people involved in manufacturing are declining as automated systems take over. For those who are left, their jobs have changed significantly (see Figure 3.3.5). For instance, as we have seen, CNC machines can be controlled directly from a central computer system. This means that some shop floor jobs such as those in a large print room consist of nothing more than 'machine minding'; correcting faults that arise or replacing materials or resources that have run out. With developments in automation, robotics and

Figure 3.3.5 A manufacturing cell at a Ford plant

artificial intelligence it is likely that there will be even less human involvement in the production processes of the future.

Employment trends on the design side

CAD systems reduce the need for large and labour-intensive drawing offices where historically drawings were hand-drawn by skilled technicians. However, the effective use of CAD requires workers who have high levels of computer and visual literacy alongside creativity and problem-solving skills. These desk-based workers do not necessarily need to be office based. With the growth in information communication technology they can work almost anywhere and if they do so they are classified as remote out-workers.

The increased use of CAD and CAM systems has also raised significant issues concerning initial and on-going training. For instance, the effective introduction of these new ways of computer-based working relies on the willingness of employees to adopt more flexible approaches to work and the recognition that they will need to retrain or update their skills as systems continue to develop. It also requires that people coming into this sector are more highly trained with much higher levels of basic skills than previously expected.

New opportunities for employment also arise with the adoption of CAD/CAM systems. For many small to medium-sized manufacturers, employing a specialist bureau service to provide professional CAD/CAM services to support a number of different design disciplines is an effective and flexible way to cater for the specific needs of their different customers.

Employment trends on the processing side

CAM operators now have to be trained to operate different machines and carry out many processing operations within their manufacturing 'cell' or work centre. They have to be good 'team players' offering help to others as required. Gone are the days of rigid demarcation and narrow specialisation when, for instance, lathe operatives were not allowed to set up and use another machine in the factory. This method of redeploying workers to the area of greatest need is a feature of modern production systems. These systems reduce queues for machines, remove potential bottlenecks on the production line and so reduce processing times. They are not viable without a skilled, trained and flexible workforce.

In the UK there are severe shortages of these multi-skilled, multi-function operators in many

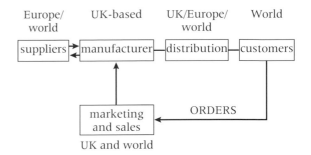

Figure 3.3.6 *The worldwide distribution of a UK manufacturing company*

areas of manufacturing. This, combined with the flexibility offered by computer-based systems, often means that companies can relocate their manufacturing operation to another part of the world. Design and development can remain in one country while manufacturing operations move abroad. This has a devastating impact on local and regional employment prospects. Until recently, large Japanese manufacturing companies offered 'jobs for life' but Japanese workers are now also feeling the pressure on their jobs from emerging manufacturing centres in Taiwan and other Pacific Rim countries. Figure 3.3.6 shows how the distribution of a UK manufacturing company has been spread worldwide.

The impact of CAD and CAM technologies on future industrial innovation

The globalisation of manufacturing industry means that considerable research and development time is being spent on developing innovative manufacturing systems that fully integrate the use of CAD/CAM with artificial intelligence or 'expert' systems, information systems and databases. Innovation is a process involving three types of technologies:

- **Critical technologies** are the 'building blocks' from which products develop. CAD/CAM approaches will be involved in the continued development of computerised sensing and control systems, materials handling, storage and retrieval systems and the development of industrial robots.
- **Enabling technologies** such as CNC machines are needed to make use of critical technologies.
- **Strategic technologies** are concerned with the decision-making process. In

143

manufacturing this can range from decisions about capital investment in new products, the most effective factory layouts and facilities, to the future use and potential benefits of systems based on artificial intelligence.

The drive for profitability

There is much research and development activity into data manipulation systems because a company's continued existence in both low- and high-volume manufacturing relies on the establishment of efficient, cost-effective and 'quick response' systems. Effective CIM systems that include CAD and CAM reduce the 'lead time' from product development to market; time to market is a key factor in profitability. The development of more sophisticated and

powerful DTP technology has improved the effectiveness and output of individuals, while developments in information and communication technologies now allow for the rapid and effective deployment of vast amounts of electronic data to virtually anywhere in the world. The development in mobile phone technologies is one area that will also contribute to the drive to profitability.

Task
Explain how DTP technologies contribute to increased profitability in both a small- and large-scale graphics-based industry.

2. Computer-aided design

A computer-aided design system is a combination of hardware and software that enables designers working individually or in teams to design everything from promotional gifts such as pens or desk tidies to the International Space Station. CAD provides increased flexibility for designers and allows them more control over the quality of the finished product. In situations such as the production of a computer-animated feature film it allows digital image manipulation and the facility to combine computer-generated images with filmed images from real life.

In many manufacturing situations a CAD system will automatically cross-reference many drawings to one another to create a product inventory. This not only details the materials and components required to assemble the final product, it also informs the stock control systems in the manufacturing facility. The successful application of CAD is dependent on the amount of computing power that is available to the user.

In general terms, CAD requires a powerful central processing unit (CPU) and a large amount of memory. The images on a computer screen are made up of pixels. The term bit depth refers to the amount of data used to describe each pixel on the computer screen. Black and white line work is 1 bit deep. Greyscale is 8 bits deep. RGB is 24 bits deep. Images to be printed as CMYK separations should be 32 bits deep. In graphic terms, images can now be created and manipulated in 2D, 3D and video forms creating a need for greater processing speeds and increased amounts of computer memory.

Until the mid-1980s, all CAD systems operated on specially constructed 'dedicated' computers. The processing power and the speed of processors and graphics interfaces on personal computers has improved dramatically in recent years. 'Fully featured' professional quality CAD systems can now be operated effectively on **local area networks (LANs)**, wide area networks (WANs) and general-purpose office workstations, and on personal desktop and laptop computers.

The components of a CAD system
The following are the main components of a CAD system; you will consider them in more detail later in this option.

Hardware
This term describes all the physical components of a CAD system including the input and output devices. There are three main categories of computer that can be used in a manufacturing situation:

- Mainframe computers have high processing speeds and the memory required for handling and storing the large amounts of data that are generated by high-volume manufacturing. A mainframe computer is accessed via a network of computer terminals connected to it. The networks can be local area networks, **intranets** or wide area networks.
- Minicomputers are smaller versions of the mainframe system used by large organisations mostly in various network configurations.

- Microcomputers, desktop or laptop personal computers (PCs) are used for individual computing needs or as machine control units (MCUs) for controlling a range of CNC machines.

Data storage devices include hard and floppy disks, CD-ROM and DVD, and other external storage devices such as Zip drives.

Input devices include keyboard, mouse and tracker-ball, graphics tablet and stylus, digitiser, puck (cursor) and mouse, digital camera and video, and 2D and 3D scanners.

Output devices include monitor, printers of all types, plotters and cutters, CNC machines, and increasingly Rapid Prototyping systems (RPT) for modelling 3D product ideas.

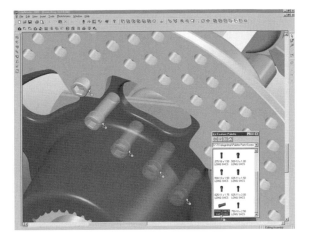

Figure 3.3.7 *Example of a CAD window*

Software

The operating system (OS) provides the platform to run all the application programs. It is also used to manage the files in the computer. There are a number of operating systems available such as **Unix**, Windows, DOS and MAC-OS (Apple computers).

There are many CAD software applications based on these operating systems. All CAD software, whatever the operating system, provides the data for the graphic display and the other output devices such as printers and plotters.

A modern CAD program contains hundreds of functions and features that enable you to accomplish specific drawing tasks. A task may involve drawing an object, editing an existing drawing, displaying a specific view of the drawing, printing or saving it, or controlling any other operation of the computer. The software will contain a number of menus, commands, functions and features that enable you to specify exactly what you want to do and how you want to do it. For instance, the edit feature is a convenient electronic drawing aid that enables designs to be changed easily. The software may also have a number of specialised features such as providing animated 3D views or printing design information on different layers. This feature is useful in areas such as desktop publishing or the screen printing of a graphic design.

Drawings can also be plotted with specific colours, pen thickness and line types. In some CAM programs lines on a CAD design can be drawn in different colours to indicate the different depths of cut that a cutting tool has to make. This is a particularly useful feature when making folded card or 'creased' plastic products (such as folders and wallets to hold promotional leaflets) where one colour is used to indicate the folds or creases that are required to make the 3D product.

Graphical user interface (GUI)

This is a two-way link between the user and the computer. It provides an on-screen display giving visual clues to help the user communicate with the computer. The graphical user interface (GUI) allows the user to enter data through commands or functions that are selected from menus; by keyboard stokes (e.g. Ctrl+P); from toolbars, buttons and on-screen icons and from text or dialogue boxes. GUIs also allow a range of users with different operational requirements to set up their own preferred 'on-screen' working environment such as specifying dimensions and text that match industry standards such as **ANSI**, ISO and BSI. An example of a typical CAD window is shown in Figure 3.3.7.

Standards for CAD systems

A format, programming language, or operating **protocol** only becomes an officially recognised **standard** when it has been approved by one of the recognised standards organisations. However, you will find that many of the CAD/CAM standards are, in fact, accepted standards because they are widely used and recognised by the industry itself as being the standard. Some examples of such standards that you will come across include:

- Hayes command set for controlling modems
- Hewlett Packard Printer Control Language (PCL) for laser printers
- PostScript (PS) page description language for laser printers
- Data Exchange File (DXF), a graphic file format created by AutoDesk and used on many CAD systems.

Managing CAD data

All CAD applications have a range of data management features and options. Drawings can be stored either on a hard disk, tape, CD-ROM or on a central network **server**. Drawing files are often large and take up a lot of memory. Some applications allow files to be compressed (zipped) to take up less of the system memory. Files are managed in directories and subdirectories. The software should also be able to translate drawings created by other CAD programs. Data Exchange Format (DXF) is one of the common data translation formats used by CAD programs. There are a number of other data formats available. Initial Graphics Exchange Specification (IGES) is an international standard, which defines a neutral file format for the representation of graphics data across different PC-based CAD systems.

The common formats for storing CAD data are:

- Windows Metafile Format (WMF)
- PICT – the standard format for storing and exchanging graphics on Apple computers
- Tagged Image File Format (TIFF)
- Hewlett Packard Graphics Language (HPGL)
- **Virtual Reality Modelling Language (VRML).**

Tasks

1 Find out and record what the initials ISO, ANSI and BSI stand for.
2 Give three examples of standards that apply to CAD/CAM resources, equipment or software.
3 Give three examples of standards that apply to graphic products.

The basic operating characteristics of CAD programs

- The user is able to interact with and manipulate images on screen.
- Displays can be divided into two or more windows, sometimes referred to as tiles or panes.
- Each window can contain a different view of the product or the different parts that make up the product.
- Models can be repositioned or edited independently in each window, the effects of one action affecting all relevant views.
- When working with multiple windows only one of the windows remains current or active; other windows are activated by clicking on them.
- The coordinates that make up the multiple views to create a 3D model or product are calculated automatically.
- Models or products can be displayed or viewed from any direction or viewpoint.
- Standard projections are supported including orthographic (1st and 3rd angle), oblique, isometric and perspective.
- Pull-down or pop-up menus contain the various commands that allow text, dimensions, surface finishes, textures and labels to be added. Features such as line styles and use of colour can be adjusted. Most CAD packages have keyboard shortcuts for drawing and other features such as Ctrl+G for grouping a collection of individual elements in addition to the standard word-processing shortcuts, for example Ctrl+P for printing.
- Primitives are predefined graphic elements. Most CAD packages have a library of 2D and 3D vectored objects that can be drawn, stretched and resized. Typical primitives are:
  – 2D: lines, polylines, arcs, polyarcs, circles, ellipses, splines, Bezier curves and polygons
  – 3D: cones, cylinders, prisms, pyramids and spheres.
- Predefined specialist libraries of graphic shapes can be provided to support specific design disciplines such as architecture, interior or textiles design.

Task

Logos often make use of primitives like the recently introduced British Farm Standard logo (see Figure 3.3.8). It is designed to help UK farmers compete with imported food supplies and it can be found on fresh produce lines in supermarkets. Analyse the shapes in the British Farm Logo. Create a display of advertising logos or slogans that are based on predefined geometric shapes.

Figure 3.3.8 British Farm Standard logo

Computer-generated images and models

All 2D and 3D CAD systems are based on the fundamental need for a designer from any specialist discipline to work with a 'model' of a product design. These systems are classed as interactive graphics systems because the user has control of the image on screen; data can be added, edited, modified and deleted. The generation of digital models or virtual products removes the pressure to produce a physical model too early in the product development process. The virtual product becomes the designers' main means of communicating and talking about their ideas with others in their creative and manufacturing team. Creating a digital product model requires a system that is capable of processing mathematical functions and making complex calculations in order to produce models and manipulate graphic images in 2D and 3D. These images and models are either generated by vector or raster graphics.

Using vector and raster graphics in CAD systems

In systems using vector and object-oriented graphics, geometrical formulae are used to represent images as a series of lines. Raster graphics represent images as 'bitmaps' and the image is composed of a pattern of dots or picture elements (pixels). In design terms, the way a graphic image is generated or displayed affects the impact it will make or the message it conveys. Draw programs create and manipulate vector graphics. Programs that manipulate bitmapped or raster images are called paint programs. Vectored images are more flexible because they can be resized and stretched. In addition, images stored as vectors look better on screen or paper, whereas bitmapped images always appear the same regardless of a printer or a monitor's resolution or picture quality. Another advantage of vector graphics is that representations of images often require less memory than bitmapped images. CAD programs employ a combination of these two graphics. Vector graphics are used to draw lines and produce 3D shapes; raster graphics are used for the rendering of surfaces and textures.

The uses of CAD-generated images and models

Designers use CAD systems in a variety of ways depending on the properties of the image or the product to be modelled. They are concerned with how the image will sell the product, or in the case of a product, how it will function, how the parts go together (structure), its form (shape), as well as the materials, surface finishes, textures and functional dimensions. The design models that are generated on screen can be sent directly to a manufacturing centre. For instance, for folding box production CAD data can be used to produce the cutting dies for the carton and to prepare the labels in different colour variants and in different language versions if required.

Computer to plate (CTP)

Traditionally, printing companies would receive the camera ready artwork from a graphic product designer and then use it to create the film and printing plates necessary for the printing process to proceed. Increasingly, a new technique called computer to plate (CTP) is being introduced that removes the need to create films from the production workflow. Using computer-to-plate equipment designers can now use the digital artwork to output a set of printing plates directly from the desktop. There are no films, no chemical film processing and no operators tied up exposing plates so that make-ready times are significantly reduced. The quality of the final output is improved because a laser is used to cut the plates and the images produced have much greater printing definition.

CAD-generated images on printed plastic are ideal for sales promotion and merchandising. Figure 3.3.9 shows plastic items that have been thermoformed, folded or die cut. The images are digitally printed directly on to the plastic, and this offers the opportunity to personalise marketing materials, such as mouse mats, by adding individual names. Each type of plastic has its own unique properties. In terms of visual design, plastics are often far more versatile and effective than either paper or card. It is possible to digitally print directly from a computer on to mirror-effect plastic, plastics that are transparent or tinted, plastics with integral patterns and plastics with different textures.

Task

Polypropylene, PVC, polystyrene and polyester are all used in packaging and promotional materials that have images generated in a CAD program and that are digitally printed. Using the Internet and other sources, research the process of digital printing on plastic, illustrating your answer with examples of commercial products that use plastics.

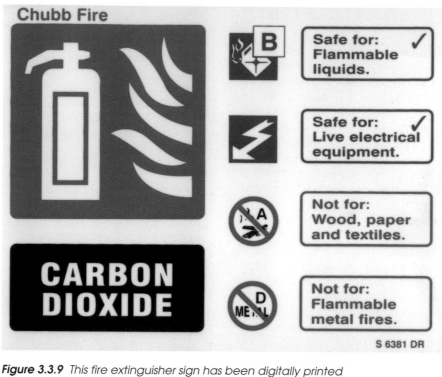

Figure 3.3.9 *This fire extinguisher sign has been digitally printed*

Comparing the benefits of 2D and 3D drawings

Historically, 2D CAD drawings have been used to develop, share and exchange product shapes and designs by electronic means such as email or local and global networks. The benefit of a 3D drawing is that it provides additional visual information about important things such as the form of the product. The development of affordable 3D CAD systems that do not rely on powerful mainframe computers is providing a means to communicate and exchange much more detailed design information. 3D facilities allow complex screen images or virtual products to be rotated, sectioned, measured and annotated to create a range of digital data. These data can be used via CAM to support an increasing range of CNC and automatic processing operations.

Using design data

In modern CIM systems design data can also be shared with the suppliers of materials, components and sub-assemblies, as well as the workers and machines on the production lines. Product data can be used directly or converted into other file formats for a variety of other uses, such as product and sales presentations, company reports, marketing materials and brochures that are made available at a company website on the Internet. The websites are designed to boost brand awareness and electronic retailing presence in home and international markets. Virtual Reality Modelling (VRM) and the use of 'knowledge-based' expert systems via the Internet are growing in importance for all the purposes described above. Multi-media approaches that create more flexible mer-chandising formats for retailers are covered in more detail later in this unit and in Unit 4.

The importance of CAD modelling

CAD modelling is now a key part of the industrial and creative design process for the following reasons:

- Designs can be developed and electronically shared with others which enables a fast turn-around of ideas. A team of specialists often design products such as point-of-sale displays. The first stage in the process of designing a 'shelf wobbler' or moving display (see Figure 3.3.10) might involve drawing the overall layout in 2D including all the crucial features of the brand or product being promoted. The design constraints can then be applied along with the way the different components that make up the display need to move (kinematics). The basic solid model properties such as colours and textures are added and

Figure 3.3.10 *Examples of 'shelf wobblers' and other kinetic displays*

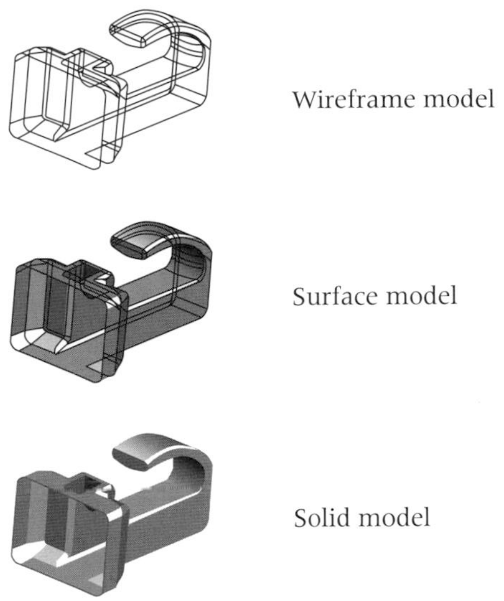

Wireframe model

Surface model

Solid model

Figure 3.3.11 *Applications of 3D modelling techniques*

the top-level layout is complete. The specialist teams such as those responsible for 'die cutting' the final shapes work on their particular components or features. If there are any design changes, these take place on the top-level layout as the reference point and all the design teams receive updated versions automatically.

- Ideas can be tested, evaluated and modified at all stages at any point in the process.
- The need to produce a range of costly prototypes or samples is reduced. Photorealistic images can be created and modified without the need for physical models and expensive photography.
- Products and processes can be simulated or animated and then evaluated on screen. This means that development time, design costs and the use of resources is significantly reduced.

2D and 3D geometric models
Geometric modelling is concerned with describing an object mathematically (algorithm) in a form that a CAD or graphics program can display visually. Geometric models are subdivided based on the amount and kind of information they store. The three divisions are wireframe, surface and solid models (see Figure 3.3.11).

Wireframe models
Wireframe models in 2D or 3D are most effective for sheet metal products and simple frame constructions without a great thickness of material. In a wireframe model an object is represented as a collection of points, lines and arcs. Wireframe models can be ambiguous and difficult to 'read' or interpret. A realistic form is only achieved by the generation of a lot of data, which increases image-processing times. Dimensions, annotations and other 'attributes' of the object may be stored but there are no visible surfaces. This means that surface or solid properties cannot be computed and rendered images cannot be generated. Additionally, 3D-wireframe objects lack information about points inside the object and the geometric data that are produced are incompatible with the requirements of CNC programs.

Surface models
Surface models in 3D can provide more machining data than wireframes and generate a more realistic 'picture' of the model. This technique is an alternative to solid modelling but provides fewer data. Surface models, either flat or curved, are created by 'patches'. Polygons define contours and surfaces. As with the wireframe, the surface model contains no data about the interior of the part.

Solid models
Solid models in 3D can produce full digital mock-ups and a comprehensive dataset including product assemblies. Solid models are clear; there are no visual confusions as with the other models described. They provide complete representations of the properties of the solid.

149

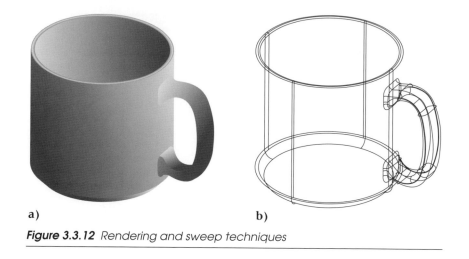

a)   b)

Figure 3.3.12 *Rendering and sweep techniques*

Rendering

Rendering is the process of adding realism to a computer model by adding visual qualities (see Figure 3.3.12a). These include colour; patterns and textures; surface shading with or without light sources; hidden line removal and hidden surface removal. Hidden line removal is an important drawing function as it removes lines from the drawing that would normally not be visible from the chosen viewpoint making the model less ambiguous. The semi-hidden function displays 'hidden' lines in the 'dashed' line style (hidden detail), which you will be familiar with in engineering drawings. Hidden surface removal is a technique for filling shapes on the model with colour to improve visual understanding.

Rendered images are also used in advertising, sales literature, assembly illustrations, operational instructions and other information sources. The degree of realism that can be achieved is dependent on the quality of the available software and hardware, the designer's creative and visual abilities and the time available.

Shading

CAD software has a range of available shading options. The three most common are as follows:

- *Flat shading* is quick and simple. The surface of the object is divided into small polygons that are all shaded uniformly. This type of shading gives the object a faceted appearance. The curved surfaces are represented as a series of flat surfaces rather than a smooth curve.
- More realistic effects are achieved by *graduated shading*, which removes the sharp edges created by flat shading and replaces them with a gradually changing shading pattern.

- *Phong shading* is the most accurate as it incorporates 'highlights'. Each pixel on the shaded portion can be assigned a brightness value. As a result, the rendering quality and visual realism are very good but Phong shading is time consuming and slows down the processing of images.

Sweep techniques

Sweeping refers to a class of techniques used for creating curved or twisted solids (see Figure 3.3.12b). Sweep techniques involve drawing a profile along a path. The profile is usually a closed geometric form such as a 'D' shape. The path indicates how or to where the profile will be 'swept'. Moving the profile along the path that can be linear, circular, radial, spiral or some other configuration then creates the solid. Handles on a cup or a threaded part are examples of profiles that can be generated by sweeping.

Textures

Textures can have the 2D qualities of colour and brightness and they can have the 3D properties of transparency or reflectivity. Textures can be mapped electronically around any 3D model, a technique known as 'texture mapping'. Textures are an important part of creating ever-more realistic images but they use lots of memory and image processing can be slowed down.

Task

Produce a series of drawings for your project folder to demonstrate your capability in using the techniques described above using a CAD program that you have access to.

Constructing accurate drawings within a CAD system

To describe an object for manufacture accurately all the appropriate 2D orthographic views and 3D visualisations must be drawn. Production drawings communicate all the information necessary for the production of products and assemblies. All production drawings, whether CAD or manually generated, can be classified into two major categories:

- *Detail drawings* are drawings of single parts and include the additional information such as dimensions and notes relating to materials, finish, weight or calliper of paper, or standard colour options such as Pantone numbers that are required to produce the parts.
- *Assembly drawings* document all the necessary parts needed to assemble a product and how they fit together. The dimensions in an assembly drawing, such as a point-of-sale display, usually refer to the spatial relationships of different parts to each other rather than the size of the individual parts. An assembly drawing may be a multiview drawing or a single profile view. Ballooned letters or numbers are attached to leader lines to 'reference' or identify the parts in the assembly. The letters or numbers also identify the part in a list which is usually placed to the bottom right of the drawing. The 'parts list' provides information regarding the name of the part, what material it is made from and the minimum number of each part that is required. Standard 'off-the-shelf' parts and components like fixings and fastenings are also included in the parts list.

Dimensioning and annotating a drawing

Dimensioning and annotating a drawing provides accurate information about the size of the product and its component parts. Different types of features require the use of different dimension formats. CAD systems can provide linear, angular, cylindrical and radial dimensioning. Notes are added to drawings to provide additional information about the project. They are used to indicate specific surface finishes or materials; or other special manufacturing requirements such as the size or depth of holes.

The importance of dimensioning standards

In any manufacturing system, it is important that all the people reading a 'drawing' interpret it in exactly the same way. However, dimensioning practices may vary from company to company or from country to country. A set of international standards (ISO) has been developed to specify acceptable dimensioning practices.

CAD systems offer standard dimensioning formats allowing the fundamental principles of dimensioning to be followed. Some are listed below:

- Each feature of an object is only dimensioned once in the view in which it is most clearly seen.
- Each dimension should include an appropriate tolerance.
- Dimensions should be located outside the boundaries of the object wherever possible and there should be a visible gap between the object and the start of a dimension line.
- Crossing of dimension lines should be avoided wherever possible.
- Dimensions should refer to solid rather than hidden lines.
- Dimensions should be placed as close as possible to the feature they are describing.
- When dimensions are 'nested', the smaller dimension should be placed closer to the object.

Figure 3.3.13 shows some standard dimensioning formats.

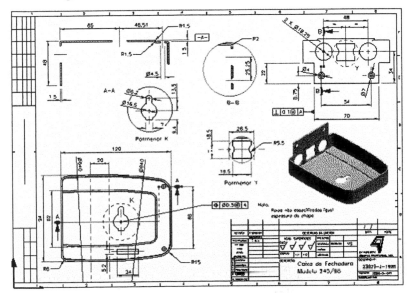

Figure 3.3.13 Standard dimensioning formats

Sections

Sections are an essential aid to understanding the complexity of a product. They should make a drawing easier to understand. Standard views show all the exterior features of objects, but if the interior features are shown just as series of dashed lines (hidden lines) it can cause confusion to less expert readers of the drawing. In a CAD drawing, as in manually produced drawings, sectioning cuts the object with an imaginary plane (cutting plane), making interior features, which were hidden, visible. The generation of sectioned views is quick and relatively easy when using a CAD program. The solid parts of the object in contact with the cutting plane are cross-hatched. CAD systems will allow many types of section view to be drawn. The choice of method depends on the internal complexity of the object.

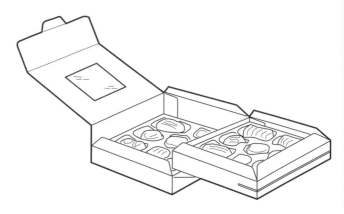

Figure 3.3.14 *An exploded view of a box of chocolates*

Tasks

1 Using a CAD program, generate some simple block shapes with holes, recesses and cavities.
2 Using either the views you generated above, or views supplied by your teacher, investigate how to produce:
   a) a full-section view, i.e. show an entire orthographic view as a section view, with half the object removed
   b) a half-section view, i.e. show one-half of the orthographic view as a section view
   c) an offset-section view, i.e. a type of full section using two or more cutting planes that meet at 90° angles
   d) a removed section view – this is similar to a revolved section, but it is not drawn within the view containing the cutting plane; it is shown displaced from its normal projection position.

Exploded views

Manufacturers in all design disciplines use exploded views to provide a visual explanation of how a product is assembled. A typical example is shown in Figure 3.3.14.

Virtual reality (VR) techniques

Virtual reality is an emerging technology that combines computer modelling with simulations in order to enable a person to interact with or be immersed in an artificial 3D visual or other sensory environment. Three-dimensional 'virtual products' can be created and viewed from different angles and perspectives.

Virtual Reality Modelling Language (VRML) is a specification for displaying and interacting with 3D objects on the **World Wide Web** using a **web browser** with a VRML plug-in or one that supports it. The development will have a significant impact on all industrial sectors but especially manufacturing. For instance, potential customers will be able to download virtual products and examine all aspects of them offline anywhere in the world. There are many other exciting developments that are beyond the scope of this book, but here are two examples for you to find out more about:

• Virtual manufacturing is a rapidly developing technology being pioneered by research teams all over the world. One of these teams, based at the University of Bath, produced this definition in 1995: 'Virtual manufacturing is the use of a desktop virtual reality system for the computer-aided design of components and processes for manufacture.' In their system, a user wearing a helmet with a stereoscopic screen for each eye views animated images of a simulated manufacturing environment. The illusion of being there (telepresence) is caused by motion sensors that pick up your head movements and adjust the view on the screens accordingly, usually in real time. Real time is the actual time during which something takes place. Simulations of manufacturing processes are usually accelerated to save time.
• Denford Ltd has created an innovative 3D website that uses virtual reality worlds to create a tour of its Professional Training and Development Centre where visitors can control a robot and try out trial versions of the company's CAD/CAM products (see Figure 3.3.15).

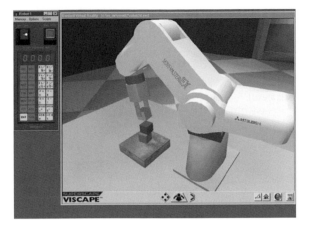

Figure 3.3.15 *3D website*

Input devices used in CAD systems

As we have seen, CAD systems have to be both interactive and graphical in use. In addition to the keyboard such systems need input devices and a user **interface**, operating as an input/output device, that allows interaction with the computer-generated model. There are two types of interface in common use:

* *Command-driven interfaces* require specific commands or codes to be used in order to make something happen. These interfaces process data quickly and are flexible in use but they rely on operators trained in a particular code or command set. Incorrectly entered codes make it difficult to edit or revise the drawing. These types of interface are being gradually replaced by graphic user interfaces.
* *Graphic user interfaces* (GUIs), first developed in the 1970s, take advantage of the computer's graphics capabilities making the CAD program easier to use. Most GUIs use WIMP format (Windows, Icons, Mice and Pull-down/pop-up menus operating environment).

The user does not need to learn complex command codes. GUIs are effective in allowing the user to:

- control the system by using set commands; by selecting functions via a series of windows; by a menu system; by screen icons or by direct actions such as 'clicking and dragging'
- receive information and feedback relating to what the system is doing, for example displaying an hourglass icon or progress meter
- enter data that will be used by the system in constructing the model
- select relevant data or parts of the model for the system to manipulate.

Input devices position or locate, point or pick or combine these functions. To signify that an action has to take place the user presses a button or switch that is provided on the input device. Many input devices produce an **analogue signal** (A) that needs to be converted into a **digital signal** (D) in order for it to be processed by computers. This (A to D) process is completed using a device called a digital signal processor (DSP) which can also be used to produce an analogue signal for use by an output device (D to A).

The mouse

Invented in 1963 at the Stanford Research Centre, the mouse is the most common input device. It operates either mechanically or optically to control the movement of a cursor or a pointer on the graphic display. It allows the user to 'point' to a function and 'click' to 'execute' it. All mice contain at least one button and sometimes as many as three, which have different functions depending on what software is used. Some also include a scroll wheel for 'scrolling' through large documents. The big disadvantage of using a mouse is that a positional error will occur if the mouse is lifted from the surface that it is running on. It is also virtually impossible to trace a drawing from a paper sketch or drawing (see digitiser).

The trackball or tracker-ball

The trackball or tracker-ball is basically a mouse lying on its back. The cursor on the screen moves as a thumb or fingers, or even the palm of the hand, rotates the ball. As with the mouse there are one to three buttons, which are used like mouse buttons. Unlike a mouse the trackball remains stationary so has the advantage of not requiring much space and it will operate on any type of surface. For both these reasons, trackballs are popular pointing devices for personal and laptop computers.

Table 3.3.2 Types of mice and computer connections

There are three basic types of mice:	Mice and tracker-balls connect to computers in different ways:
Mechanical – has a rubber/metal ball on its underside that can roll in all directions. Mechanical sensors within the mouse detect the direction the ball is rolling and move the screen pointer accordingly	Serial mice connect directly to the RS-232C serial port or a PS/2 communication port. This is the simplest type of connection
Optomechanical – similar to a mechanical mouse, but uses optical sensors to detect motion of the ball	PS/2 mice only connect to a PS/2 communication port
Optical – uses a laser to detect the mouse's movement along a special mat with a grid that provides a frame of reference. Optical mice have no mechanical moving parts. They respond more quickly and precisely than mechanical and optomechanical mice, but they are also more expensive	Cordless mice are more expensive and do not physically connect to the computer. They rely on infra-red or radio waves to communicate with the computer. On Apple Macintosh computers the mouse connects through the ADB (Apple Desktop Bus) port

Table 3.3.2 shows the three basic types of mice and identifies the different ways in which mice and tracker-balls are connected to computers.

Digitiser

Digitising tables have a large working area, typically over A0 paper size, enabling users to enter large-scale drawings and sketches into a computer. They operate to a great degree of accuracy and avoid the positional problems described previously when using a mouse. A digitising table consists of a reactive electronic surface and a cursor (also called a puck) that has a window with cross hairs for pinpoint accurate placement and it can have up to 16 buttons to execute various functions. Each point on the table represents a fixed point on the computer screen. To determine the exact position of the puck the surface may have a grid of embedded wires, each carrying a coded signal. The puck has an electronic device that can pick up these signals that are then digitally translated to give an exact position on the screen.

Graphics or digitising tablet

A graphics tablet is a tabletop digitiser consisting of a rectangular board (tablet) and pen (stylus) electronically connected to a computer (see Figure 3.3.16). The board contains electronics that detect movement of the pen and in the more sophisticated tablets a pressure sensitive stylus enables the user to simulate the marks that a pen or brush would make when it is pressed harder on the paper. The advantage of this device is that the designer can use hand-drawing techniques in order to 'draw' ideas electronically. The drawing can then be manipulated in the normal way by the CAD software.

Figure 3.3.16 A graphics tablet

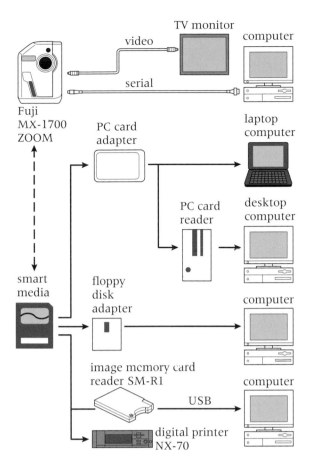

Figure 3.3.17 *Connecting a typical digital camera to a computer system*

digital camera allows an image to be put in a CAD system for processing and image manipulation within minutes of shooting. Digitally enhanced images now appear in a whole range of media ranging from product presentations such as photo montages on product labels to the Internet and print-based product catalogues. If the requirement is to get images in electronic form in the fastest possible time, then a digital camera is the best choice.

PhotoCD™

PhotoCD™ is a useful way of getting images from an ordinary camera into a computer fitted with a CD-ROM drive. The film to be developed and converted is sent to an authorised Kodak PhotoCD™ developer who provides a CD-ROM containing the compressed digitised images. The quality of the images is relatively good and the CD often contains several sizes of each individual image for various uses, the largest size being 2048×3072 samples per inch (spi). One of the advantages of this method is that CD-ROMs are an effective way of keeping image files organised without occupying expensive disk storage space on the computer system. However, in some cases, a dedicated slide scanner, although a more expensive and less convenient option, provides a higher quality image that has not been compressed.

Digital camera

Digital cameras store images digitally rather than recording them on film. Once a picture or image has been 'captured', it can be downloaded as data into a computer system for manipulation within a CAD program, stored on a photo CD or printed on a dedicated digital printer (see Figure 3.3.17). The quality of a **digital image** is limited by the amount of memory in the camera, the optical resolution of the digitising mechanism, and by the resolution of the final output device. Established printer technologies have limitations but there are three printer technologies, 'thermo autochrome', 'dye sublimation' and digital printers that produce better images. We shall look briefly at all three in the section on output devices.

The main advantages of digital cameras for designers are the operating costs; the speed of data conversion (there is no film processing at the start of the process); and the image manipulation and editing that is possible. A

Tasks

1 Investigate the use of a digital camera and how the images are transferred from the camera to a computer. Once your image is on screen how does your software allow the image to be edited? Produce a range of digital product drawings and other graphic images for use in your project portfolio.

2 An example of a product package that incorporates text, graphics and photographic images is known as a composite image (see Figure 3.3.18). Find a similar example of your own and write a brief description of the key merchandising properties and types of images and text that have been incorporated into it. Using a product that you have designed, or are in the process of designing, develop a promotional leaflet to create an identity for your product. You must incorporate images from a digital camera alongside computer-generated text and graphics.

Figure 3.3.18 *A **sleeve** for a food product that combines text, graphics and photographic images*

Digital video (DV)

A new generation of video cameras that are entirely digital are now on the market. In a digital video (DV) camera the output from the camera is already in a compressed digital format, so no analogue to digital conversion is required as with analogue video cameras. Images can be taken straight to a PC-based video editing system. Digital video systems allow the communication, control and interchange of digital, audio and video data. The benefits of applying these DV systems within CAD other than for video conferencing are an area of considerable research.

Scanner

A scanner is a device that converts analogue data into digital data that can be read by computer software for desktop publishing (DTP), CAD and other programs that combine text and graphics. The data can be edited, stored and used to develop design ideas and specifications.

The quality of a scanned image depends on the resolution of the scanner. The more dots per inch (dpi), the sharper and more detailed the image on screen. As is the case with most computer-generated text and graphics, the final appearance of the image will depend on the quality of the output device – a laser printer produces better quality images compared to a low-quality inkjet printer. High resolution scanning is important for line artwork to ensure that the lines retain their smoothness when printed on a commercial high-resolution output device such as a 2500 dpi imagesetter.

Storing scanned images requires large amounts of random access memory (RAM) on the hard disk or other storage space. A colour image measuring 200 mm x 250 mm at a low resolution of 300 dpi would require in excess of 7 Mb (megabytes) of RAM. The higher the resolution required, the more memory intensive the image. Graphic designers use scanners mainly for:

- scanning in line artwork for reproduction or positioning on a graphic image
- scanning images to be used as positional guides on a page or other graphic layout
- inputting text copy by means of optical character recognition (OCR) software
- scanning images that will be digitally manipulated to achieve a particular visual effect.

Scanners are able to scan both 2D surfaces and 3D objects (see Figure 3.3.19). Developments in the light-sensitive scanning head called the charge coupled device (CCD), the control technologies and the processing software mean that scanners are now much easier to operate. The images produced range from simple black or white (line images) to greyscale and full-colour images with in excess of 256 colours.

Types of scanner in common use in CAD

A scanner may be monochrome, greyscale or colour and it may be one of the following:

- *Flatbed scanner* – the most versatile and popular desktop device for scanning reflective, line and greyscale artwork. It looks and works like a photocopier as the image is placed face down on a flat glass screen and a scanning head passes across the image. There are three types of flatbed scanner:
  - entry-level flatbed scanners generally share the following specifications: 216mm × 280mm scanning area, low to medium resolution and low cost. They often come with entry-level image manipulation software. These machines frequently offer excellent price/performance ratio
  - mid-level flatbed scanners differ in three important ways: they cost much more; they are targeted at a professional market and often come with more sophisticated image manipulation software. They also have significantly better technical specifications resulting in scans of correspondingly higher quality. Some mid-level scanners may also offer a larger scanning area compared to flatbeds
  - high-end flatbed scanners are often seen as a practical alternative to drum scanners offering sophisticated design features that graphic design requires. They are expensive

but offer a noise-free design, large scanning areas and very high image resolution.

- *Sheet-fed* or *edge-feed scanners* – similar to flatbed scanners, they operate differently as the original is moved under a fixed scanning head. They are used for high-volume work.
- *Handheld scanners* – portable, low cost, low resolution and small capacity devices with restricted scanning widths.
- *Large drum scanners* – capable of scanning both opaque documents and transparencies at high resolutions of over 400 dpi; expensive.
- *Transparency/slide scanners* use a scanning camera to produce high-quality colour images when the transparency or slide is placed between the camera and a light source. Slide scanners cost a lot more than the relatively inexpensive flatbed transparency option. For those who may need only an occasional transparency scanned, a flatbed with transparency adapter is the way to go. But if you scan a lot of transparencies, then the only equipment that offers the best quality scans are dedicated transparency scanners. Expensive, multi-format transparency scanners aimed at the professional market can scan everything from 35 mm slides to 100 × 130 mm high-quality transparencies. As scanning technologies improve these high-end transparency scanners are presenting an alternative to drum scanners by offering more features, better software and faster scanning time.
- *Overhead scanners* – scanning cameras mounted above a copy table on which the original such as a photographic image is placed. They can also be used for scanning transparencies that have been mounted on a lightbox on top of the copy table.
- *Drum scanners* – desktop scanners in which a laser beam acts as the light source to capture the image from the original attached to a spinning glass drum.
- *Video Digitisers* are mostly used for multimedia purposes, especially in the creation of QuickTime™ movies, but they can be used to capture still images for print. Video cameras utilise the same digital CCD arrays found in flatbed scanners. These CCD arrays produce an analogue signal (50 or 60 MHz) that either drives other analogue devices such as VCRs and television sets, or is captured on to videotape. Video cameras are not technically scanners in the truly digital sense of the word, but the analogue video signal can be digitised using specialised hardware and software. Video image capture software is very similar to traditional scanning software, while the hardware is usually a board that fits inside a computer. Video cameras provide an inexpensive way to get images into a computer, but the image resolution will be low (only 640 × 480 pixels) and the colour accuracy is often poor.

3D scanners

3D scanning or digitising creates a series of profile curves that define the surface characteristics and the physical geometry of a three-dimensional object. The digital data collected and recorded allow a 3D representation or model to be created within a computer. CAD and graphics software can then be used to blend and render surfaces to add colour, reflections, texture and other visual techniques to add visual realism. There are two different methods of scanning a 3D object in order to produce a digital file: contact or non-contact.

Contact scanning systems make physical contact with the object with a probe that can be passed around and over the surface of the object. The tracking is done by a probe that is machine driven or manually operated. A typical manual system uses a mechanical arm that has digital sensors in each joint (see Figure 3.3.19). A typical use of these devices is in converting a concept model made in any suitable medium such as clay into a set of electronic data that can be used by a CAD program or a CNC machine. An operator moves a probe over the surface of the object clicking and recording the positional data. These points generate the required electronic profile curves. A machine-driven device common in UK schools and colleges is the Roland model,

Figure 3.3.19 *An arm scanner*

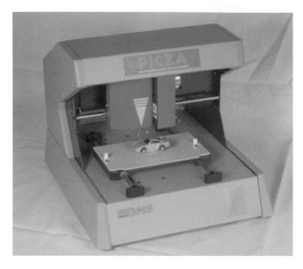

Figure 3.3.20 *A Roland scanning machine*

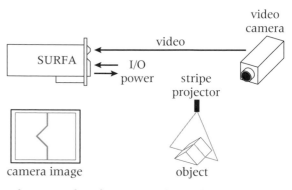

The SURFA board uses Digital Signal Processing of video data to capture surface shapes in real time at over 14,000 points per second.

Figure 3.3.22 *Laser stripe triangulation*

which can operate as a low-cost 3D scanner or as a small capacity CNC milling machine (see Figure 3.3.20). The scanned data can be exported in 3D DXF or VRML format for use in other software.

Non-contact scanning systems mostly use a geometric technique known as triangulation to create the 3D shapes. Laser scanners are non-contact and high-speed devices but they are expensive. These devices scan by either directing beams at various points to create a profile curve or by generating a 'laser stripe profile' of the object (see Figures 3.3.21 and 3.3.22). The beams or stripes are reflected back to a series of sensors or into a video camera. In both cases an accurate surface representation is built up by scanning from different planes and angles and mathematically processing the digital data.

Figure 3.3.21 *Laser scanner in operation*

Ultrasonic scanners are also non-contact devices that 'bounce' sound waves off the surface of the object and triangulate the points in 3D. The devices are not portable or easy to set up. They are noisy in operation and do not provide a high level of accuracy.

Magnetic tracking scanners work on a similar principle to ultrasonic devices using magnetic fields to create the triangulated points. The major drawback of such systems is that they cannot scan any object with a metal part.

Common output devices

An output device is what makes a computer capable of displaying and manipulating pictures (graphics) or machining a variety of materials. Laser printers and plotters are 'graphic' output devices because they permit the computer to output pictures. They produce 'hard copy' on paper or film. A monitor is an output device that can display pictures. A CNC milling machine is an output device that can produce physical objects in a range shapes and materials.

Linking CAD and CAM

CNC machines provide an interface between CAD and CAM. Numerically controlled (NC) machining processes were first developed in the USA in the 1940s. The motors and machining tables on the NC machines were originally driven from instructions provided on a punched card system and then from magnetic tapes. Programs generated and transferred from computers now control the operation of many NC machines, hence computer numerically controlled or CNC machines. There are a host of CNC machines including milling machines, lathes, drilling machines, press engravers and die cutters used for producing cartons and packages. New classes of 'tool-less' CNC machines are in use in industry, for example CNC laser devices are increasingly being used in cutting and machining applications.

Transferring data to output devices

Both CNC machines and graphic output devices are generally connected to the computer by either 'series' or 'parallel' cables, but some devices such as printers can be operated by infra-red sensors and are therefore 'wireless'. When cables are used they are connected into the computer via communication ports, the size, shape and number of connecting pins varying according to the specific make of computer. Almost all graphics systems, including CAD systems and animation software, use a combination of vector and raster graphics. Most output devices are raster devices.

Displaying graphics

The graphic capabilities of a monitor or display screen make it an important output device in any CAD system. There are many ways to classify monitors. The most basic is in terms of colour capabilities, which separates monitors into three classes:

- Monochrome monitors actually display two colours, one for the background and one for the foreground. The colours can be black and white, green and black, or amber and black. They are often used on the displays of a CNC machine where a colour image is not so important.
- A greyscale monitor is a special type of monochrome monitor capable of displaying different shades of grey. They are also often used on the displays of CNC machines.
- Colour monitors can display any number from 16 to over 1 million different colours.

The importance of screen size for CAD work

The choice of screen size, or viewable area, is particularly important in a CAD system. Monitors that measure 400 or more millimetres from corner to corner are often called full-page monitors and are considered more suitable for graphics work. Large monitors can display two full pages, side by side.

The resolution of a monitor indicates how many pixels are on the screen. In general, the more pixels (often expressed in dots per inch – dpi), the sharper the image and the better the resolution.

Printers

A printer is an output device that prints text or graphics on to paper, card and film. Not all printers can produce high-quality images. Printers are classified by the following:

- The quality of type they produce – either letter quality (as good as a typewriter), near letter quality, or draft quality. Inkjet and laser printers produce letter-quality type.
- The speed at which they print. Printing speeds are measured in characters per second (cps) or pages per minute (ppm). The speed of printers within a particular class can also vary widely. Laser printers print at speeds ranging from four to 20 text pages per minute.
- The effectiveness of the methods used to create colour and more realistic images by employing techniques such as dithering, colour matching between the screen and paper, half toning and continuous tones.

The following printers are considered below:

- inkjet
- laser
- dye sublimation
- thermo autochrome
- digital
- offset.

Inkjet printers

These non-contact printers produce low-cost, high-quality text and graphics in colour by combining cyan, magenta, yellow and black (CMYK). A typical commercial use of a larger inkjet printer would be in the production of exhibition panels or short-run or one-off posters and photographic enlargements that require clear and colourful images at a comparatively low cost. The images are drawn on-screen to produce digital artwork for printing on to A2, A1, A0 and custom paper sizes paper before being encapsulated in clear plastic, for added durability, and mounted on foamboard or some other suitable substrate.

Inkjet printers all use some sort of thermal technology to produce heated bubbles of ink that burst, spraying ink at a sheet of paper to form an image. As the ink nozzle cools a vacuum is created and this draws in a fresh supply of ink for heating and spraying. The print head prints in strips across the page, moving down the page to build up the complete image with a resolution that can range from 300 to 1200 dpi. They are sometimes referred to as 'bubble jet' printers, which is actually the trade name of Canon's own inkjet technology.

Disadvantages of the inkjet printer are that the ink cartridges need to be changed more frequently and expensive specially coated paper is necessary to produce really high-quality images, which significantly raises the cost per page. Choosing the right paper for inkjet printing is important. Some images, especially those with large areas of colour, can 'bleed' if the paper is too absorbent or too much ink has been applied. Bleeding causes images to blur as the colours merge and run together.

Another disadvantage of this system of printing is that images are easily smudged as some types of ink take a little time to dry. A further drawback can be that the diameters of the nozzles used in inkjet printers are very small and can easily become clogged.

Laser printers

These printers use the same technology as photocopier machines. The printer receives data from the computer, which is processed and used to control the operation of a laser beam directing light at a large roller or drum. Altering the electrical charge wherever laser light hits the drum creates the required image. The drum then rotates through a powder called toner. The electrically charged areas attract the powder and the print is made when it is transferred on to the paper by a combination of heat and pressure. Laser printers produce very high-quality text and graphics.

Dye sublimation printers

These low-speed devices produce relatively expensive but high-quality graphic and photographic images. The four coloured inks or dyes (CMYK) are stored on rolls of film. A heating element turns the ink on the film into a gas. The amount of ink that is put on paper correlates to the temperature of the heating element. The temperature varies in relation to the image density of the original drawing or artwork. The reason that it produces images of such high quality is that the ink is applied as a continuous tone rather than as a series of dots and special paper is used that allows the dyes to diffuse into the paper to mix and create precise colour shades.

Thermo autochrome (TA) printing

This method is used to print high-quality images generated by digital photography. The process is more complex than either inkjet or laser printing. TA paper contains three layers of coloured pigment, cyan, magenta and yellow, each of which is sensitive to a particular temperature. Three passes are needed to get the three colours to show. The printer is equipped with both thermal and ultraviolet heads; the heat from the thermal head 'activates' the colour in the paper, which is then 'fixed' by the ultraviolet light.

Digital printing

Digital printing is a system of printing based on rapidly evolving digital technology, which involves linking state-of-the-art printing presses and computers, bypassing the traditional route of making printing plates. High-speed digital printers use laser technology to provide 600 dpi print quality for text, photographs and graphics on a wide range of media. These printers combine ultra high resolution with over 4000 adjustable levels of shading and can print at a rate of 250 copies per minute. For printing from hard copy the original is scanned in once reducing wear and tear on the master copy. If a digital file is used every print is a 600 dpi original. There is no deterioration in print quality. The printers come with a full range of image handling capabilities including image clean-up, reduction and enlargement, photo-screening, cropping, masking, rotating and the moving of text and graphics. The advantages of digital printing include:

- digital job preparation and processing is controlled from the end user's desk without having to save data on disks, CDs or other traditional media for later delivery to the printer
- reliability as end users can download their printing jobs via a modem, **ISDN** or network without worrying that they will output incorrectly; the software allows users to specify finishing and media types and sizes from the desktop
- what you see is what you get, as documents can be seen on-screen as they will be printed
- print proofs can be made at the end user's terminal for checking
- the ability to select media finishing such as stapling and gluing
- less operator stress as different printing jobs can be sorted so that those requiring the same media can be printed together
- more productive print operators thanks to streamlined job flow and lower document production costs
- printing on demand as digital technology allows images to be scanned or transmitted electronically into memory to be stored for future use
- the smoothing out of production runs which is important when market demands are fluctuating; for example, you can print 700 jobs one week and then increase production to

1000 copies the next with no additional set-up charges as all the files and processing settings are stored digitally.

In your school or college you may come across specialist types of digital printers that are used to produce photographic quality images from a digital camera without having to transfer data to a computer. The printers can be connected to a monitor for viewing and editing images and layouts. The widespread introduction of digital printers is limited at the time of writing because there are no agreed standards for them. Manufacturers of digital cameras will supply digital printers that match their range of cameras.

Offset or offset lithography (litho) printing

Offset printing is currently the most common commercial printing method, in which ink is offset from the printing plate to a rubber roller, then to paper. This type of printing technique can be divided into two parts: duplicating or standard printing, and commercial quality or special printing.

Standard printing is a cheap alternative to photocopying. Any reasonably typed, hand-drawn or computer-generated original can be used to produce long copy runs. It is often offered as a 'same day' or an 'over the counter' service for all small- to medium-run jobs, i.e. 1–20 originals with 100 copies of each.

Special printing describes high-quality and multicolour commercial type printing. In addition to paper and card, a wide selection of materials can be printed such as plastic, foils and other laminated paper finishes. Increasingly, a graphic artist provides the designs, producing digital data that can be sent directly for computerised phototypesetting. These computer-based production methods provide considerable savings and can be made with no loss of quality when compared with the traditional commercial printing methods.

The use of computers in specialist printing situations

Screen process printing, sometimes referred to as silk-screen printing or serigraphy, is very common. The introduction of computer technology in the 1980s, in the pre-press side of screen printing, particularly in image capture and manipulation, was a major step forward in the scope and range of images that could be applied to objects. With this process it is possible to print on to a great variety of surfaces such as plastics, wood, metal, glass, textiles, paper and board.

The process is typically used to place images, many of which are digitally produced, on to products such as binders, document wallets, mouse mats, window and car stickers as well as posters and all types of point-of-sale and promotional materials. Other uses of screen printing include displays on computers and the badges and control panels on electrical equipment. In the home you will find that many textiles and items of clothing, sports bags and T-shirts have all been screen printed.

Artists have also used silk-screen printing, especially since the days of 'Pop Art' in the sixties; Andy Warhol, Rauschenberg and Hamilton are a few examples. Some artists and print makers who want to produce multiples of their work, limited editions and other projects often like to provide their own colour separations prepared on film or as camera-ready artwork, while others prefer to supply working drawings, sketches, together with their instructions and colour specifications. There are no set rules to producing a print and every artist works differently so printing houses have to be able to operate flexibly. Typically, they might offer a photo stencil-making service to printmakers wishing to screen print their own work. For those that are now using digital imaging techniques the company must be able to offer the production of computer-cut stencils.

Continuous stationery covers products such as invoices and statements and involves the original being created in a word processing or desktop publishing application before direct or indirect transfer into the printing process.

Flexographic printing (flexography) provides economical printing on unfinished surfaces, wrapping paper, cardboard, plastic film, and it is also used for printing self-adhesive and fabric labels. Digital images can be drawn from a range of sources and then transferred to an image setter or to a plate setter for production of the printing plate.

Other uses for computers in the commercial print process

Computers are playing an increasing role in each of the three stages of the commercial printing process pre-press, press and post-press:

Pre-press includes all the various printing related services performed before ink is actually put on the printing press and includes image capture and manipulation, scanning and rasterising, colour separating and proofing. A typical pre-press department concerns itself with pasting-up camera ready artwork to produce final films of a specified area, usually a page, which contains all the images, text and tints in position, necessary for the production of a printed plate. Images and text produced on a computer are sent directly to a Raster Image Processor (RIP) which converts the electronic information into a format that an image setter can understand and process. An image setter is a high-resolution device that prints directly to film or bromide significantly improving print quality and reducing preparation time. Increasingly, pre-press also involves electronic publication direct to devices such as a plate setter which is a device to create printing plates directly from a digital page, missing out the need for a film stage.

Press is the stage of printing the image on to the chosen media in which automated, CNC and microprocessor-controlled presses are being increasingly used. The market for short four-colour runs is one of the fastest growing and dynamic of all. It is characterised by short runs with extremely fast delivery and the need to be able to receive and act on last-minute changes. With changing market requirements, new approaches have to be taken and digital technology is increasingly at the core of workflow. The technology enables companies to become complete service providers from digital data to the final product. In the new digital printing presses the printing plates are imaged using data direct from a computer which reduces the number of production steps, increases speed and improves reliability. Computers provide profitability with a high degree of automation, user-friendly operation and enormous flexibility for job changes.

Post-press/finishing is the stage at which the printed item is cut to size, folded as necessary and finished as appropriate by binding and stitching.

Plotters

A plotter is a high-quality impact-printing device that draws images on paper or any other suitable medium directed by commands from a computer. There are two classes of plotter: vector and raster. Vector plotters produce an image as a set of straight lines, fill patterns are clearly visible and they operate comparatively slowly. All raster plotters generate an image as a series of points. The way that the points are printed varies and the methods are beyond the scope of this book. These plotters produce very large, full-colour drawings with a high degree of quality.

XY plotter

The term XY refers to the axes along which the

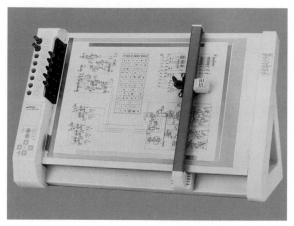

Figure 3.3.23 *An XY plotter*

plotting pen can travel. Plotters differ from printers in that they draw lines using a pen. As a result, they can produce continuous lines, whereas printers can only simulate lines by printing a closely spaced series of dots. Multicolour plotters use different-coloured pens to draw different colours. Pens can be picked from a bank of penholders or changed individually as the different colours are called for. In general, plotters are considerably more expensive than printers. They are used in engineering applications where precision is essential. An XY plotter is shown in Figure 3.3.23.

Plotter-cutters

Plotter-cutters can plot drawings in the same way as the XY plotter described above but they can also produce cut shapes in card, vinyl and other sheet materials using thin blades that can be adjusted for depth and pressure of cut. This allows the plotter-cutter to undertake finely controlled cutting techniques such as 'scoring' in which a card that is to be folded is only cut to a certain depth to make the subsequent folding easier. Computer-controlled vinyl cutters are used in the production of advertising and promotional products as well as in exhibition screens, sign and banner making and many other similar processes. Figure 3.3.24 shows a typical plotter-cutter.

Engraving machines

Engravers also have a variety of uses ranging from sign and print plate making to the production of jewellery, medals and 3D reliefs (see Figure 3.3.25). They can be relatively small devices that sit on a desktop or larger floor-standing machines. The machines can operate in

Figure 3.3.24 *A plotter-cutter*

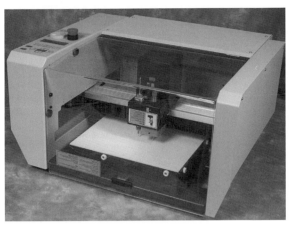

Figure 3.3.25 *A CNC engraver*

the x, y and z axes. This allows the engraving of 3D surfaces and curves as well as lettering. Cheaper versions of CNC engravers for schools and colleges have less control over the z axis of the cutting tool. This reduces the overall flexibility of the machine because it has to engrave at a fixed depth. One of the characteristics of an engraver is the high spindle speed that is needed because of the very small diameter of the V-point engraving tools. We shall return to CNC cutting speeds and feeds later in this unit.

3. Computer-aided manufacture

The production benefits of using CNC machines

CNC machines are widely used in a range of industries. Because they can operate in more than one axis they can generate shapes ranging from straight lines to very complex curves. The technology and functionality of CNC machines has continued to evolve since they were first used in the 1970s. CNC machining centres allow a single machine to carry out more than one manufacturing operation, for example a plotter-cutter can perform two main manufacturing functions. It can be used for high-quality, large-scale colour plotting on a range of media. If the cutting knife is fitted it can be used to cut and perforate card and other media such as vinyl to produce 2D shapes, such as packaging nets, screen printing templates, exhibition screens and advertising banners.

Processing flexibility and improved output quality is the key to the future development and profitability of computer-based manufacturing systems. Where they can be used, computer-controlled machines offer significant reductions in processing times. In volume production, such as commercial printing, set-up or make-ready times are significantly reduced, allowing greater flexibility supporting 'quick response' production capable of handling production runs ranging from under 100 to print runs in the thousands.

CNC prototyping

The development of new 'tool-less' cutting technologies such as the use of computer-controlled lasers is further extending the range and complexity of products that can be prototyped or manufactured such as new designs for shampoo or perfume bottles. Rapid Prototyping (RPT) is an emerging CNC application that creates 3D objects using laser technology to solidify liquid polymers in a process called stereolithography.

Boxford Ltd in the UK has produced a relatively low-cost RPT system for schools and colleges based on a prototyping system called layered object modelling (LOM) (see Figure 3.3.26). Models are imported from a CAD program as in the industrial systems but the

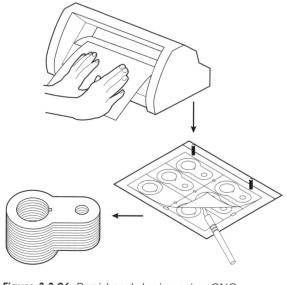

Figure 3.3.26 *Rapid prototyping using CNC machines*

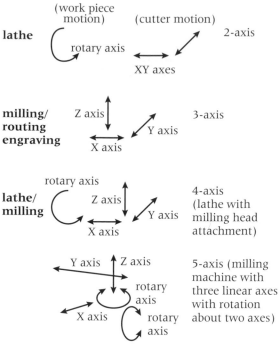

Figure 3.3.27 *CNC machines can move and operate in up to five axes*

models are assembled from thin sticky backed paper. A CNC vinyl cutter cuts the slices (in the industrial process a laser is used). The layers of sliced paper are built up on a pegged jig to make the 3D prototype.

Task

Use the Internet and other information sources to find out more about RPT. Produce an information sheet to record your findings. How could you use an RPT system in your designing and manufacturing?

Operating characteristics of CNC machines

On CNC machines such as plotter-cutters and embroidery machines either the tool or the work piece is able to move in up to five axes to generate the required point-to-point, straight or contoured tool paths, as shown in Figure 3.3.27.

On larger-scale manufacturing machines such as plotting and cutting devices a computer-generated cutting or plotting program is fed into a machine control unit (MCU) which includes a manual control keypad and in some cases a display screen. The MCU reads and converts the digital data it receives in order to control the analogue machining movements that the machine has to make. Movement is controlled by a series of servo systems or, in low-cost devices, stepper motors, called actuators. As the machine is operating the MCU is constantly receiving

feedback information from sensing devices (**transducers**) or encoders on the machine. This information is used to correct any errors of spindle speed, feed rate or cutter position. This sensing and response mechanism is called a **closed-loop control** (see Unit 4, Section B, Option 2). On smaller desktop CNC devices such as engravers, plotters and vinyl cutters the cutting program is transferred from the computer by cable or infra-red to the CNC machine. The machine has the circuitry and computing capacity to generate the codes needed to move the machine in each of its three axes. Figure 3.3.28 shows a block diagram of a typical CNC machine.

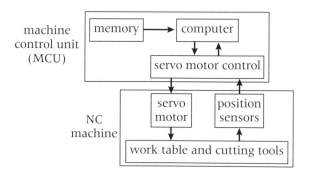

Figure 3.3.28 *A block diagram of a typical CNC system*

Toolpaths, cutting and plotting motions on CNC machines

There are three type of motions employed on CNC machines:

- The cutter or plotter moves from 'point to point' as it moves between two specified positions. However, the path between the points does not have to be a straight line; it can be an arc or a series of curves.
- Moving a cutting tool or plotting pen parallel to one machine axis is known as 'straight cutting'.
- Contouring allows both point-to-point and straight movement in more than one axis, allowing complex curves and shapes such as spirals to be generated (see Figure 3.3.29).

On an engraver or another cutting machine the depth that the tool cuts in any one pass depends on the amount of material to be removed and the surface finish required on the finished article. The inside of a mould for a precision plastic product to be made by injection moulding has to have a finer surface finish than is needed on the outside. Roughing cuts remove large amounts of material, set out the basic profile and cut close to the different depths that are required. Finishing cuts provide the required final shape and surface finish. Finishing cuts are made in a number of ways. Depending on the functions available the cutter can:

- move in one direction, constantly lifting back from the end of the cut to the start
- constantly cut as it moves backwards and forwards
- follow the contours on the surface
- follow parallel paths over the surface of the object (see Figure 3.3.30).

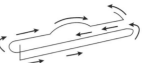

a) The cutter moves in one direction constantly lifting back from the end of the cut to the start

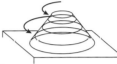

b) The cutter cuts constantly as it moves backwards and forwards

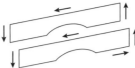

c) The cutter follows the contours on the surface

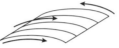

d) The cutter follows parallel paths over the surface of the object

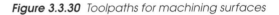

Figure 3.3.30 *Toolpaths for machining surfaces*

Task

Determine what types of movement the CNC machines that you have access to in your graphics and practical work are able to make. Compile a table showing the results of your investigations using one of the four toolpath models to describe the movements you observe.

Types of CNC machine

- Lathes for operations in metal and plastics used to produce 3D product shapes and moulds for plastic products.
- Milling machines for operations such as mould making, profiling, pocketing and surface milling used to produce the cutting dies used in the print industry.
- Routers and engravers for producing printing plates and 2D and 3D graphic designs.
- Flatbed cutting and perforating machines used in the post-press stage of commercial printing.
- Pressing, punching, bending and die cutting machines for processing sheet materials such as card and plastic sheet to produce products such as flat carton packs and document wallets.

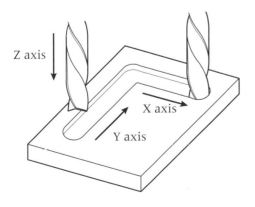

Figure 3.3.29 *Point-to-point cutting of a printing plate in three axes*

Speeds, feeds and rapid traversing

The speed at which the cutting tool or device such as a plotting pen or the work piece moves in relation to one another on a CNC machine is critical to its efficient operation. The plotting pen has to have time to make its mark on the paper. On most XY plotters it is usually automatically programmed. The feed or cutter moving rate is also programmed and is measured in millimetres per second (mm/s). It determines the rate of progress along a cutting or a plotting path. The feed rate has to be precisely controlled on CNC devices such as embroidery and knitting machines to avoid machining problems such as snagging and jamming. On engraving machines metals are machined with slow feed rates and small depths of cut in comparison to engraving plastics. Rapid traversing describes the way the cutter, plotting pen or printing head moves when it is not engaging with the material being worked. On a plotter, for instance, different lines are drawn apparently at random but, in fact, the plotting software has calculated the most efficient way of drawing and building up the image and when different pen changes need to occur.

On a CNC engraver or other cutting device the cutter revolution or spindle speed is measured in revolutions per minute (rpm) and will vary according to the material being machined and the diameter of the cutter. As a general rule, the bigger the cutter diameter, the slower the spindle speed and the softer the material, the faster the spindle speed. On a typical engraving machine spindle speeds usually range between 5000 and 15,000 rpm. To give an example, acrylic is engraved at a spindle speed of 10,000 rpm with a depth of 0.2 mm at an XY feed of 15 mm/sec and a Z feed of 5 mm/sec.

The types of cutting tools on CNC machines

CNC machines such as engravers and millers use cutting tools that are similar to those used in their manual equivalents. The tools are held in collets. The choice of cutter depends on the geometry or shape required and the cutting loads it has to withstand. Cutters are typically made from high-speed steel (HSS), tungsten carbide or ceramic materials. These materials can operate at high cutting speeds without losing their hardness. They can resist the heat generated as a result of friction during the cutting process. Tungsten carbide and ceramic materials are relatively expensive so these materials are only used on the cutting edges of a 'tipped' tool. Some cutters have removable blades for ease of maintenance and for sharpening purposes.

The geometry of the cutting tool determines what shape the tool generates and how the waste materials are moved away from the cutting edges. On a twist drill or a milling cutter 'flutes' force the materials up and away from the cutting operation. On a lathe tool the angled surfaces combine to move the cut material away. Routing cutters are small in diameter and can be half round or three-quarter round in shape. Engraving cutters are V-shaped.

Task

Slot cutters, end mills, bull nose cutters and face mills are some of the cutters you might find on a CNC milling or routing machine.

a) What features do they have in common?
b) What are their respective advantages and disadvantages?
c) What are the benefits of a collet rather than a chuck for holding the cutting tools?

Scale of production and the use of CAM

The choice to use CAM will depend on the intended scale of production, the number of identical components to be produced and the cost benefits that it can bring. There are essentially three categories of production:

- one-off production (sometimes called 'jobbing')
- batch production
- mass-production systems that include high-volume 'runs' or continuous operation.

One-off production

This means that just one product is 'made to order', such as an architectural model or a proto-type. Whatever the product, the company making it has to have a highly flexible production facility and in most cases a highly skilled team of workers. Products made by this process are invariably more expensive to buy because they are individually designed and made.

Batch production

This is a flexible production system that is used to make a relatively small number or batch of components or products. It is found in areas such as commercial printing where print runs are of different lengths and degrees of complexity. A key feature in batch production is the ability to respond quickly to demand. In modern commercial printing houses once a batch run has been completed the computer-controlled machines can be easily reprogrammed and quickly made ready for another printing job.

Mass production

Set-up and break-even costs can be higher than in the other two systems but the high volumes of product generated in continuous operations such as those in the printing industry significantly reduce manufacturing costs. In an industry such as carton making the products are mass produced and though they vary in their complexity they all are highly standardised, which means that processes can be highly automated, are less labour intensive and more cost efficient.

Task

Use the information you have been given about different scales of production to describe how they might be applied in different types of printing operation. Give examples of products that incorporate graphic elements to support your answer.

Recent developments in production systems

Mass-production companies, such as carton manufacturers, are looking to develop more flexible systems based on the batch production principle. Earlier in this unit we looked at the pressures for change, including the demand for a greater variety of products and the move towards products that are 'customised' to meet the demands of a particular market sector. In the advertising and manufacturing sectors both print and carton manufacturers are under pressure to reduce the unit costs of their products but at the same time they are expected to offer their customers a greater range of options as 'standard' rather than as 'optional extras'. To meet these demands all manufacturers, whatever their sector, need flexible production or manufacturing systems to allow products that can be 'made to order'. Flexible manufacturing systems (FMS) are particularly important for companies like those involved in merchandising, promotional or point-of-sales display when the scale of production cannot justify a fully automated production line but where there is a need to have a 'quick response' system. Outsourcing becomes an option in an FMS as it allows a company to take advantage of the IT revolution, reduce and control costs by freeing up capital that would otherwise be invested in costly specialist equipment.

An example of outsourcing is in the growth of CAD bureau services. These are specialist companies that have a vast range of specialist equipment aimed at the design and reprographic needs of the graphic market. Typically, these bureaux provide large format plotting and printing of electronic files created from a variety of CAD applications. These files can usually be plotted or printed in monochrome and full colour on various media including bond paper, tracing paper, drafting film, vinyl for banners and signs and textiles in much greater widths than is possible with most common output devices. For smaller companies, a bureau service offers the most cost-effective way to use expensive specialist equipment for jobbing or small batch production such as high-resolution full-colour printing up to 1440 dpi or digital scanning to copy, enlarge or reduce hard copy originals into an electronic file format. Additionally, bureau services might offer a range of computer design services such as slide and multimedia presentations, Internet websites, logo design, business stationery including business cards, letterhead and envelope design, brochures, flyers and posters and advertising design.

Advantages of CAM

Control of the production process to remove risks

CAM allows production processes to be automated so that manufacture can be more precisely monitored and controlled. Raw materials such as corrugated packaging materials and additional components such as card dividers in composite packages can be moved safely so that they are at the right stage of production at the right time. Automated lifting, handling and carrying systems allow people to be taken off tasks that are potentially hazardous or dangerous to health.

Flexible scale of production

CAM systems allow the scale of manufacturing to range from batch and mass to continuous production. For example, in the carton manufacturing industry the advantage of CAD/CAM is a reduction in make-ready times. The CAD programs not only generate the carton graphics and labelling, they also generate cutting programs that are transferred to the die cutting machines online leading to extremely fast and flexible order processing. The introduction of computer-controlled systems of manufacture such as 'just in time' (JIT) means that:

- CNC machines can be operated more flexibly allowing manufacturers to develop quick response systems to meet changes in design or marketing strategy such as the addition of information on an existing package design as part of a sales campaign for the product

- production levels can be linked directly to the size of the customer order book because manufacturing can be speeded up by a reduction in make-ready or set-up times
- the company is less likely to be damaged by sudden changes of demand in its particular market
- orders for raw materials can be generated automatically as and when required, just in time, so avoiding large and costly stockpiles of materials and components that are not adding value to the process
- the inventory or levels of finished products ready for distribution to other manufacturers or retailers can be kept to a minimum which keeps costs to the minimum
- production increases and quality is improved.

Reduced manufacturing times

CAM can reduce the manufacturing time from days to just a few hours. Complex 2D and 3D shapes and forms are easily developed, edited and tested for ease of manufacture through on-screen simulation. As an example, computer-based design teams in a promotional merchandising company can now provide advice on all areas including structural design, they can provide rough visuals and graphics including coloured samples, complete camera ready artwork and cutter drawings. The technology now available also allows the production of computer-cut samples and direct access to suppliers and customers via modem and ISDN links.

Operational reliability

CAM improves levels of operational reliability and the finished quality of products. Modern CAM software can generate and simulate cutter and plotting paths that highlight potential production problems. In specialist areas like carton packaging, the bending allowances for producing 3D forms from 2D sheet materials using creasing and scoring rules can be calculated by the CAD software and applied directly to the cutting or creasing presses.

Consistency in repetitive situations

CNC machines only need to be 'trained' once and they can do repetitive tasks rapidly with fewer errors than human operators. Machines can operate 24 hours a day, do not need rest breaks or holidays and are not affected by other human constraints.

Improved productivity

Production 'throughput' is improved with lower processing costs and less waste of materials and resources. Manufacturing costs can be estimated with a greater degree of certainty, as production rates are more consistent than is possible with human-operated machines.

Disadvantages of CAM

Costs

The cost of buying and installing computer-controlled machines is high when compared to manually operated machines. For really high-volume mass production such as carton making a purpose-built automatic machine may be a more cost-effective solution to the use of CNC machines.

Employment

With developments in automation, robotics and artificial intelligence it is likely that there will be even less human involvement in the design and production processes of the future.

Worker involvement

For some people CAM can create emotional and other psychological problems at work. When machines and processes such as those found in the printing industry are controlled directly from a central management system, some jobs consist of nothing more than 'machine minding' leading to poor job satisfaction and reduced productivity. Many companies are having to work hard to devise systems and develop employee schemes to maintain the interest, enthusiasm and cooperation of their workers to maintain high levels of productivity.

Task

You might want to refer back to the first section of this unit to consider the employment issues surrounding the introduction of CAM in more detail. Explain what you consider the key issues to be.

Exam preparation

You will need to revise all the topics in this unit, so that you can apply your knowledge and understanding to the exam question.

In preparation for your exam it is a good idea to make brief notes about different topics, such as 'Global manufacturing'. Use sub-headings or bullet point lists and diagrams where appropriate. A single side of A4 should be used for each heading from the Specification.

It is very important to learn exam skills. You should also have weekly practice in learning technical terms and in answering exam-type questions. When you answer any question you should:

- read the question carefully and pick out the key points that need answering
- match your answer to the marks available, e.g. for two marks, you should give two good points that address the question
- always give examples and justify statements with reasons, saying how or why the statement is true.

Practice exam questions

1 a) Outline three reasons why the use of computers has enabled changes in design and production methods. (6)
   b) The use of computer-aided design (CAD) has changed the way designers work. Explain **two** of the following terms related to CAD:
   - 2D modelling
   - 3D prototyping
   - accurate drawings. (4)
   c) Describe the benefits to the manufacturer of using CAD in the design process. (5)

2 a) Describe what is meant by a CNC machine. (2)
   b) Outline **two** reasons why CNC machines improve the manufacture of products. (2)
   c) Describe the following terms:
   - computer-integrated manufacturing (CIM)
   - concurrent engineering
   - flexible manufacturing system (FMS). (6)
   d) Explain the role of computers in a flexible manufacturing system. (5)

3 a) Describe how the following are used:
   i) hardware
   ii) software
   iii) graphical user interface (GUI). (9)
   b) Explain how virtual reality techniques are helping designers to develop products. (6)

4 a) Explain and give an example of **three** input devices used in CAD systems. (9)
   b) Outline the meaning of an output device. (2)
   c) Describe the benefits of using **two** of the following output devices:
   - laser printer
   - digital printer
   - XY plotter. (4)

5 a) Describe the benefits of using a CNC machine for prototyping a product. (4)
   b) Describe the following terms:
   i) computer-aided manufacture (CAM)
   ii) just in time (JIT). (6)
   c) Explain the benefits of use of CAM in mass production. (5)

Total: 75 marks

Part 3
Advanced GCE (A2)

UNIT 4A

Materials, components and systems (G401)

Summary of expectations

1. What to expect

Unit 4 is divided into two sections:

- Section A Materials, components and systems
- Section B consists of two options, of which you must study the same option that you studied at AS level.

Section A is compulsory and builds on the knowledge and understanding of materials, components and systems that you gained in Unit 3.

2. How will it be assessed?

The work that you do in this unit will be externally assessed through Section A of the Unit 4 examination paper. You are advised to spend **45 minutes** on this section of the paper.

3. What will be assessed?

The following list summarises the topics covered in Section A and what will be examined:

- Selection of materials
  - the relationship between characteristics, properties and materials choice.
- New technologies and the creation of new materials
  - the creation and use by industry of modern and smart materials
  - the impact of modern technology and biotechnology on the development of new materials and processes

- the recycling of materials
- modification of the properties of materials.
- Values issues
  - the impact of values issues on product design, development and manufacture
  - the responsibilities of 'developed' countries in relation to production and the environment.

You should apply your knowledge and understanding of materials, components and systems to your Unit 5 coursework.

4. How to be successful in this unit

To be successful in this unit you will need to:

- have a clear understanding of the topics covered in Unit 4A
- apply your knowledge and understanding to a given situation or context
- organise your answers clearly and coherently, using specialist technical terms where appropriate
- use clear sketches where appropriate to illustrate your answers
- write clear and logical answers to the examination questions, using correct spelling, grammar and punctuation.

5. How much is it worth?

This unit, together with the option, is worth 15 per cent of the full Advanced GCE.

Unit 4A + option	Weighting
A2 level (full GCE)	15%

1. Selection of materials

Everyone involved with either designing new products or improving existing ones must have an understanding and working knowledge of materials in order to be able to select, process and finish the material. For example:

- A graphic designer must have a working knowledge of paper and board and printing processes in order to produce a flyer. Full-colour printing, metallic inks or special printing effects, such as spot varnishing, may give dramatic effect but may not be economically viable. Such effects must therefore be explored with the client in order to produce a flyer that will attract the attention of the potential customer.
- A product designer may have to compromise on the aesthetics of a product due to manufacturing processes and current technologies. Injection moulded casings, for example, are far more cost effective than pressed aluminium casings which may create a greater aesthetic appeal for the product.

When choosing any material for a specific application, full consideration must be given to its characteristics and properties.

Quality

The quality of the material relates to its fitness-for-purpose for the function and aesthetics of a product. For example, newsprint is a low cost, low quality type of paper that is only required to last for a short period of time. On the other hand, cast-coated board is a high quality, high cost material used for luxury products requiring expensive looking effects. Naturally, higher quality materials place a premium on the product and therefore demand a higher price.

Manufacturing processes and level of production

The manufacturing processes and level of production required to produce a product have an immediate effect upon materials choice. For example, the designer of a new mobile phone will produce several concept models, using easy-to-shape-and-finish Styrofoam, to determine aesthetic values and ergonomic factors. In consultation with engineers, several working prototypes are then made using a range of possible materials. The most cost-effective method of manufacture is determined at this point and an appropriate final material decided upon; for example ABS plastic for the casing as this can be injection moulded and is shock resistant.

Once the actual product is finalised and tooling and machinery is set up for high-volume production, a package designer develops the product packaging. A well designed package is not only a protective container, but a powerful

Table 4.1.1 *Materials and production processes used in different levels of production for graphics with materials technology products.*

	Level of production		
	One-off	**Batch**	**High volume (mass)**
Products	Product development, e.g. prototype and concept models, mock-ups, etc. Response to individual client needs, e.g. custom-made vinyl graphics and signage Preparation for manufacture, e.g. mould, pattern and plate making	Specified quantities of products up to 1000 items, e.g. business stationery, limited edition products, etc.	Large quantities of products sometimes on a 24-hour basis (continuous production), e.g. metal drinks cans, plastic drinks bottles, etc.
Production process	CAD/CAM, e.g. plotting/ cutting, engraving, laser cutting, etc. Laser/inkjet printing Screen printing	Injection moulding Blow moulding Vacuum forming Sand casting Printing processes (short runs), e.g. offset lithography, letterpress, screen printing	Injection moulding Blow moulding Vacuum forming Die casting Printing processes, e.g. offset lithography, letterpress, screen printing
Materials	Paper, card and board Woods, e.g. MDF moulds and models Plastics, e.g. polystyrene for vacuum forming, Styrofoam for modelling Metals, e.g. light-sensitive metal for printing plates	Paper, card and board Plastics Metals, e.g. aluminium and brass for casting	Paper, card and board Plastics Metals

advertising tool. Successful package designers are skilled at working three-dimensionally with a range of materials including paper and board, plastics, glass and metal. They must combine this ability with a knowledge of graphic design and printing processes. A number of card and board mock-ups are produced using CAD/CAM techniques to plot and cut out appropriately sized and shaped nets. Once finalised, commercial printing and finishing methods can be utilised using the correct type of cartonboard for high-speed processes.

Limitations

The limitations of the material determine the effectiveness of the product. Mechanical properties such as compressive or tensile forces, and environmental properties such as temperature and humidity can cause materials to fail. Steel drinks cans will corrode so a layer of tin plate is added, whereas the surface of aluminium cans will oxidise so a protective lacquer is applied. Corrugated board is an ideal choice for an industrial package. The specific type of corrugated board chosen, however, is based upon the limitations of that specific material through the results of testing.

In the example below, double-wall corrugated board was used to package a computer system. The tests carried out on the board determined its very good stacking strength and resistance to shock and set the limitations for its use.

Wear and deterioration

The wear and deterioration of a material is an important factor when designing for the intended life-span of a product. Some materials will naturally wear and deteriorate with prolonged use. For example the book you are reading now will be well thumbed over your two-year course. It is therefore important that

Figure 4.1.1 *Box certificate stamped onto a corrugated board industrial package*

the publisher selects an appropriate weight of paper for the printed pages and card for the cover. When producing vinyl signs for shops or vans, the correct vinyl must be used (usually 3- or 5-year vinyl) depending upon the operating conditions. Outdoor signage must incorporate a vinyl that is resistant to weather conditions such as UV light from natural sunlight, rain/sea water, and oils and hydrocarbons (diesel, petrol).

Maintenance

Issues of maintenance do not usually apply to the graphics element of a product. A menu in a restaurant may be laminated (encapsulated) to increase its durability and maintained simply by wiping it clean. 3D products made from resistant materials do, however, require maintenance and suitable materials must be considered. For example, modern mountain bike frames can be made of carbon fibre instead of the tradition steel or indeed aluminium. If a carbon fibre frame breaks, the structural integrity of the whole frame is compromised due to the make up of the fibres used in the production process. If a steel frame breaks at the join of two tubes then at least it can be re-welded or brazed. In extreme conditions the suitability of materials is extremely important. For example, the NASA space agency invests millions of dollars in developing suitable modern materials that can operate and be maintained in the vacuum of space.

Life costs

The life costs of the material are not simply economic factors, but consist of a wide range of factors affecting its entire life cycle from the 'cradle to the grave'. Paper, card and board require trees to be felled, wood pulp to be produced, and processing into products. These products are used and discarded, or recovered and recycled. The entire life cycle requires energy to be consumed, involves social factors with human involvement and environmental effects such as pollution. All of these factors contribute to the choice of material for a specific application.

> **SIGNPOST**
> 'Values issues' Unit 4A page 194

Selecting materials for product packaging

The design and manufacture of packaging for products has become almost as important as the product itself. The presentation of a product reflected in its packaging allows the manufacturer to target a specific market for its goods. Therefore, the packaging of a product must fulfil two major functions:

- The aesthetic presentation of a product for **marketing** purposes:
  - to influence the consumer to notice and buy the product
  - to identify the product and **brand**.
- Meet the functional requirements of its contents:
  - to contain and protect the product during distribution and storage
  - to maintain the hygiene and safety of the contents
  - to communicate information and usage.

These requirements will be important factors in the design of the product packaging. The designer will have to select a suitable material for the product and its target market. The cost of the material and the production process are also determining factors in the selection of materials as they will have to be incorporated in the retail price (see Figure 4.1.2).

As packaging is designed to be disposable, it is a major environmental concern. It contributes around a third of household waste in Europe and North America. This has led to European Union legislation on waste management, targeting packaging as a specific area for improvement.

The use of **non-renewable energy resources** in the production process and the raw material itself, along with excessive packaging used for some products, have been targeted by environmental groups campaigning for reform. Environmental groups have applied pressure to manufacturers to change how their products are packaged and the type of materials used. This awareness of environmental responsibilities has changed the way product packaging is designed

and manufactured. Manufacturers have placed a greater emphasis in selecting materials and production processes to improve their environmental record.

As well as environmental concerns, the function of the packaging in relation to the product inside is a major consideration. The overall cost of producing the final item, along with the materials innate properties, are the two determining factors in material selection. The following, along with glass, are the three main materials used for packaging.

Card and paper

These materials are commonly used for packaging as they are inexpensive, easily formed and offer a wide range of properties. When card is shaped in a 2D form, as a net, it has benefits such as higher volume storage and transportation. It also enables common commercial printing methods to be used such as **offset lithography** without the need for specialist machinery.

To reduce costs card used for packaging is often produced from low-grade unbleached pulp. One side is then coated or a layer of higher grade paper is applied for better results during printing.

The weight of card used ranges from 120 gsm to 220 gsm. Card that exceeds this weight would not be used because of the high cost. Corrugated board would provide a cheaper alternative – this is used mainly for products that are not purchased from off the shelf such as white and brown goods that do not require such a high level of printed presentation. Corrugated board also offers the product greater protection because it can be produced in a greater thickness and, due to its construction, offers a slight padding.

As card and paper are naturally absorbent they are susceptible to water and grease. A polyethylene coating can be applied to the

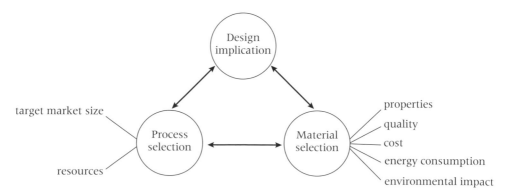

Figure 4.1.2 An overview of the relationship between the manufacturing processes, design implications and the choice of material

surface of the card to give it a resistant finish. In doing this, the card cannot be recycled.

Card packaging can also be produced in a 3D shape. Egg cartons and electrical product packaging are examples of pulp formed in a one- or two-part mould. The initial tooling cost for this production method is high and provides a crude surface finish. However, it does allow for intricate shapes to be produced and delicate products can be effectively supported. This type of packaging is an environmentally friendly alternative to polystyrene and thermoplastic.

Plastic

Thermoplastics are widely used in packaging, particularly for food. This is because they provide a non-absorbent, hygienic surface and an air-tight seal can be created to prolong the shelf life of the food inside. These materials also allow the customer to view the product before purchasing it without having to open the packaging. Two common forms of thermoplastic used are PET (polyethylene terephthalate) and HDPE (high density polythylene), which are both recyclable materials (see Figure 4.1.4). The two main production methods used are as follows:

- **Vacuum forming/blow moulding** – an inexpensive method that can be used for both mass and batch production. Production costs of the moulds are low in comparison to injection moulding. This is especially true when large, intricate, departmentalised forms are required for products such as toys and food. The resulting form may lack rigidity and durability so card may be used for support.

Figure 4.1.4 *The Body Shop uses bottles made from HDPE and also offers a refill service*

- **Injection moulding**. The tooling costs for injection moulding can be very expensive, especially for packaging inexpensive products like food. Items such as bottles and tubs have to be produced in high volumes to warrant outlay costs. Companies will often use generic products produced by packaging manufacturers and apply their own labels. The advantages of injection moulding is the creation of more rigid and accurate shapes.

Metal

Metal is used predominately in packaging food and drink. Aluminium and tin-coated steel are the two main materials used because of their non-corrosive properties when vacuum sealed. This permanent seal maintains freshness, longevity and protection for the product.

In the manufacture of cans a high proportion of reclaimed and recycled material is used. Mechanical magnetic sorting at waste sites can detect and withdraw the steel to be melted down. Reclaiming aluminium is more difficult so manufacturers have offered economic incentives for people to recycle. This has resulted in the use of aluminium in drinks cans becoming more common. Extracting aluminium is an expensive process with high energy costs so recycling is an economically viable alternative. The properties of aluminium are ideal for packaging as it is a lightweight, malleable material which can be easily formed and shaped.

Selecting materials for product design: kettles

Kettles are good examples of how materials have changed and brought about new developments in manufacturing technologies (see Figure 4.1.5). The Victorian kettle made from cast iron pushed the material to its limit as a hollow product at the time. As a metal, it would naturally have the required thermal conductivity properties where it could be heated over a fire or on a range. The kettle would require careful handling with gloves or some insulating material because the handle would also get very hot.

As materials became more readily available, copper replaced cast iron kettles. Copper kettles could be fabricated by soldering or formed from a single piece or spun. The spouts would be made from a single piece and soldered into place. Copper has a much higher coefficient of thermal conductivity than cast iron and so is much more efficient in terms of the amount of energy required to heat the water inside. Great care still needs to be taken though, as the handle gets even hotter due to the thermal conductivity of the copper.

Copper kettle

Victorian cast iron kettle

'Graves kettle' designed by Michael Graves and produced for Alessi, 1985

Plastic injection-moulded kettle, 1980s

'Hot Bertaa' aluminium kettle, designed by Philippe Starck, 1991

Figure 4.1.5 *Kettles through history depicting how materials have allowed designers to challenge both shape and form*

With the advent of plastics and injection moulding, it became possible to mass produce components quickly and cheaply. Plastics are an ideal material for use in kettles as they are excellent insulators of heat and electricity. Due to new manufacturing processes, new and improved shapes were designed with enhanced ergonomic performance. It also became possible to build in new features such as water-level indicators.

Designers such as Michael Graves and Philippe Starck challenged the concepts of shape and form in the 1980s and 1990s with the 'Graves kettle' and the 'Hot Bertaa' aluminium kettle respectively. These have become modern-day icons of design and yet they still retain the basic function of the kettle – that of heating water.

Task

Working in small groups, consider the following aspects of mobile phone design:

- How will mobile phones develop in the future?
- What will customers want?
- Are there any new technologies that can be incorporated at present?
- Are there any new or different materials that could be used?
- Produce some concept sketches for a new generation mobile phone.

Explain your choice of materials and technologies to support your ideas. Present your findings to the others in your class in a formal presentation (key skills).

2. New technologies and the creation of new materials

LCD displays

Liquid crystal displays (LCDs) are used as numerical and alpha-numerical indicators and displays. As they require much smaller currents they have replaced LED (Light Emitting Diode) displays because they prolong the life of batteries by using microamperes rather than milliamperes.

Liquid crystals are organic, carbon-based compounds, which exhibit both liquid and solid characteristics. When a cell, containing a liquid crystal, has a **voltage** applied across its terminals, and on which light falls, it appears to go 'dark'. This is caused by the molecular rearrangement within the liquid crystal. A liquid crystal display has a pattern of conducting electrodes that is capable of displaying the numbers 0 to 9 via a seven-segment display. The numbers are made to appear on the LCD by applying a voltage to certain segments, which go dark in relation to the silvered background. LCD technology is now commonplace in a wide range of modern products from mobile phones to microwave ovens.

Advanced LCDs make use of cholesteric liquid crystals, making it possible to design smaller pixels resulting in higher resolution displays with much sharper and brighter colours. These crystals, widely used for laptop computer screens, are more stable, eliminating the screen flicker that makes laptop screens wearisome to read when used for prolonged periods of time. Their main advantage, however, is that they use far less power, running about ten times longer on batteries and being far less expensive to purchase.

Smart materials

Smart materials have been developed through the invention of new or improved technologies. Smart materials respond to differences in temperature or light and change in some way as a result. They are classified as smart because they sense the conditions in their environment and respond to these conditions. Smart materials appear to 'think' and some have a 'memory' as they revert back to their original state.

Thermochromic liquid crystals

Thermochromic liquid crystals are used in a number of applications including forehead thermometers, battery test panels and special printing effects for promotional items. In the case of a forehead thermometer, a layer of conductive ink is screen printed on to the reverse of the thermometer strip – this area makes contact with the forehead. On top of the conductive ink is a layer of normal ink that conveys the temperature gauge colour bars. Finally, there is the thermochromic layer which is black when cool. By pressing the thermometer to the forehead, the temperature generated will turn the thermochromic ink translucent. This reveals the temperature gauge colour bars that are printed in normal ink. Depending upon inner body temperature, most or all of the thermochromic ink will heat to the temperature needed to become translucent. The same process applies to battery test panels where the electrical charge of the battery generates the heat required.

Other special printing inks are available to enhance printed materials especially for promotional use. Thermoreactive inks can be used to reveal graphics if a warm hand is placed over them or, conversely, if they are placed in a fridge – an ideal device for revealing a lucky winner on a promotional pack. As well as inks that react to changes in temperature, some inks react to UV radiation in natural sunlight; these are known as photochromic inks.

Future developments using this technology may involve thermocolour displays as seen in the Sci-Fi film *Minority Report*. Here products such as cereal boxes could comprise a plastic film that is overprinted with thermochromic liquid crystal ink (currently being developed as electronic ink). Once a voltage is applied to the display, perhaps activated by the opening of the box, surface designs could be animated and sound activated. Imagine this technology applied to side of an articulated lorry where the whole side of the trailer could become a moving commercial. At present piezo-electric actuators are used in greetings cards that play tunes when opened. They produce a sound from an electrical signal as a result of the card being opened.

> ## Task
> Discuss the future applications and products that could incorporate thermochromic liquid displays and the advantages they will possess over current products.

Smart labels

The Electronic Point of Sale (EPOS) system has been used with great success for a number of years in modern business. Information collected by scanning a product's unique bar code enables businesses to supply and deliver their products faster by reducing the time between placing of an order and the delivery of the product. For example, once the barcode has been scanned at a supermarket checkout, the electronic information not only checks the price against the retailer's database, but also automatically deducts the quantity from stock levels for automatic reordering from the manufacturer. This system, however, relies on the checkout assistant individually scanning each item which can cause long queues at the checkout.

In the future, bar codes will be replaced by smart labels known as Radio Frequency Identification (RFID) tags. RFID tags are intelligent bar codes that can talk to a networked system to track every product that is purchased. This will enable the customer to fill up their shopping trolley with RFID tagged products that communicate with an electronic reader which

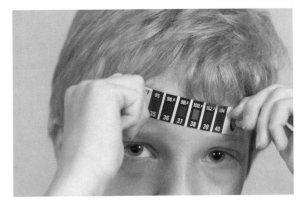

Figure 4.1.6 Forehead thermometer

detects every item and rings up the bill almost instantly. The reader will be connected to a large network that will send product information to the retailer and product manufacturers. The customer's bank will then be notified and the amount of the bill will be deducted from their account, avoiding lengthy queues.

Existing bar codes use read-only technology and cannot send out any information after the product had been purchased. RFID tags will enable business to track the product throughout the entire supply chain, location of consumer use and even disposal, making it an invaluable marketing tool. At present, inductively coupled RFID tags are used in the USA to track cattle and airline baggage using a silicon microprocessor, a metal coil that acts as an antenna and an encapsulating material to hold the tag together. These are quite expensive per unit and capacity coupled RFID tags are being developed in order to reduce costs. These tags use a silicon microprocessor capable of storing 96 bits of information, conductive carbon ink that acts as the tags antenna, which can be printed onto a paper label (currently being developed as power paper). The only limitation at present is the tag's range – just several feet.

Once the range is increased it will lead to a ubiquitous network of smart packages that track every phase of the supply chain. For example:

- The customer picks up a carton of milk at the supermarket. The carton contains a smart label that stores the expiry date and price, which is communicated wirelessly to the customer's personal digital assistant or mobile phone.
- The milk and other items on the shopping list are automatically tallied as the customer walks through the doors, which have an embedded tag reader.
- The information from the purchases is sent to the customer's bank, which deducts the amount of the bill from their account. The supermarket deducts the quantities from stock levels and can reorder from the manufacturers.
- At home the milk is placed in the fridge, which is also equipped with a tag reader. This smart appliance is capable of tracking all of the products contained in it – the foods you use, how often you restock and can let you know when that milk or other food will expire.
- Once the milk has been consumed and disposed of, the smart fridge could add milk to a shopping list, or could be used to reorder the product automatically over the internet.

Composite materials

In general, reinforced plastics are best suited for large structural items such as boat hulls and storage tanks. The first reinforced plastics were made with glass fibre strands as the reinforcing material and this is still the most common kind today. Glass reinforced plastic (GRP) is available in several forms: a loosely woven fabric, a string of filaments wound together, a matting of short fibres, or loose short strands.

More recently, we have seen the advent of carbon fibres and usually used in the same format as glass fibres. They are, however, much stronger composites and are used in the aerospace industry, for making sports equipment and for body armour and protection. Composites based on carbon fibres or with a special polymer known as Kevlar have excellent protective applications and are used by the police and military for bullet-proof vests and helmets. The added advantage is that great weight reductions have also been achieved, making the vests 3 kg lighter and helmets 1 kg lighter.

The carbon-fibre strands are held in a matrix in a similar fashion to GRP. Once the liquid resin has been applied it is cured or hardened around the fibres. The polymerisation reaction that follows converts the resin into a rigid cross-linked thermoset solid. The chemical reaction is started by the addition of a catalyst or hardener and in some cases an accelerator is used to control the speed of the reaction.

Pigments, filters and other additives are also used where necessary to change the colour or to increase buoyancy. Polyester resin and epoxy resin are most commonly used as the matrix material in composite fibre structures. Polyester resins are used with GRP in the production of boat hulls because they are much cheaper than epoxy resins. In the production of storage tanks and pipework, epoxy resins are used because of their excellent resistance to chemical attack.

Large products such as boat hulls have traditionally been produced by hand lay up, which is a relatively laborious method. The process has undergone some mechanisation recently. The basic and simplest method of making a reinforced structure is by laying sheets of matting over a wooden mould and wetting it with resin. Additional layers are added with resin to build up extra thickness where required. The production method of boats and car body panels is now more automated and, when produced with carbon fibres, they are lighter and mechanically stronger.

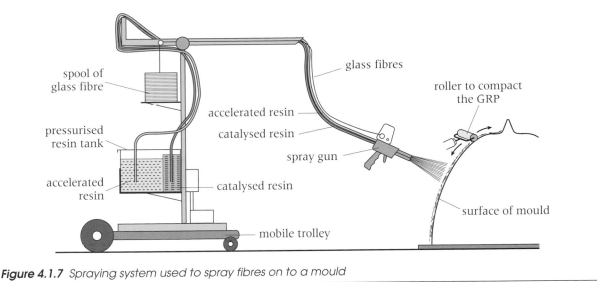

Figure 4.1.7 Spraying system used to spray fibres on to a mould

One method now in use is illustrated in Figure 4.1.7, which shows a mixture of fibres and resin sprayed on to a mould. A continuous spool of fibres is wound off a drum, mixed in a spray gun with the resin, catalyst and hardener, and sprayed on to the mould. The process is obviously much quicker, but great care is needed to ensure that an even application of material is made across the whole mould, avoiding building up excessive thickness in one area.

Task

Composites are now widely used in the production of racing and mountain bikes. Detail the benefits of this new technology compared with the more conventional and traditional metal-framed bikes.

New materials used in the computer and electronics industry

Semiconductors are a special type of material that have an electrical resistance which changes with temperature, i.e. they conduct electricity better as the temperature rises. The resistivity falls as the temperature rises. Semiconductors include both compounds and elements. Perhaps the most common type of element is silicon.

Silicon is the second most abundant and widely distributed of all the elements after oxygen. It is estimated that about 28 per cent of the Earth's crust is made up from silicon. It does not occur in a free elemental state but it is found in the form of silicon oxides, such as quartz, sand and rock, or in complex silicates such as feldspar.

Silicon is used in the steel industry as an alloying element but its main use is in the computer technology industry. In particular, it is used in the manufacture of silicon chips and other electronic components such as transistors and diodes. It is also used in the manufacture of glass, enamels, cement and porcelain.

Like other semiconductors, silicon behaves as an electrical insulator or conductor depending on the temperature. At absolute zero, the material is an electrical conductor. When heat is applied to a piece of silicon, some of the electrons gain enough energy to jump from a valence band to a conduction band. The movement of these electrons can carry a charge through the material when a potential difference is applied across it. The higher the temperature becomes, the more electrons jump across the bands, and the resistance falls making it an electrical conductor.

The production of integrated circuits (ICs) involves cutting single crystals of silicon into thin wafers. The surface of each wafer is then coated with a photosensitive polymer so that it can be etched by exposing light to predetermined areas. Several masks of different layouts are needed in order to construct the integrated circuit. As the accuracy of photolithography (the process used to transfer the patterns) has improved, accuracy has improved to within less than one micrometre. Due to the individual devices being so small, the integrated circuits are produced on a larger wafer up to 20 cm in diameter that can contain up to one thousand million circuit elements (see Figure 4.1.8).

Computers have their entire central processing unit (CPU) made from a single

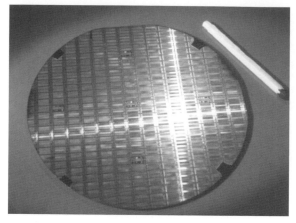

Figure 4.1.8 *Silicon wafer containing many smaller ICs*

integrated circuit containing more than one million transistors. The same computers also require memory chips and their capacity continues to increase at an ever-increasing rate.

Use of high-wattage lighting for projecting images onto buildings

The large-scale projection of images on to buildings using long-throw lenses has been used by the advertising industry for a number of years – often referred to as 'ambient' advertising and more recently as part of 'guerrilla' marketing campaigns. Ambient advertising encompasses all of the devices used by advertising agencies to promote a product to its target market wherever they are, from the back of bus tickets, the nozzles of petrol pumps to the sides of airships or blimps. The use of high wattage lighting for projecting images onto buildings became an ideal marketing device for the emerging UK club scene during the 1990s. The top clubs used such images to promote club nights, usually operating the equipment from the back of open vans so that promoters could change the location and reach larger areas of a city. At night, the sight of these vast graphic images incorporating their brand identity could be seen by thousands of potential customers against the backdrop of a normally dull office block.

Perhaps one of the most successful corporate uses of ambient advertising was made by Adidas during the 1998 World Cup finals in France. The images of several of England's football stars, including a young David Beckham, were projected on to the white cliffs of Dover with the strap line *England expects every man to do his duty*. This immediately roused a spirit of patriotism within the England supporters and was an excellent vehicle for promoting the Adidas brand.

On a less grand scale, projections are often used in shopping centres. Potential customers may walk the same journey every day through a crowded street or shopping centre and completely bypass a store. When an image is projected onto the street ahead of them their attention is grabbed and they are more likely to investigate the source.

The impact of modern technology and biotechnology on the development of new materials and processes

Genetic engineering in relation to woods

Recent developments in science and technology have given rise to a whole new culture of genetically modified foods, plants, materials and animals.

Biotechnology is now at the forefront in the production of new materials and research, and experimentation into genetically engineered and modified timber is growing rapidly. Genes are currently being investigated with a view to providing faster growing trees. This would enable forests to be managed more efficiently in terms of replacing trees at the same rate at which they are cut down. The new trees would reach maturity much quicker in terms of the useful bulk of timber that can be obtained from it.

Investigations are also taking place as to how wood can be engineered to be more resistant to wear, rot and animal infestation. Obviously, there are benefits to be had in relation to these improvements. What though are the implications beyond this as a result of timber not rotting? If it were not for decaying and decomposing timber millions of years ago, we would have no coal or oil today. If timber does not rot, then maybe it will have to be burnt to be disposed of. The consequences of burning the timber will result in emissions causing damage to the ozone layer and to acid rain.

One major area of development as far as genetically engineered timber is concerned is in the production of paper. The process of making paper is an environmentally damaging one, which involves the use of some very toxic chemicals. The chemicals have to be used to remove lignin from the wood pulp. Lignin is a natural tough polymer-like material that gives the tree its strength and rigidity. In the USA, scientists have discovered a way of reducing the natural lignin content and also producing trees that grow faster. It is thought that this will lead to the reduction of toxic chemicals used in the paper making industry.

The same scientists, when working on aspen trees (the trees traditionally used for pulp production), made some remarkable discoveries. As they peeled back the bark from the saplings, they found that the timber was not its usual whitish colour. The timber was in fact a salmon reddish colour. Each sapling exposed was slightly different in colour and some in fact were spotted in appearance. Its use is now being considered beyond that of pulp and paper production. It is not inconceivable to think that we may end up with coloured timbers being used in furniture, panelling or even external cladding that would require no maintenance.

The scientists involved in this work are now considering the study of these aspens in a natural environment. As with any work of this nature, however, they are having to apply to the United States Department of Agriculture for a licence to do so.

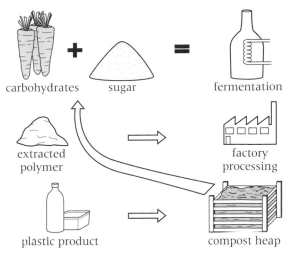

Figure 4.1.9 *The production of Biopol*

Task

What benefits might the impact of genetically engineered timber bring to the following?

• The construction of exterior furniture.
• The long-term future of a managed timber supply.

Environmentally friendly plastics

A plastic is basically any material that can be heated and moulded so that it retains its shape once it cools. Today, a wide variety of plastics exist as polymers and resins. Synthetic plastics are derived from oil and hydrocarbons form the basic building blocks. The carbon-based molecules bond together through a chemical reaction and the bonds made determine the physical properties and characteristics.

One disadvantage of plastics generally has been the issue of how they are disposed of. Recent advances in biotechnology and bioengineering have seen the development of environmentally friendly plastics. Sixty years ago the world's first biodegradable polymer, Biopol, was created (see Figure 4.1.9).

Biopol has the advantages of both synthetic and natural polymers. It is biodegradable, can be produced in bulk and the raw materials are readily available. Scientists discovered that a natural polymer produced from food stuffs could be mass produced by fermentation. Once the polymer had been extracted it was substituted into the production process to produce plastic products. The major advantage of the 'plastic'

once it had reached the end of its life was that it could be disposed of by biodegradation.

'Green' credit cards made from Biopol have now started to find their way into the market place as a genuine replacement for some of the 20 million credit cards in circulation today.

Task

Discuss the environmental benefits, in terms of recycling plastics, that materials such as Biopol have brought.

Special effects on television

Television production companies are beginning to incorporate the computer technology that has been developed by the film industry to create sophisticated special effects. Two main types of visual effects used are:

• blue screen
• computer-generated image.

Blue screen

This is a popular and widely used effect as it enables actors or scale models to be imposed on a separate filmed or computer-generated background. This technique can be used to create scenes that would be impossible (or too dangerous!) to film or to impose actors in a futuristic computer-generated scene.

The background and foreground of a scene are shot separately as two pieces of film and then combined in a process known as compositing. This can be achieved by projecting the two films simultaneously, frame by frame on to a third film. However, with the advancement of digital technology it is more likely that the two films will be digitised and the composite image made up on a computer before being processed as film.

First, the background is filmed or created on computer – this is called a background plate. Then the actor or model is filmed against a blue background or screen, which can be passed through a red filter to make it appear black. Silhouettes are created of the model/actor from the blue screen footage; one is black on a white background and the other is white on a black background. There are now four pieces of film – two originals and two mattes. They are:

- the background
- the actor/model
- the actor/model in black silhouette on a white background
- the actor/model in white silhouette on a black background.

These pieces of film are layered over one another and combine to make a composite image. The black silhouette is placed on the background plate and this creates a 'hole' into which the footage of the actor/model can be accurately placed. For each movement of the actor/model a separate frame must be produced.

Computer-generated image

Computer-generated image (CGI) is a development from blue screen technology in which computer-generated images are combined with film sequences in layers. The techniques used to manipulate existing images are part of a stage called post-production.

The scenes are filmed and scanned into a machine at a resolution of 12,750,000 dots per frame. Once this has been done, the scene can be manipulated in a number of ways.

- Rotoscoping – elements within the scene can be outlined and lifted out so that they can be replaced by other images.
- 2D painting – this technique can be used to add separately produced computer-generated images or elements to the scene. In addition to enhancing elements, this technique can also be used to remove unwanted elements from the scene, i.e. wires, safety equipment.
- Compositing – the layering the separate images together to produce a final sequence.
- 3D tracking – a 3D model of the scene is created in computerised form that incorporates the position of the camera and where it will move. This computer-generated replica of the scene allows the 3D elements that are added to be positioned accurately and realistically in relation to the rest of the scene and the camera.

Figure 4.1.10 *Motion capture is used in 3D modelling*

- 3D modelling – figures and objects are computer generated in a 3D model form. The movement of any figure or object will have to be incorporated to make it appear realistic in the context of the film. For example, human movements can be replicated from motion capture data. An actor is fitted with a suit with light reflective markers positioned on every joint. A number of 3D cameras capture the light from the actor's movements from a variety of angles and the information is transferred into digital data, processed by the computer and used to develop the model.

Digital photography

Digital cameras employ a 4.4 mm × 1.6 mm sensor that converts light into electrical charges. This sensor is either a charge coupled device (CCD) for high-quality cameras or a complementary metal oxide semi-conductor (CMOS) for more basic models.

The CCD is a group of miniature light sensitive diodes called photosites. These diodes convert light (photons) into an electrical charge (electrons). The brighter the light that hits the photosite, the greater the electrical charge will be. The charge is converted from analogue to digital as each pixel value is recorded as a digital value.

The photosites do not register colour so to obtain a full multicolour image the sensor filters light into its three basic primary colours. This process can be achieved in a variety of ways depending on the camera. The most practical method to capture a full-colour image is the installation of a permanent filter positioned over each individual photosite. Most cameras use a Bayer filter pattern, which alternates a row of red

Figure 4.1.11 *A digital camera*

and green filters with a row of blue and green filters. There is twice as much information from the green filters as the red and blue combined. This pattern is used because the human eye is not equally sensitive to the colours blue, red and green, more green pixels are required to create an image that will be perceived as true colour by the human eye. In effect, four separate pixels determine the colour of a single pixel by forming a mosaic.

Digital cameras use 'demosaicising algorithms' to convert the mosaic of separate colours into true colours. A true colour is formed for a pixel by averaging the colour value of the pixels that are closest to it.

To enable the camera to store the vast amount of information needed for these images, files have to be compressed. This can be achieved in two ways; repetition or irrelevancy.

Repetition
This process relies on the fact that certain colour patterns develop on a digital photograph and some shades of a colour will be repeated. The basic information needed to reconstruct the image is stored. This technique may only reduce the file by 50 per cent or less.

Irrelevancy
A digital camera records information in such detail that not all of it is detectable by the human eye. Therefore, any unnecessary information is discarded; the camera will offer different levels of compression by allowing the user to opt for varying grades of resolution. Higher resolution equals less compression and vice versa.

The information for photographic images is stored internally to be **downloaded** on to a computer via a serial or parallel port. Removable flash memory devices such as memory sticks are used to store files but some models store the file on a standard floppy disk.

As with standard film cameras, digital cameras use aperture and shutter speed to control the amount of light reaching the sensor. Most digital cameras are set for optimal exposure and focus automatically. However, some models do offer adjustment options allowing the user more creative control over the final image.

Internet website design
All websites are constructed using **HTML (hyper text mark-up language)**. This language or code forms the common structure of the **World Wide Web**; therefore, to place a site on-line all the information will have to be coded as such. For large commercial sites HTML code will be written for quicker navigation and downloading. Software applications for creating web pages will automatically apply HTML code, allowing web pages to be easily formed.

When creating web pages a combination of text and image is applied. The layout can be formed in a similar way to **desktop publishing (DTP)**, but with the added advantage of animation, sound and scrolling pages. A number of pages create a site so the design should allow for easy navigation between them, as well as consistency in appearance. Connections from one page to another can be achieved using a:

- hyperlink – text
- hotspot – picture, graphic, designated area.

When the mouse pointer passes over a hyperlink or hotspot it will change into a different icon such as a pointing hand. This indicates a link to the user. Hyperlinks and hotspots can be created in the website design software with the appropriate tool. Once a link has been created, it must be assigned. This can be to:

- another page on the site
- another section on the same page
- another website on the Internet
- an email address.

When creating a website consideration must be given to file sizes. The inclusion of large files such as scanned images will extend downloading time. This may prove tedious for users, so compression formats such as JPEG can be used to minimise size.

Colour printing in newspapers
Colour printing in newspapers is achieved by an offset lithographic printing method (see page 100). The **layout** of images and text is prepared on a desktop publishing (DTP) application; and a film negative is created from these digital files in order to produce the lithographic plates.

Colour images are separated into tonal black, cyan (blue), magenta (red) and yellow (see Figure 4.1.12) for individual plates to be produced. This separation of colours takes place

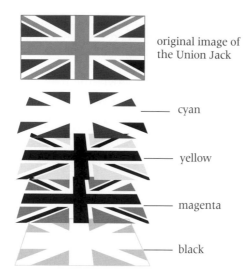

original image of the Union Jack

cyan

yellow

magenta

black

Figure 4.1.12 *The four colour layers*

at the film output stage. Registration marks are placed on the image to denote each colour area. These marks enable the image to be aligned accurately and prevent it from becoming blurred. A blue line print is made from the negatives to check the position before printing begins.

During printing, a continuous roll of paper sheet is passed through a bank of rollers that add the colours separately. The process begins with black, then cyan, magenta and yellow are added in sequence to build up the full-colour image. Registration alignment can be adjusted during this operation. The register control system (RCS) works in conjunction with a strobe light and camera to read the alignment of the registration marks. Adjustments are made to the colour rollers throughout the process to ensure the colour registration is correct. The amount of ink released into the units is also controlled, blending the colours to achieve the desired look. Prior to being placed on the press, the plates are scanned and the data are transferred to the main control unit. These data provide information regarding the levels of ink to be released to maintain the integrity of the colours. The registration marks are removed when the product is cut, folded and bound.

Holographic images

Holography is the science of producing 3D pictures. Photographic images and television pictures are viewed on a planar medium and only capture the irradiance of the subject. Holographic images in contrast include the irradiance and phase of the object.

Holographic science is constantly evolving and inventing new methods of application. It can be utilised for a variety of purposes such as embossed surface images on credit cards or cockpit instruments for planes.

To produce a basic holographic image a photosensitive material needs to be used to record the image. This can be specific holograph film or other materials such as photochromatic thermoplastic. The holographic image is not recorded like a camera image. In photography differing intensities of light are reflected by the object and imaged by a lens. In holography the phase difference of the light waves are captured after bouncing off the object to give them depth.

A reference beam or laser light emits a 'plane' wave. By using a beam splitter two beams are formed. The one (reference) beam is spread with a lens and aimed at the film. The other beam (object) is spread with a lens and aimed at the object. When the beam hits the object it is changed from a plane wave. The wave is modulated according to the physical dimensions and characteristics of the object. This light, which deviates with intensity and phase, hits the film along with the reference beam (see Figure 4.1.13).

The two beams interfere with one another as they pass through each other. The crest of the one plane wave meets either the crest of another (constructive interference) or a trough (destructive interference). Both types of interference are needed to visualise the image.

The film records the wavefronts. It is not a point-to-point recording like standard photography but a recording of the interference between light that hits the object and light that does not. It acts like a complex lens, reconstructing the image so it is perceived as if the object were really there.

Holographic images are used extensively where security is required, for example on credit cards and tickets, because they are almost impossible to counterfeit. VideoClip Holograms are currently available, but can, at present, only capture and display approximately 3–6 seconds of motion. To get to a point where we can access 'Holodecks', as seen on *Star Trek: The Next Generation*, current communications technology will have to be developed. Tele-immersion will bring videoconferencing to the next level where a 3D hologram could be projected instead of viewing on a computer screen. Merge tele-immersion with virtual reality, and interaction inside a simulated environment is possible. Another future use of holograms will be in optical data storage, currently limited to the surface area of a CD or DVD. If storage can go beneath the surface and use the volume of the recording medium, then 3D data storage enable the storage of more information in a smaller space and offer faster transfer times.

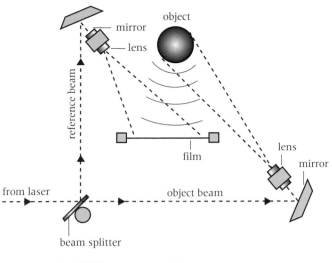

Figure 4.1.13 *How holographic images are produced*

Mail merge software

Mail merge software is available with most word processing packages. It is used to create personalised documents, letters and electronic mail from a standard template. There are two key files needed to perform a mail merge:

- Data list – name, address or any specific information to be included in the template.
- Document letter template – the standard document to be sent to individuals on the data list.

Special place holders are inserted into the template to mark where the information is to be placed, for example 'Dear Sir/Madam' will be replaced by 'Dear [name]'. This is then assigned to the correct field in the data list. Mail merge will replace this generic entry with the actual names of the recipients on the list. This process can be repeated throughout the document for various fields of information.

Non-carbon reproduction

Non-carbon reproduction (NCR) or carbonless copy paper permits multiple copies to be made without intervening layers of carbon paper. The paper translates pressure, such as that made by a pen, into a dye reaction which transfers the image to the copy. Carbonless copy papers are mainly used for continuous form sets, covered pay slips, delivery invoices and payment receipts.

The recycling of materials

Essentially, recycling takes waste materials and products and reprocesses them to manufacture something new. Some materials, such as paper and boards, can be made into the same products, while others can be made into something completely different, such as plastic vending cups into pencils. Recycling is an important aspect of a modern consumer society with millions of tonnes of waste being disposed of in landfill sites or incinerated, causing environmental concerns.

Recovery and recycling of paper and board

The cellulose fibres found in a variety of plants can be used to manufacture paper and board, and it can generally be reused between 4–6 times. New fibres can be processed in pulp mills to produce wood pulp for virgin paper manufacture. This virgin paper and board is of a high quality and strength due to its new fibre, so is used for specialist applications. However, almost any type of paper and board can be recycled using an elaborate grading and sorting system, which ensures that maximum value is obtained from the recovered fibre. As these recycled fibres are of a lower quality – weaker and may be contaminated – the products they are made from are designed to take into account the particular characteristics of recycled fibre.

Corrugated board can be made from recycled paper for single – or multi-ply kraft liners and fluting, provided liquid starch is added to provide stiffness and extra strength. UK manufacture of materials for corrugated cases uses almost 100 per cent of recycled paper and board, but the strength of this product depends on a proportion of new kraft fibre which comes from abroad.

Newsprint contains the weakest fibres because a newspaper has a short life span. Following discussions between newspaper publishers, newsprint manufacturers and the Government, an agreement was reached for UK newspapers to contain 40 per cent recycled content by the year 2000. This target was achieved four years early primarily because of a new reprocessing plant in Kent. New voluntary targets have been agreed for 65 per cent recycled content by the end of 2003 and 70 per cent by the end of 2004. These targets may prove challenging because of the growth in consumption of newsprint and the economics of investing in new reprocessing plants as opposed to using imported virgin fibres.

Although the recycling of paper and board may make economic and environmental sense, waste paper cannot be used in all paper grades or be used indefinitely. Three factors affect the use of recovered paper waste:

- *Strength* – every time a fibre is recycled it loses some of its strength. For papermaking purposes fibres can only be reused about six times.
- *Quality* – some paper grades make little or no use of recycled fibres because they require certain qualities only provided by the use of new fibres.

• *Utility* – it is not possible to recover all types of paper. Some are put to permanent use in books, etc. and some are disposed of when flushing the toilet. Others are not recoverable because they are part of a laminate or have a glossy surface finish.

Waste paper, or recovered paper as it is now often known as, is the most important raw material for the UK paper and board industry. Paper waste is collected by either the Local Authority or by a waste paper merchant. It is then sorted, graded and sent to the paper mill for recycling. The paper mill uses a hydrapulper filled with water to make the paper waste into a 'slush' and large contaminants are removed. The paper slush is then filtered and screened through a number of cycles to make it more suitable for papermaking. Depending upon the quality of paper being produced, quantities of virgin pulp may be added. Four broad categories can be used to identify grades of waste paper, based upon criteria such as the type of material the waste contains and how it will be used in the recycling process (see Table 4.1.3).

Lyocell

Lyocell is the generic name for a high-performance, staple viscose fibre produced from renewable sources of wood pulp. It is an extremely versatile fibre that can be used on its own or as a blend with cotton or synthetic fibres. Applications for special papers include tea bags, air/smoke/oil/coffee filters, printing papers and high-strength envelopes. Lyocell is made using an environmentally friendly process, which recycles the non-toxic solvent amine oxide used in its manufacture. Products made from lyocell can be recycled, incinerated, land-filled or digested in sewage by anaerobic digestion. In a sewage farm, the fibre degrades completely in just eight days to leave only water and carbon dioxide, which can power the sewage plant itself.

Remarkable Pencils

Remarkable Pencils Ltd has successfully managed to recycle plastic vending cups into pencils and other stationery. The plastic contained in one High Impact Polystyrene vending cup is enough to make one Remarkable Pencil. Some people may argue that plastic vending cups are themselves unnecessary,

Figure 4.1.14
Remarkable pencils

but daily production rates of 20,000 pencils means that 20,000 vending cups are being saved from disposal on landfill sites every day. The company has developed a strong, contemporary brand identity with its products, so that consumers are more likely to purchase an environmentally aware alternative to the traditional wooden pencil.

Tasks

1 Choose an electronic product such as a games console and describe how parts of it may be recycled.
2 Discuss some ways in which the general public can be encouraged to undertake more recycling.

Table 4.1.3 Recycling paper

Paper grade	Characteristics	Sources	Recycled applications
Pulp substitute grades	Top quality waste which can be used, with little need for cleaning	Unprinted trimmings and offcuts from printers and converters	Printing and writing papers
De-inking grades	Grades from which the ink is removed before recycling begins	Office waste, newspapers and magazines	Graphic and hygienic (tissue) papers, newsprint
Kraft grades	Long, strong fibres that generally come from unbleached packaging materials	Paper sacks	New packaging, including corrugated cases
Lower grades	Consist of mixed papers that are uneconomic to sort, due to either the small quantities of each type or the level of non-recyclable material being too high	Junk mail	Middle layers of packaging papers and boards

Modification of properties of materials

Paper and boards

The properties of paper and boards can be modified significantly with surface treatments and lamination. Here, two vastly different applications of paper and board are explored which demonstrate its important future use.

Tetra Pak aseptic cartons

Aseptic packaging means filling a sterilised package with a sterile food in a hygienic environment. This packaging system is used to maintain the high quality of the product for the length of its intended shelf life.

Tetra Pak aseptic cartons are made of three basic materials that are laminated to provide improved properties, resulting in a very efficient, safe and lightweight package. Each material provides a specific function:

- **Paper** (75 per cent) – provides strength and stiffness.
- **Polyethylene** (20 per cent) – to make packages liquid tight and to provide a barrier to micro-organisms.
- **Aluminium foil** (5 per cent) – to keep out air, light, and other flavours – all the things that can cause food to deteriorate.

Combining each of these three materials in a six-layer laminate produces a packaging material with optimal properties and excellent performance characteristics:

- Higher degree of safety, hygiene and nutrient retention in foods.
- Preserves taste and freshness.
- Increases shelf-life with no need for refrigeration or preservatives.
- Efficient – uses a minimum quantity of materials necessary to achieve a given function

(a filled package is 97 per cent product and only 3 per cent packaging material).
- Lightweight (as opposed to plastics, metals and glass packages).

Paper and board as a building material

Cardboard as a building material is relatively inexpensive, especially when made from waste paper. The only significant costs are in the processing required to make the tubing, panels and any of the complex shapes needed to ensure the aesthetics and structural strength of the building. Naturally, if such buildings were subject to investment and mass production, the potential for a cost effective building is possible.

In the case of the Westborough School, Westcliff-on-Sea, the design of the building reflects some of the properties of cardboard, particularly corrugated cardboard and origami. The zigzag shape of the south wall and the roof not only adds an aesthetic value, but provides structural strength and stability.

There are two basic cardboard components to the building – panels and tubes:

- Solid board can be used for some applications; however, multi-layer panels are often required. These consist of solid board, honeycomb card for strength, a timber framing if necessary, plastic-coated or aluminium foil layers for water resistance, and fire-treated outer board layers. Miracle board and Quikaboard are two trade examples. As a cost indication, a panel 2.5 m by 1 m and 160 mm thick would cost approxiamtely £150 per panel for an order of around 1000, or about £60 per square metre.
- Cardboard tubes are made from multiple layers of spirally wound paper plies, glued together with a starch or PVA glue. Up to 22 plies can be combined, giving thicknesses up to 16 mm. Because of the way in which the layers are wound together, the top or bottom layer of paper can be made of a different paper to the rest. This allows a treated, stronger or

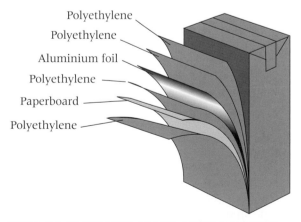

Polyethylene
Polyethylene
Aluminium foil
Polyethylene
Paperboard
Polyethylene

Figure 4.1.15 Tetra Pak aseptic carton packaging laminate

Figure 4.1.16 A cardboard classroom at Westborough School

coloured paper to make the surface, but avoids the cost of using these throughout the tube. As an indication of cost, for orders in 100s, tubes of 20 cm diameter with 5 mm thick walls cost around £2 per metre.

It is also possible to make a form of concrete, with paper as the aggregate mixed with normal cement. This is known as papercrete and has been successfully used in the USA in a dryer climate. There are limitations to the thickness of the board that can be made in this way, because of the time to dry out and it may not be suitable in the British climate.

The main potential problems for any cardboard building are strength, fire and water.

Strength – deflection of the cardboard tubes and panels may occur due to drying out, therefore tubes and panels should be left for a period of drying out before use as a structural element.

Fire – paper as a product obviously burns unless treated. Cardboard has the tendency to char rather than play an active part in a fire, with the charring protecting the surface. Fire treatments to the outer layers restrict flame spread.

Water – cardboard becomes a pulp when wetted and also absorbs moisture from the air, which can lead to the collapse of the material. Three levels of protection can be implemented:

- use of water-resistant cardboard
- use of an external poly-coated layer and internal aluminium foil layers
- use of over-cladding by a wood fibre and cement panel.

The potential for paper and board buildings in the future is extensive, both as a low-cost and 'greener' alternative to traditional building materials and as temporary disaster relief housing that can be constructed in kit form wherever it is urgently required.

Task

Discuss the possible future uses of paper and board as a building material including its advantages and disadvantages over traditional building materials.

Woods

Seasoning

Natural air **seasoning** is the traditional method of removing the excess moisture from newly felled trees. Although it is a very cheap system to operate it is dependent upon the weather conditions. The timber slabs are stacked with sticks between them allowing free air to circulate. This results in the evaporation of moisture, but this method has two major disadvantages:

- It is slow and inaccurate, taking on average one year to season 25 mm of slab thickness.
- The moisture content can only be reduced to that of the surrounding atmosphere (15–18 per cent).

Kiln seasoning, however, results in a much quicker, more reliable method. This time, the timber is stacked with sticks between, and is placed into a sealed chamber. Steam is pumped into the chamber and is absorbed into the timber. The humidity is then drawn out by an extractor fan, the temperature is raised and hot air is circulated. Very careful monitoring and recording takes place in order to attain the precise moisture content level. Moisture content can be calculated using the following formula:

Percentage of moisture content =

$$\frac{\text{Initial weight} - \text{Dry weight}}{\text{Dry weight}} \times 100$$

Overheating can result in a form of a case-hardened timber which is brittle on the outside. Kiln seasoning has a number of advantages:

- It only takes between one and two weeks per 25 mm of slab thickness.
- Less space is required as the process is quicker.
- There is improved turnover of stock.
- It kills insects and bugs in the process.
- Accurate moisture content can be achieved.

Laminating

The development of plywood was an important process that improved the physical properties of timber. Plywood is made from thin layers of wood, and veneers, about 1.5 mm thick, called laminates. They are stuck together with an odd number of layers, but with the grain of each layer running at right angles to the last. This means that the two outside layers have their grain running in the same direction.

The interlocking structure gives plywood its high uniform strength, good dimensional stability and resistance to splitting. It is also available with a variety of external facing veneers and it can be made with waterproof adhesives, which means it can be used externally.

These veneers or laminates can also be stuck together over formers to produce curved shaped forms. This process is known as laminating and since different shapes can be formed, the strength of the material can be further enhanced by the shapes into which the material is formed.

Other types of manufactured boards are also available that possess qualities of strength: block

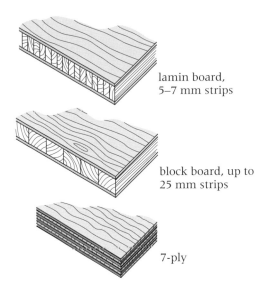

Figure 4.1.17 *Laminated forms of manufactured sheet timber*

lamin board, 5–7 mm strips

block board, up to 25 mm strips

7-ply

board, lamin board and batten board all use the laminating, joining together, approach to increase the overall strength (see Figure 4.1.17). These three boards are also clad on their external surfaces with thin laminates for decoration and strength reasons.

Particle boards such as hard board, chip board and medium density fibreboard (MDF) also exhibit properties of strength although they are made in a different way. Again, they can all be prepared so that they are waterproof and can be used externally. In particular, due to its uniform structure, MDF is particularly strong as well as having an excellent surface, which is capable of taking a variety of finishes.

Plastics

It seems somewhat strange to be thinking how plastics can be changed and modified but it is possible. In the same way that separate metals can be alloyed, separate monomers can be altered. Two or more monomers can be combined to form a new material and this process is known as co-polymerisation (see Figure 4.1.18). The new material is known as a co-polymer. An example of this would be a mixture of vinyl chloride and vinyl acetate. The combined co-polymer is known as polyvinyl chloride acetate

and is shown in Figure 4.1.18.

Cross-linking is another way of increasing strength. In the vinyl chloride monomer, one atom of hydrogen has been removed and replaced by an atom of chlorine. The new links have formed to create a new polymer, polyvinyl chloride (PVC), also shown in Figure 4.1.18.

Additives can also be used to increase the mechanical properties of plastics. Plasticisers are added as liquids to improve the flow of plastics when being used in moulding processes. They also lower the softening temperature and generally make them less brittle.

Fillers and foamants are added to plastics in an attempt to reduce bulk and overall costs because they are less expensive than polymers. They improve strength by reducing brittleness which also makes them more resistant to impact from shock loading.

One of the early problems with plastics was their inability to resist deterioration and exposure to ultra-violet light. The addition of stabilisers has made them more resistant to ultra-violet light and they no longer yellow or become transparent when exposed for long periods.

Metals

An alloy is a combination of one metal with one or more other metals and, in some cases, non-metals. Steel is an alloy of iron with carbon; brass is an alloy of copper and zinc. There are thousands of alloys in existence, but they are strictly controlled by the International Standards Organisation (ISO) and similar bodies. There are very stringent guidelines and specifications laid down that dictate the maximum and minimum limits of composition and the mechanical and physical properties required by each.

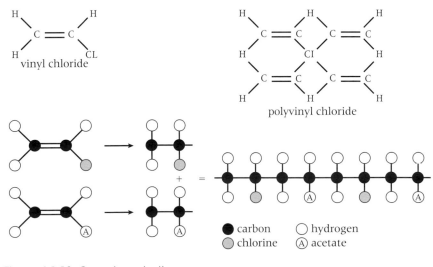

Figure 4.1.18 *Co-polymerisation*

An average car may contain as many as ten different alloys, including:

- mild steel for the body panels
- heat-treated steel alloys for gears
- cast iron aluminium alloys for the engine block
- lead alloys used in the battery, etc.

Alloys are prepared by mixing two or more metals in the molten state where they dissolve in each other before they are paired into ingots. Generally, the major metal is melted first and the minor is then dissolved into it.

Steel has been covered in some detail in Unit 3 but it is worth considering a few of the details once again. Iron as a pure metal is very soft and ductile, and carbon is very brittle. Yet when the two are combined they form a metal that exhibits none of the properties that the two original ingredients possessed.

As would be expected, the ratio in which the two materials are combined also affects the mechanical and physical properties of the new material. Figure 4.1.19 gives an indication of how the increase in the percentage of carbon affects the hardness of steel. It should be noted, however, that as a direct result of increase in

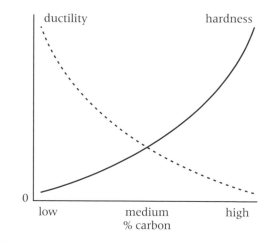

Figure 4.1.19 *The effect of carbon content on hardness and ductility*

hardness, the ductility (the ability to be drawn out) is decreased.

Engineers have to make compromises at times by having to balance how the increase in one property decreases another. Other properties can be further enhanced by alloying. Resistance to corrosion can be increased by the introduction of chromium into the alloy mix.

3. Values issues

The impact of values issues on product design, development and manufacture

Values issues incorporating technical, economic, aesthetic, social, environmental and moral factors can be encompassed in one word: sustainability. The Earth's resources are finite, large-scale industry causes pollution and the actions of a mass consumerist society are causing both social and economic problems. Designers therefore have a crucial role to play in achieving a more sustainable economic and social order. A more holistic approach to design must be adopted to determine the impact of values issues upon all aspects of a product's life.

Life Cycle Assessment (LCA)
It is the case with any design decision and solution that an optimum is looked for and a balance drawn between cost and benefit. Balancing the needs against the environmental

impact is becoming increasingly difficult for manufacturers as they strive to develop new products and processes. Life Cycle Assessment (LCA) is a technique now widely used to assess and evaluate the impact of the product 'from the cradle to the grave' through the extraction and processing of raw materials, the production phase, distribution, use and finally the disposal.

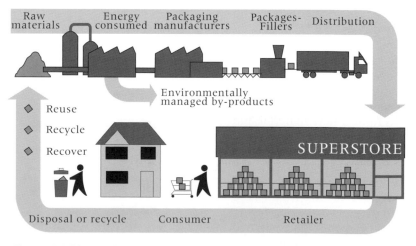

Figure 4.1.20 *Life Cycle Assessment (LCA) for packaged products*

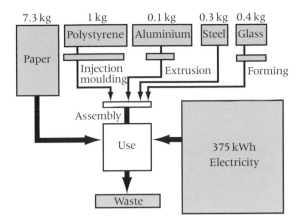

Figure 4.1.21 *This simple LCA inventory of a coffee machine clearly indicates design priorities of minimising the use of electricity and paper filters*

Life Cycle inventory

Customers simply expect that companies pay attention to the environmental properties of all products. However, British Standards and the ISO 14000 series of standards demand continuous improvement in a company's environmental management systems. The complex interaction between a product and the environment is dealt with in a Life Cycle Assessment (LCA) method called life cycle inventory. There are two main steps in an LCA:

Life Cycle inventory – the basis of an LCA study is an objective inventory of all the inputs and outputs of industrial processes that occur during the life cycle of a product, including:

- **Environmental inputs and outputs** of raw materials and energy resources and emissions.
- **Economic inputs and outputs** of products, components or energy that are outputs from other processes.

The life cycle inventory can be expressed as a process tree, where each box represents a process with defined inputs and outputs, which forms part of the life cycle.

> **Task**
>
> Write up a simple LCA inventory for a familiar product including materials, components, manufacturing processes, end use, waste production and energy used.

Built-in obsolescence

Built-in or planned obsolescence is a method of stimulating consumer demand by designing products that wear out or become outmoded after limited use. In the 1930s an enterprising engineer working for General Electric proposed increasing sales of flashlight lamps by increasing their efficiency and shortening their lifespan. Instead of lasting through three batteries he suggested that each lamp last only as long as one battery. By the 1950s built-in obsolescence had been routinely adopted by a range of industries, most notably in the American motor and domestic appliance sectors. Nowadays products such as computers are obsolete as soon as they are loaded into the car.

There are four forms of obsolescence:

1. *Technological obsolescence* – occurs mainly in the computer and electronics industries where companies are forced to introduce new products, such as mobile phones, with increased technological features as rapidly as possible to stay ahead of the competition.
2. *Postponed obsolescence* – occurs when companies launch a new product even though they have the technology to realise a better product at the time. For example, it is not unrealistic to imagine that when Sony launched its PS2 games console it knew what its next generation games console would look like – the PS3.
3. *Physical obsolescence* – occurs when the very design of a product determines its lifespan; for example, light bulbs or ink cartridges for printers.
4. *Style obsolescence* – occurs due to changes in fashion, where products seem out of date and force the customer to replace them with current 'trendy' goods.

Perhaps a fifth and more cynical form of obsolescence occurs when products are changed or discontinued simply in order to justify a higher and more profitable price.

Consumer trust

Generally, consumers understand that the only constant in a technological world is change. Society constantly demands newer and better ways of doing things – obsolescence due to technological innovation may be acceptable to most, but to deliberately design a product to fail is a serious abuse of consumer trust. Built-in obsolescence is in danger of weakening the bond of trust between customer and business. Today, when protecting the environment is such a priority goal, the question of product life and durability is a critical question. Yet current economic systems still seem to rely on the profits generated by over-consumption. Consumers are becoming increasingly aware of these issues and more pressure is being applied to business to address this with more ethical design. Designers must now consider how components can be upgraded, reused, recycled or maintained for extended product life, rather than simply thrown away.

Task

Make a list of products that have built-in obsolescence. Discuss ways in which a designer may address built-in obsolescence when developing new products from your list.

Sustainable product design

In his book *The Total Beauty of Sustainable Products* Edwin Datschefski spells out five factors that make up product sustainability:

1. *Cyclic:* products made from compostable organic materials, or from minerals that are continuously recycled therefore decreasing levels of waste and pollution; for example, products made from Biopol.
2. *Solar:* products in manufacture and use that consume only renewable energy that is cyclic and safe; for example, products made using renewable energy sources, i.e. wind power and products that operate using solar-powered (photovoltaic) cells.
3. *Safe:* all releases to air, water, land or space are food for other systems; for example, products that do not emit unnecessary pollutants or chemicals during their manufacture. For example, the redesigned Nike shoe box uses no glues or solvents in its construction and no heavy-metal inks for printing.
4. *Efficient:* products in manufacture and use that require 90 per cent less energy, materials and water than equivalent products did in 1990; for example, the reduction of materials in packaging a product.
5. *Social:* product manufacture and use supports basic human rights and natural justice; for example, Fairtrade products that help producers in developing countries receive a fair share of profits, so eliminating exploitation of the workforce.

Minimising waste production

Perhaps the most important economic factor for a designer of sustainable products to consider is that waste is lost profit. There are some simple options to consider when deciding how to minimise waste production. They are referred to as the four R's:

- Reduce
- Reuse
- Recover
- Recycle.

The UK Government's Envirowise programme has drawn up a table of ten environmental considerations for improving the environmental performance of a product covering the various stages of its life cycle:

Reduce

For all designers, one of the first priorities for sustainability should be to reduce the quantities of any material chosen whenever possible. Therefore, packaging designers must optimise the amount of materials needed to package a product in order to minimise the consumption of resources which will in turn achieve significant cost savings and improve profit margins.

Manufacturers are obliged to reduce packaging use under the UK Producer Responsibility Obligations (Packaging Waste) Regulations 1997. The Government's Envirowise programme suggests that manufacturers:

- consider the materials and designs they use
- examine ways of eliminating or reducing the packaging requirement of a product – changes in product design, improved cleanliness, better handling, just-in-time delivery, bulk delivery, etc.
- optimise packaging use, e.g. match packaging to the level of protection needed.

In one Envirowise case study, Ambler of Ballyclare implemented an environmental policy focusing upon the minimisation of packaging waste. The financial benefits were considerable (£103,002 savings per year) as well as the environmental benefits including diverting waste

Life cycle stage	Key environmental considerations
Raw materials 	• Use less material • Use materials with less environmental impact
Manufacture 	• Use fewer resources • Produce less pollution and waste
Distribution 	• Reduce the impacts of distribution
Use 	• Use fewer resources • Cause less pollution • Optimise functionality and service life
End-of-life 	• Make reuse and recycling easier • Reduce the environmental impact of disposal

Figure 4.1.22 Key environmental considerations for cleaner design

from landfill, reducing environmental impact and disposal costs. The reduced use of vehicles for the transportation of waste also saw a 42 per cent reduction in carbon monoxide emissions.

Reuse

Reusing products minimises the extraction and processing of raw materials and the energy and resources required for recycling. A number of companies adopt returnable or refillable containers for some of their products; for example, the delivery of milk in glass bottles. Refillables appear to offer environmental benefits yet they often require greater use of resources in their manufacture and distribution to enable them to withstand the rigours of repeated use. If reuse is to be economically viable then the cost of collection, washing and refilling should be less than producing a new container.

Recover

The manufacture of any product obviously requires the use of energy. If the product is simply discarded and landfilled then all of this energy is lost. Waste that cannot readily be recycled but can burn cleanly can be incinerated in specialised power stations to generate electricity and provide hot water for the local area. This is not an ideal solution, but by adopting such technology fewer finite fossil fuels are needed to generate electricity in conventional power stations. In Sweden 47 per cent of waste is recovered in energy from waste plants.

Recycle

Designers and manufacturers should be increasingly aware of the possibilities of using recycled materials as illustrated earlier in this unit. However, it is now fundamentally important for a designer to consider *design for recycling*. Here the designer must consider the LCA of a new product in order to fully appreciate how it will be disposed of when it has come to the end of its useful life. By incorporating components in new products or entire products that can be recovered easily and recycled after use then waste generation will decrease.

Eco-labelling

Labels are increasingly used by manufacturers to display claims about the nature of the product concerning its properties and social and environmental attributes. Such claims can often be misleading or overly general and only add to the confusion. Effective eco-labelling should be an objective indicator of the environmental impact of a product from cradle to grave using LCA. There are three main types of label:

- Single issue/mandatory – where a Government requires information to be displayed about either the hazardous nature of a product, or some other attribute.
- Single issue/voluntary – where one aspect of the environmental impact, of a product or one stage of the life cycle, is targeted; for example, percentage recycled content or totally chlorine-free bleaching for paper products.
- Multiple issue/voluntary – where impacts at various stages of the life cycle are considered, both socially and environmentally, e.g. Fairtrade Mark.

At present there are several eco-labelling schemes (for example, EC Ecolabel) using their own criteria for assessment. For these schemes to be truly effective they must be part of a global eco-labelling scheme using an international system for accreditation. Only then will the consumer be able to make an informed choice.

Task
Investigate the various eco-labelling schemes currently used and identify their logos on a variety of products.

Responsibilities of developed countries

Global sustainable development

The challenge for people in developed countries is the need to reduce their use of scarce resources and reduce the production of pollution, which is associated with over-consumption. This is a shift towards sustainable consumption whereby each person in the world has access to the same 'environmental space'.

The challenges for developing countries are different. In many instances people in developing countries need to consume more, needing, for example, greater access to clean water, electricity and health care. One method may be to trade more with developed countries to bring in much needed foreign reserves to invest in domestic economies. Developing countries will need to have access to markets in developed countries in order to expand. However, developed countries need to shrink their markets to address over-consumption and to take up less 'environmental space' thereby creating tighter and more impenetrable markets for developing countries to sell into.

The UN Earth Summits are the global forum for sustainability, where representatives of all nations meet to discuss sustainable development. Most countries have established some mechanism for implementing Earth Summit agreements, such as global trade or the reduction in greenhouse emissions. All countries are invited to speak at these summits. Norway has stated several practical steps toward sustainable consumption that would include:

- Improving analysis, public awareness and participation.
- Providing incentives for sustainable consumption.
- Energy – sustainable use, efficiency and renewable sources.
- Implementing new strategies for transportation and sustainable cities.
- Accelerating the use of more efficient and cleaner technologies.
- Strengthening international action and cooperation.

Extract from the Report of the Symposium: Sustainable Consumption, Ministry of Environment, Norway, 1994.

If global sustainable development is to succeed then all countries must firstly agree on the terms and conditions and secondly implement the changes needed.

Task
Organise a mini Earth Summit within your group with each student representing either a developed or developing country. Identify the main issues for global sustainable development in that country and try to agree upon resolutions to tackle these problems.

UK responsibilities: forest products
Britain relies heavily upon the import of tropical hardwoods, accounting for 8 per cent of the global trade. The developing countries that produce this timber benefit little from this trade, with only 10.5 per cent of the revenue from timber production benefiting the producing country.

Timber has been the focus of considerable efforts over the past decade to establish more sustainable production and trading systems. The main problems associated with forests are:

- *Deforestation* – the full-scale removal of forest to make way for farming, settlement, infrastructure and mining. Global deforestation is currently taking place at a rate of approximately 17 million hectares each year.
- *Environmental degradation of forest areas* – deforestation can cause soil erosion, watershed destabilisation and microclimate change. Industrial air pollution also reduces forest health.
- *Loss of biodiversity* – deforestation and environmental degradation contributes to a rapid reduction in ecosystems, species and genetic diversity in both natural and planted forests. Some scientists estimate that 1 per cent of all species are being lost each year.
- *Loss of cultural assets and knowledge* – any indigenous peoples' lives are destroyed by deforestation, because their lives depend upon the forest.
- *Loss of livelihood* – for forest-dependent peoples, particularly in poor countries.
- *Climate change* – both regional and global climate change may lead to global warming. Forests play a major role in carbon storage and with their removal more carbon dioxide enters the atmosphere.

The UK has a responsibility to encourage the development of sustainable production and trading systems in order to minimise the amount of deforestation and its effects upon the environment. This should include:

- No longer importing from sources that involve deforestation.
- Moving to sources of supply from areas of ecological surplus, e.g. the high-yielding plantations of Brazil, Chile and New Zealand.
- Certification systems that ensure that forests producing goods for the UK are sustainably managed.
- Timber-tracing systems to ensure that products from certified forests can be identified as such.
- Reducing consumption through education and advisory approaches that show how to produce the same benefits from less timber.
- Encouraging exporting countries to make the necessary policy changes required for the transition to sustainable forest management.
- Supporting international efforts to control the trade in unsustainably produced wood
- Improving aid to poor communities involved in current deforestation methods.

Offshore manufacturing of multinationals
Offshore manufacture is a driving force in the global marketplace. There is an increased awareness by multinational companies based in developed countries of the value of offshore manufacturing as a vital strategic tool. Many companies will draw upon the individual expertise of other countries in order develop new products, especially in the field of technology.

The new global job shift

'Corporate downsizing' is part of the ebb and flow of modern business practices, but in recent years unemployment is not simply occurring because demand has dried up. Companies are relocating to less-developed countries such as India, China and former Soviet nations and outsourcing their work. Modern corporate buildings and industrial estates are sprouting up in these countries to supply the new demand for outsourcing and offshore manufacturing. Initially jobs in developing countries were created through the manufacture of shoes, cheap electronics and toys. Now all kinds of knowledge, work and manufacturing can be performed almost anywhere. For example, there is an increasing trend for call centres dealing with the UK public to be based in India.

The driving forces are digitisation, the Internet and high-speed data networks that cover the entire globe. Why do multinationals manufacture offshore or outsource? The answer is quite simple: it costs less. It is now possible to receive the same quality of work at a fraction of the cost than if Western companies manufactured in their own country. For example, the cost of manufacturing injection moulds is typically 50 per cent lower in China than in the West. In addition, by having bases in developing countries it is possible to gain greater access to expanding overseas markets. Obviously this calls into question certain ethical issues such as large-scale unemployment in developed countries and exploitation of labour in developing countries. For instance, why would a British-based multinational company continue to pay the minimum wage to its UK employees when they could employ Indian or Chinese labour for 50–60 per cent less? Workers in developing countries are often not given the same opportunities for promotion, pay rises, company benefits, union membership and working conditions that their Western colleagues demand as basic human rights.

Task

Discuss the effects of offshore manufacturing and outsourcing in relation to:
a) multinational companies
b) workers in developing countries
c) workers in developed countries.

Exam preparation

You will need to revise all the topics in this unit, so that you can apply your knowledge and understanding to the exam question. In preparation for your exam it is a good idea to make brief notes about different topics, such as 'The impact of modern technology and biotechnology on the development of new materials and processes'. Use sub-headings or bullet point lists and diagrams where appropriate. A single side of A4 should be used for each heading from the Specification.

It is very important to learn exam skills. You should also have weekly practice in learning technical terms and in answering exam-style questions. When you answer any question you should:

- read the question carefully and pick out the key points that need answering
- match your answer to the marks available, e.g. for two marks you should give two good points that address the question
- always give examples and justify statements with reasons, saying how or why the statement is true.

Practice exam questions

1 A manufacturer is developing a new range of packaging for a cosmetic container and its packaging box.
   a) Describe **two** criteria that would need to be considered when selecting an appropriate material for the perfume container. (4)
   b) Suggest and justify a specific material for the packaging box. (3)

2 Liquid crystal displays (LCDs) are used in a wide range of modern products.
   a) Explain why LCDs have replaced light emitting diodes (LEDs) in displays. (2)
   b) Briefly describe how an LCD works. (3)

3 Modern materials enable the development of innovative products.
   a) Explain the benefits to the consumer of the use of **one** modern composite material. (3)
   b) Explain the term 'smart material' and give **one** example of its use in a product. (3)

4 Biotechnology is at the forefront in the development of new materials and processes. Briefly discuss the role of biotechnology in the production of the following:
   a) coloured timber (3)
   b) paper. (3)

5 Explain what is meant by **two** of the following terms:
   - digital photography
   - holographic images
   - mail merge software. (6)

6 Computer technology is used to create special effects on television. Describe how 'blue screen' is used to create scenes that would be too dangerous for actors. (5)

7 Explain how Life Cycle Assessment (LCA) is used to evaluate the impact of products on the environment. (5)

Total: 40 marks

UNIT 4 B1 Design and technology in society (G402)

Summary of expectations

1. What to expect

Design and Technology in Society is one of the two options offered in Section B of Unit 4. You must study the same option that you studied at AS level.

2. How will it be assessed?

The work that you do in this option will be externally assessed through Section B of the Unit 4 examination paper. The questions will be common across all three materials areas in product Design. You can therefore answer questions with reference to **either**:

- a Graphics with Materials Technology product, such as a 'green' credit card
- a Resistant Materials Technology product, such as a carbon fibre racing bike
- **or** a Textiles Technology product, such as a Kevlar bullet-proof vest.

You are advised to spend **45 minutes** on this section of the paper.

3. What will be assessed?

The following list summarises the topics covered in this option and what will be examined:

- Economics and production
  - economic factors in the production of one-off, batch and mass-produced products.
- Consumer interests
  - systems and organisations that provide guidance, discrimination and approval
  - the purpose of British, European and International Standards relating to quality, safety and testing
  - relevant legislation on the rights of the consumer when purchasing goods.

- Advertising and marketing
  - advertising and the role of the design agency in communicating between manufacturers and consumers
  - the role of the media in marketing products
  - market research techniques
  - the basic principles of marketing and associated concepts.
- Conservation and resources
  - environmental implications of the industrial age
  - management of waste, the disposal of products and pollution control.

4. How to be successful in this unit

To be successful in this unit you will need to:

- have a clear understanding of the topics covered in this option
- apply your knowledge and understanding to a given situation or context
- organise your answers clearly and coherently, using specialist technical terms where appropriate
- use clear sketches where appropriate to illustrate your answers
- write clear and logical answers to the examination questions, using correct spelling, grammar and punctuation.

There may be an opportunity to demonstrate your knowledge and understanding of Design and Technology in Society in your Unit 5 coursework.

5. How much is it worth?

This option, together with Section A, is worth 15 per cent of the full Advanced GCE.

Unit 4A + option	Weighting
A2 level (full GCE)	15%

1. Economics and production

Figure 4.2.1 shows the UK's manufacturing sectors.

Economic factors in the production of one-off, batch and mass-produced products

The sequence of activities required to turn raw materials into finished products for the consumer is called the **production chain**. It is the aim of all manufacturing companies to undertake such activities in the most cost-effective manner. The purpose of this is to maximise profit.

The economic factors that combine in order to produce profit include:

- variable costs
  - the costs of production, such as materials, services, labour, energy and packaging
- fixed costs
  - related to design and marketing, administration, maintenance, management, rent and rates, storage, lighting and heating, transport costs, depreciation of plant and equipment.

In order to remain profitable, a manufacturing company must calculate accurately its total costs and set a suitable selling price. This calculation must allow for variable costs, fixed costs and a realistic profit.

The production chain

The production chain includes the following:

- The primary sector is concerned with the extraction of natural resources such as mining and quarrying.
- The secondary sector is concerned with the processing of primary raw materials and the manufacture of products. Although this sector supplies a large proportion of exports from developed countries (such as those of Western Europe and the USA), it employs a decreasing proportion of the workforce, as changes in technology and the global economy occur.
- The tertiary sector industries provide a service and include employment in education, retailing, advertising, marketing, banking and finance. This sector employs the most people in developed countries.

In your design and technology course you are mostly concerned with the secondary and tertiary sectors. When you design and manufacture a coursework product, your main concern should be product viability in terms of its cost of manufacture and market potential. For any industry in the secondary sector, however, product viability relates to market potential and profit. Product viability is essential to its very existence and to the employment of that industry's workforce.

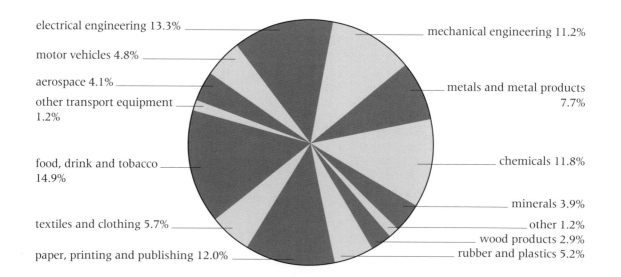

electrical engineering 13.3%
motor vehicles 4.8%
aerospace 4.1%
other transport equipment 1.2%
food, drink and tobacco 14.9%
textiles and clothing 5.7%
paper, printing and publishing 12.0%
mechanical engineering 11.2%
metals and metal products 7.7%
chemicals 11.8%
minerals 3.9%
other 1.2%
wood products 2.9%
rubber and plastics 5.2%

Figure 4.2.1 *Different sectors of manufacturing are located in every region of the country. The sectors of manufacturing that use resistant materials include 'engineering and allied industries', which is the largest group*

Productivity and labour costs

A company's productivity is a measurement of the efficiency with which it turns raw materials (production inputs) into products (manufactured outputs). The most common measure of productivity is output per worker, which has a direct effect on labour costs per unit of production. The higher the productivity, the lower the labour costs per unit of production and therefore the higher the potential profit. Table 4.2.1 compares the weekly wage, output per worker and labour cost per unit of production for two companies. It shows how an efficient company with high output per worker can keep labour costs per unit of production low.

Table 4.2.1 *Comparison of productivity*

	Weekly wage	Output per worker	Labour cost per unit of production
Efficient company	£280	40.0	£7
Less efficient company	£250	25.0	£10

Labour costs are linked to the type and length of any production process. For example, a washing machine made from mild steel sheet will have sharp edges where the steel has been cut to size. In order to produce a good quality product these sharp edges will need to be removed or covered. Introducing an additional process such as this will increase labour costs and therefore increase the cost of the product.

Scale of production

Within the economies of the developed countries of the West and Asia, there is a responsibility for all the functions of a company, including manufacturing, to be profitable. For example, in many large companies:

- different internal departments are 'customers' to other internal departments within the same company
- each department has a budget and has to work within this budget or make a profit.

In manufacturing companies the level or scale of production is an extremely important factor for profitability, because it influences the process of manufacture, where the product is manufactured, the choice of products available on the market and their resulting selling price.

One-off products

One-off or custom-made products need to use materials, tools and equipment that are available or that can be resourced fairly easily. They are often less efficient to manufacture, because they may use different or 'specialised' materials and processes. One-off products are usually much more expensive to buy, since materials and labour costs are higher, but they do provide 'individual' or made-to-measure products, such as hand-made kitchen units.

Mass production and the development of new products

On the other hand, mass production has revolutionised the choice and availability of a range of relatively inexpensive products:

- It is increasingly difficult for a company to remain in profit without developing new products, even when existing ones are selling well. Consumer demand can be very fickle and sales of even a best-selling product tail off in time, as new or different competing models come on the market.
- The cost of developing new products is high. It includes initial manufacturing setting up costs such as the factory and its layout and the cost of the workforce.
- New products may require changes in production, which are often difficult to achieve due to the high levels of investment required.
- The development of new production methods needs long-tem planning and may need to overlap with existing production in order to keep the company running and in profit.
- There is a constant need to reduce the time-to-market of new products. The most successful companies produce the right product at the right time, in the right quantity and at the right cost – one that the market will see as providing the right image as well as being 'value for money'.

Task

Select two similar products, such as a hand-made board game in a wooden box and a mass-produced chess set in a cardboard box. For the two products compare the following:

- the type of materials and used for the product and its packaging
- the number and estimated length of time of manufacturing processes
- the selling price.

Sources, availability and costs of materials

The cost of materials depends on the type and quantity of materials required and their availability. As a general rule, it is a question of supply and demand, so that materials in short supply cost more. Different scales of production also result in different levels of costing. Materials costs are generally lower in high-volume production than in one-off manufacture. All product manufacturers require a reliable and continuous supply of raw materials at an economic price to enable profit to be made.

Softwoods and hardwoods

Most softwood is grown in the colder regions of northern Europe and North America. Careful management of these forests enables the control of supply and demand. As conifers are relatively fast-growing and produce straight trunks, they are economic to produce, with little waste. Softwoods are therefore relatively inexpensive. They are used extensively for building construction and joinery. Waste softwood is used in the manufacture of manufactured board and paper.

Hardwoods are slower growing and therefore more expensive than softwoods. They come from broad-leaved deciduous trees, growing in the temperate climates of Europe, Japan and New Zealand. The hardwoods grown in the tropical climates of Central and South America, Africa and Asia are mainly evergreen. These grow all year round and reach maturity earlier, so they are cheaper than the northern hemisphere hardwoods. They are used for furniture, kitchen utensils, flooring, toys, etc.

Manufactured board

The UK is one of the least wooded areas of Europe and has to import almost 90 per cent of its timber needs. Manufactured board is usually supplied debarked, in board form, ready for further processing. Importing the timber in board form is more cost-effective than importing raw timber. There is less waste from processing and manufactured board can travel faster in freight container ships rather than in bulk cargo ships.

Plywood and blockboard are more expensive than medium density fibreboard (MDF) because they are made from better quality timber.

- Plywood is made from European or American birch and meranti from South East Asia.
- Blockboard is made from birch and pine. Birch is used for the facing because it is more durable. The core is made from European or American pine.
- MDF is less expensive than plywood or blockboard because it is made from small section or thinned timber and reconstituted wood.

Manufactured board is made in high volume

and is widely used in self assembly furniture, shop fittings and flooring.

Task

Using suppliers' catalogues or the Internet, investigate the cost of softwoods, hardwoods and manufactured board. Explain why there are variations in the cost of different sizes.

Metals

Metal ores form about a quarter of the weight of the Earth's crust; the most common ores being aluminium followed by iron. These metals are relatively inexpensive. They have no particular pattern of distribution around the world, but some countries have larger deposits than others. Iron and steel account for almost 95 per cent of the total tonnage of all metal production. Ships, trains, cars, trucks, bridges and buildings and thousands of other products depend on the strength, flexibility and toughness of steel.

Sources of ore

- The main sources of iron ore are Europe, North America and Australia. Iron ore has a high metal content and is easily available, so the price of steel is relatively low.
- Over-production and fierce competition have led to the loss of steel making in some countries.
- Aluminium ore (bauxite) is mainly found in the southern hemisphere. Bauxite is easily accessible, so aluminium is cheaper to make than steel.
- Copper ore is found in the USA, Canada and Chile. The comparative rarity of copper makes it much more expensive than iron ore or bauxite.

Metals in common use, such as iron and aluminium are easily available and the supply problems associated with timber do not normally occur. Lower metal production costs are often achieved by smelting ore close to its source. This reduces transport costs and may make use of lower labour costs.

The importance of oil

Crude oil is an important commodity because it supplies much of the world's energy needs. It is also the principle raw material for making plastics and polymers, without which modern society could not long continue to flourish.

Unfortunately, oil is rarely found where it is needed. The largest oil-producing countries are not themselves major consumers and they are therefore able to export much of their oil. These countries are part of the Organisation of Petroleum Exporting Countries (OPEC), a cartel which sets output quotas in order to control crude oil prices. The members of OPEC are the Middle East, South America, Africa and Asia, but not the USA, the Russian Federation or European oil producers such as the UK. In the 1970s, OPEC controlled 90 per cent of the world's supply of crude oil exports. The resulting high price of oil that OPEC maintained then allowed more expensive fields such as the North Sea and Alaska to be brought into production (see Figure 4.2.2). Oil costs continue to fluctuate, depending on the control of its supply. This can result in higher petrol, energy and raw materials prices worldwide.

Figure 4.2.2 *Offshore oilfields currently produce about a quarter of the world's crude oil. Great technological skill has been required to design platforms that are stable enough to allow drilling to take place and resilient enough to withstand the harsh conditions of wind and waves*

Plastics

Most thermoplastic and thermosetting plastics are derived from crude oil so are easily available. In many industries traditional materials such as wood or metal are replaced by inexpensive plastics, which often use fewer processes, as products can be made in one piece. For example polyester (PET) bottles are safer, lighter and cheaper to produce than ones made from glass.

- Thermoplastics are available in sheet or rod form for processing into products, or as granules for injection moulding. Acrylic, polythene, polystyrene, polypropylene and polyester are widely used for a whole range of domestic products.
- The thermosetting plastic epoxy resin is used in adhesives, surfboards and motorbike helmets.

Cost of materials

As we have seen, the cost of timber, manufactured board, metals and plastics is related to their sources and availability. The following can be said about most raw materials:

- materials in short supply or in great demand cost more
- materials that are difficult to process cost more
- materials that come from isolated sources have to be transported further so cost more.

Advantages of economies of scale of production

Economies of scale are factors that cause average costs to be lower in high-volume production than in one-off production (see Figure 4.2.3). The unit price is lower because inputs can be utilised more efficiently. Economies of scale in high-volume production are brought about by:

- specialisation – the work processes are divided up between a workforce with specific skills that match the job

Figure 4.2.3 *One-off products are more expensive to produce, such as this cap and necklace by Val Hunt. They are made from beer cans, which are annealed, woven, pleated, frilled or curled to produce exciting and creative products*

- the spread of fixed costs of equipment between more units of production
- bulk buying of raw materials at lower unit costs
- lower cost of capital charged by providers of finance
- the concentration of an industry in one area – attracting a pool of labour that can be trained to have specialist skills
- a large group of companies in one area – attracting a large network of suppliers whose costs are lower, because of their own economies of scale.

Task

Devise a checklist that will make your own production more cost effective. Think about the range of materials, components and processes you might use.

The relationship between design, planning and production costs

SIGNPOST
'Efficient manufacture and profit' Unit 3B1 page 123

In order to remain profitable, manufacturers must accurately calculate their total costs and set a suitable product selling price. Target production costs are established from the design stage and checked against existing or similar products. This is done so that the design team can make sound decisions early in the design task.

All costs in a manufacturing company are initiated in the design phase. It is in the manufacturing stage that the major costs are incurred. It is often said therefore that designing for manufacture (DFM) is directly related to designing for cost. The main aims of DFM are:

- to minimise component and assembly costs
- to minimise product development cycles
- to enable higher quality products to be made.

The cost of quality

SIGNPOST
'Aesthetics, quality and value for money' Unit 3B1 page 124

Manufacturing a competitive product based on a balance between quality and cost is the aim of all companies. The costs of quality are no different from any other costs, because like the cost of design, production, marketing and maintenance, they can be budgeted for, measured and analysed. There are three types of quality costs:

- the costs of getting it wrong
- the costs checking it is right
- the costs of making it right first time.

The costs of getting it wrong

There are two ways to get it wrong: **internal failure costs** and **external failure costs**.

Internal failure costs occur when products fail to reach the designed quality standards and are detected before being sold to the consumer. They include costs relating to:

- scrap products that cannot be repaired, used or sold
- reworking or correcting faults
- re-inspecting repaired or reworked products
- products that do not meet specifications but are sold as 'seconds'
- any activities caused by errors, poor organisation or the wrong materials.

External failure costs occur when products fail to reach the designed quality standards and are not detected until after being sold to the customer. They include costs relating to:

- repair and servicing
- replacing products under guarantee
- servicing customer complaints
- the investigation of rejected products
- product liability legislation and change of contract
- the impact on the company reputation and image – relating to future potential sales.

The costs of checking it is right

These costs are related to checking:

- materials, processes, products and services against specifications
- that the quality system is working well
- the accuracy of equipment.

The costs of making it right first time

There is one way to get it right: **prevention costs**. These are related to the design, implementation and maintenance of a quality system. Prevention costs are planned and incurred before production and include those relating to:

- setting quality requirements and developing specifications for materials, processes, finished products and services
- quality planning and checking against agreed specifications
- the creation of and conformance to a quality assurance system

- the design, development or purchase of equipment to aid quality checking
- developing training programmes for employees
- the management of quality.

Costing a product

Costing is the process of producing an accurate price for a product which will make it saleable and create a profit (see Figure 4.2.4). Setting the selling price too high may reduce sales below a profitable margin, while setting it too low won't allow a profit even if vast numbers are sold. Checks against a competitor's product are often used to establish the potential price range of a new product because they can give an idea of what the market can stand.

Products are sometimes said to have a value, a price and a cost:

- Manufacturers want the income from selling the finished products, rather than keeping them in stock, so for them the product value is always lower than the selling price.
- On the other hand, consumers want the product more than the selling price, as they see the product value as being higher than the selling price.

The total cost of a product takes account of the following:

- **Variable costs** (also known as direct costs) – the actual costs of making a product, such as materials, services, labour, energy and packaging. Variable costs vary with the number of products made. The more products that are made, the greater the variable costs of materials. Variable costs may account for around 50–65 per cent of the total product selling price (SP).
- **Fixed costs** (also known as indirect or overhead costs) – the costs, for example, of design and marketing, administration, maintenance, management, rent and rates, storage, lighting and heating, transport costs, depreciation of plant and equipment. Fixed costs are not directly related to the number of products made, so they remain the same for one product or hundreds. A company's accountants will establish a way to divide up fixed costs between the various product lines made by a company, so that each product carries its share. Marketing and selling costs often account for 15–20 per cent of the total SP.
- **Profit** – the amount left of the SP after all costs have been paid. Profit is referred to as gross or net. The gross profit is calculated by deducting variable plus fixed costs from the revenue from sales. Net profit is gross profit minus tax. Net profit is used to pay dividends to shareholders, bonuses to employees and for reinvestment in new machinery or in new product development.

The break-even point

In order to cover the cost of manufacture, enough products need to be sold at a high enough price. Calculating this requirement is called 'break-even analysis'. The starting point for working out the break-even point is the relationship between fixed costs, overhead costs and the selling price. For example, if the selling price of a chess set and its packaging is £65, how many would need to be sold to cover manufacturing costs and break even? The

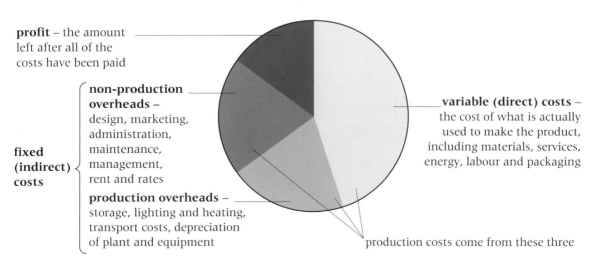

profit – the amount left after all of the costs have been paid

non-production overheads – design, marketing, administration, maintenance, management, rent and rates

fixed (indirect) costs

production overheads – storage, lighting and heating, transport costs, depreciation of plant and equipment

variable (direct) costs – the cost of what is actually used to make the product, including materials, services, energy, labour and packaging

production costs come from these three

Figure 4.2.4 *What's in a price?*

following formula can be used to work out the break-even point:

$$\text{Break-even point} = \frac{\text{Fixed costs}}{\text{Selling price} - \text{variable costs}}$$

Suppose the variable costs of making a chess set and its packaging are £35 and it sells for £65. If the fixed cost of making the chess set and packaging is £5000, what is the break-even point?

$$\text{Break-even point} = \frac{5000}{65 - 35} = 167$$

So 167 chess sets and packaging would need to be sold to reach the break-even point. Any chess sets and packaging sold over 167 would make a profit for the company.

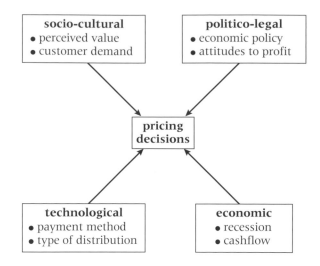

Figure 4.2.5 *The forces acting on pricing decisions*

Tasks

1 The variable costs for making a light from wood, paper and polymer are £15 and it sells for £35. If the fixed cost of making the light is £1800 work out the break-even point.
2 Explain the difference between gross and net profit.

The material and manufacturing potential for a given design solution

When manufacturers calculate the potential selling costs of a product, they have to take into account more than just manufacturing costs. Consumer expectations of the value of the product play an important part. This will depend on two things: what customers perceive as value for money and what the competition is offering.

As can be seen from Figure 4.2.5, the forces acting on pricing decisions are complex. The perceived value of a product is reflected in the likelihood of customers continuing to buy a product in the face of a price increase. This is the pattern for products like food items or services like heating. Unless they become prohibitively expensive, we continue buying them. For non-essential purchases such as a second car, demand can drop when they become more expensive. Other forces acting on price include the economic and political climate – how high are taxes, is there a recession, is there full employment?

Task

Some say that there is no direct connection between the product cost and the selling price. In the same way that materials in short supply cost more, the manufacture of a popular product in limited supply can add a large mark-up to the price. Give examples of products that prove or disprove this theory.

2. Consumer interests

The relationship between manufacturers as producers and consumers as buyers is important. Manufacturers want their products to sell well and make a profit. Consumers want to buy high-quality, attractive products that are reliable, easy to maintain, safe to use and that provide value for money at a price they can afford. A manufacturer's success in producing goods that meet consumer requirements relies on keeping consumer needs and values at the heart of the business.

Systems and organisations that provide guidance, discrimination and approval

Most large companies use market research to establish consumer needs, values and tastes. They must also take into account consumers' statutory rights when buying products. These rights are enforced and regulated by a wide range of legislation relating to consumer protection and fair trading. There are many systems and organisations that provide guidance, discrimination and approval for consumers. These include:

- the Institute of Trading Standards Administration
- British, European and International Standards organisations
- consumer 'watchdog' organisations.

Consumer 'watchdog' organisations

These days consumers are much more aware of new products. One reason for this developing consumer awareness is the profusion of 'style' sections in newspapers and magazines. Products ranging from cook-chill foods to computer games are regularly featured and their desirability and value for money evaluated.

Consumer 'watchdog' organisations and specialist magazines also provide guidance, discrimination and approval for new products (see Figure 4.2.6). These organisations are independent of product manufacturers and provide objective reviewing and testing of products. One such organisation is the Consumers' Association, which publishes the magazine *Which?*. This provides reports about product testing and 'best buys'. It has a website, found at www.which.co.uk, which provides a range of information, such as an overview of its activities, 'headlines' for daily consumer news and links to electronic newspapers.

Other groups that provide consumer support include television and radio programmes, many

DIGITAL CAMERAS	Canon Digital Ixus 400	Canon PowerShot G3	Casio Exilim EX-Z3	Fujifilm FinePix F410 Zoom	Fujifilm FinePix M603 Zoom	Kodak EasyShare LS633 Zoom	Konica Digital Revio KD-500Z	Kyocera Finecam S5	Minolta Dimage F300	Minox DC 3311	Nikon Coolpix 310
SPECIFICATION											
Price (£) 1▷	450	600	350	296	350	300	450	350	430	350	280
Largest image size in pixels	2,272x1,704	2,272x1,704	2,048x1,536	2,816x2,120	2,832x2,128	2,032x1,524	2,592x1,944	2,560x1,920	2,560x1,920	2,048x1,536	2,048x1,536
Effective megapixels 2▷	4.0	4.0	3.2	3.1	3.1	3.3	5.0	5.0	5.0	3.1	3.2
Zoom range 3▷	36-108	35-140	35-105	38-114	38-76	37-111	39-117	38-114	38-114	32-96	38-115
Size (hxwxd) (cm)	6x9.5x3.1	8x13x7.5	6x9.5x2.4	7x9x3.5	9.5x7x4	6x12x4	6x10x3.5	6x9.5x4	6x11.5x4	8x11.5x7.5	7x9x4.5
Weight (g) 4▷	231	525	147	199	288	240	229	198	239	349	219
Number of batteries required	1	1	1	1	1	1	1	1	1	4	2
Supplied with rechargeable batteries 5▷	✓	✓	✓	✓	✓	✓	✓	✓			✓
Supplied with battery charger 5▷	✓	✓	✓	✓	✓	✓	✓	✓			✓
Software included 6▷	Zoom Browser EX, ArcSoft photo suite	Zoom Browser EX, Adobe Photoshop	Photo Loader	Finepix Viewer	Finepix Viewer	Kodak Easy-Share software	none supplied	ArcSoft photo suite	Dimage Image Viewer	MGI PhotoSuite, Photo Vista	Nikon View, Adobe Photoshop Elements
FEATURES											
Manual focus setting 7▷	✓	✓	✓					✓	✓	✓	
Short video recording 8▷	✓	✓	✓	✓	✓	✓	✓	✓	✓		✓
Sound recording 8▷	✓	✓			✓	✓	✓	✓	✓	✓	
MEMORY											
Camera's internal memory (Mb) 9▷	0	0	10	0	0	16	2	0	0	8	0
Memory card supplied (Mb)	32	32	0	16	16	0	16	16	64	0	16
Compatible memory cards 10▷	CF I	CF I or II	SD or MMC	xD	xD or CF II	SD or MMC	SD or MS	SD or MMC	SD	CF I	CF I
Image quality settings 11▷	12	13	12	4	5	4	8	8	16	6	4
PERFORMANCE											
Shutter delay on auto focus (sec) 12▷	0.9	1.4	0.5	0.5	0.7	0.7	1.1	1.2	1.5	2.5	0.7
Flash	☆	★	○	☆	☆	☆	☆	○	☆	○	☆
Battery life	○	★	☆	★	○	★	☆	○	◕	★	★
Time to download pics to PC	☆	★	★	★	★	◕	◕	○	★	☆	★
Focusing 13▷	☆	☆	☆	○	☆	☆	☆	○	☆	☆	☆
Close-up rating 14▷	◕	☆	★	★	☆	☆	★	★	★	★	☆
Time between shots 15▷	★	★	☆	★	☆	★	★	★	★	○	★
Overall picture quality	☆	☆	○	○	☆	○	☆	○	○	○	☆
Overall ease of use	☆	○	☆	☆	○	○	○	○	○	○	☆
TOTAL TEST SCORE (%)	64	62	51	58	59	56	56	55	60	48	63

Figure 4.2.6 Products are regularly reviewed in consumer magazines

of which give information and guidance on consumer rights. Consumer advice and support is also available from the Citizens Advice Bureau (CAB).

> ## Task
> Investigate the work of two different consumer organisations. Compare their roles and the range of products they evaluate. For each organisation explain how their product reviews:
> - help or hinder product manufacturers
> - guide consumer choice.

The purpose of British, European and International Standards relating to quality, safety and testing

> ### SIGNPOST
> 'British and International Standards' Unit 3B1 page 132

European and International Standards organisations set national and international standards, testing procedures and quality assurance processes to make sure that manufacturers make products that fulfil the safety and quality needs of their customers and the environment. Most standards are set at the request of industry or to implement legislation. Manufacturers of upholstered furniture, for example, have to conform to established fire safety standards. The test procedures for checking fire safety have to comply with British Standards (BS) guidelines and must be carried out under controlled conditions.

Any product that meets a British Standard can apply for and be awarded a 'Kitemark'. This shows potential customers that the product has met the required standard and that the manufacturer has a quality system in place to ensure that every product is made to exactly the same standard.

The relationship between standards, testing procedures, quality assurance, manufacturers and consumers

Common to all commercial product manufacture is the need to produce a quality product. For manufacturers, incorporating quality management systems into the design and manufacture of products is therefore important. ISO 9000 is an internationally agreed set of standards for the development and operation of a **quality management system (QMS)**. ISO 9001 and 9002 are the mandatory parts of the ISO 9000 series. They specify the clauses manufacturers have to comply with in order to achieve registration to the standard. A QMS involves a structured approach to ensure that customers end up with a product or service that meets agreed standards.

Quality management systems
All industrial quality management systems use structured procedures to manage the quality of the designing and making process. The following designing and making procedures illustrate the kind of quality management process that industry adopts and uses:

- Explore the intended use of the product, identify and evaluate existing products, consider the needs of the client.
- Produce a design brief and specification.
- Use research, questionnaires and product analysis.
- Produce a range of appropriate solutions.
- Refer back to the specification.
- Refer to existing products. Use models to test aspects of the design.
- Check with the client. Use models to check that the product meets the design brief and specification.
- Plan manufacture and understand the need for safe working practices.
- Manufacture the product to the specification.
- Critically evaluate the product in relation to the specification and the client. Undertake detailed product testing and reach conclusions. Produce proposals for further development, modifications or improvements.

You will probably be familiar with many of the procedures outlined above, because they are very similar to ones you use in your coursework.

> ## Task
> Compare the designing and making procedures listed above with the ones you used in your most recent project. Comment on any similarities and differences you find.

Applying standards
Risk assessment
In any product manufacturing situation, hazards must be controlled. A hazard is a source of potential harm or damage or a situation with

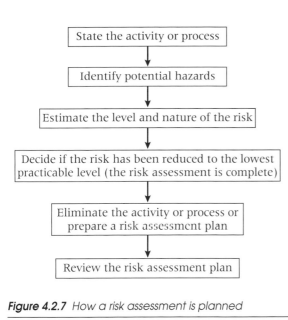

Figure 4.2.7 How a risk assessment is planned

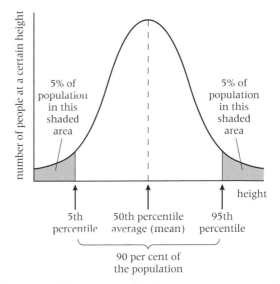

Figure 4.2.8 *Anthropometric data uses measurements representative of 90 per cent of the population (the fifth to ninety-fifth percentile)*

potential harm or damage. Hazard control incorporates the manufacture of a product and its safe use by the consumer. As part of the Health and Safety at Work Act 1974 risk assessment is a legal requirement for all manufacturers in the UK. Its use is a fundamental part of any quality management system. The main purpose of the Health and Safety at Work Act is to control risks and to enable decisions to be made about the level of existing risk control. Figure 4.2.7 shows how a risk assessment may be planned.

A risk has two elements:

- the possibility that a hazard might occur
- the consequences of the hazard having taken place.

Task
Using the risk assessment plan shown in Figure 4.2.7, draw up a risk assessment plan for your own product.

Ergonomics

SIGNPOST
'Anthropometics and ergonomics' Unit 3B1 page 130

Ergonomics is the application of scientific data to design problems. This means applying the characteristics of human users to the design of a product – in other words matching the product to the user. In order to do this, data about the size and shape of the human body is required – this branch of ergonomics is called anthropometrics.

Anthropometric data must take into account the greatest possible number of users. This data exists in the form of charts, which provide measurements for the 90 per cent of the population which falls between the fifth and the ninety-fifth percentiles (see Figure 4.2.8). Consumer products ranging from tools to storage units are generally easier, safer and more efficiently used if anthropometric data is used in their design.

Relevant legislation on the rights of the consumer when purchasing goods
Consumers are protected by a body of law called 'statutory rights'. This sets out what consumers should *reasonably* expect when buying products. The body of law includes legislation such as the Sale of Goods Act 1979, the Supply of Goods and Services Act 1982 and the Sale and Supply of Goods to Consumers Regulations 2002.

Statutory rights protect consumers' 'reasonable expectations' when buying products in a shop, market, on the doorstep, by mail order, through direct response TV, leaflets, magazines and newspaper advertisements or on the Internet. In fact, it does not matter where or how goods are bought, consumers' statutory rights remain the same.

Task

Anthropometric data is fine for the 90 per cent of people who fall between the fifth and ninety-fifth percentiles.

a) Explain what is meant by the fifth and ninety-fifth percentiles.

b) Discuss what problems the remaining 10 per cent of people might find in using products.

c) Anthropometric data was collected some years ago. Explain how changes in the sizes and shapes of people might impact on the design of products.

Statutory rights

As a consumer you expect any product that you buy to look and perform satisfactorily. In order to meet the Sale of Goods Act 1979 products must satisfy the following conditions:

- They must be 'as described', 'of satisfactory quality' and fit for any purpose which the consumer *makes known* to the retailer. For example, if you are told a computer game will work on your particular machine it must do so, or you have good reason to complain.
- Products are of satisfactory quality if they reach the standard that a reasonable person would consider satisfactory, taking into account their price and how they are described. For example, if a product is described as being scratch-proof and scratches appear during reasonable use, you have good reason to complain.
- Satisfactory quality covers fitness for purpose, appearance and finish, freedom from minor defects, durability and safety.

Under the Sale of Goods Act 1979 it is the *retailer*, not the manufacturer who is responsible. If a product is not of satisfactory quality the buyer is entitled, within a reasonable time, to return the product and get a refund. In order to build customer loyalty, many retailers exchange products that consumers decide are the wrong size, fit or colour, as long as there is proof of purchase.

Sale goods

The Sale of Goods Act 1979 still applies to sale goods, so notices that say 'no refunds on sale goods' are illegal and local authorities can prosecute traders who display them. However, you are not entitled to a refund on a sale item if you were told about some fault before purchase, did not see the fault, damaged the product yourself, bought the item by mistake or changed your mind about it.

The Sale and Supply of Goods to Consumers Regulations 2002

The Sale and Supply of Goods to Consumers Regulations 2002 became law in March 2003. It ensures that consumers enjoy a minimum level of protection when they buy goods from any country within the European Community. The new legislation states that:

- every consumer has the right to a repair or replacement for faulty goods
- if the consumer requests a repair or replacement, then for the first six months after purchase it will be for the retailer to prove that the product was not faulty when sold
- after six months and until the end of six years, it is up to the consumer to prove that the goods were faulty when sold. This does not mean, however, that a product has to last for six years as the life expectancy of products will vary.

Task

Before you buy

Choose two mid-priced products that you recently bought. For each product answer the following questions:

- Did you think through buying it?
- Is it what you really wanted?
- How much did you want to pay?
- How did you pay – cash, credit or a different way?
- Did you shop around?
- Did you compare prices?
- Did you ask about after-sales service?
- Could it have been delivered?
- Was it an 'impulse buy'?
- Did you buy it from a trader who is fair?

Write a report on the two products, summarising the above points. Compare your report with a colleague.

If things go wrong

If a product is not of a satisfactory quality when you buy it or when you try it out, you can reject it and get your money back.

- You need not accept a replacement, a free repair or a credit note.
- If you do agree to a repair, you can still claim your money back if the repair turns out to be unsatisfactory.
- Be careful if you accept a credit note as there may be restrictions on its use.
- You do not lose the right to reject a product by signing a delivery note.

If you have used the product for more than a trial and it goes wrong or is of unsatisfactory quality, you cannot reject it, but can claim reasonable compensation. This can be for loss in value of the product and for any harm caused by its use. Reasonable compensation is often repair, replacement or a price reduction.

If the product was faulty when you bought it and if it is reasonable for it to last that long, you can claim compensation for up to six years after purchase. If you bought the product on credit or with a credit card, you may also have rights against the credit company.

If you need to complain about a faulty product, you should quickly return to the shop, state the problem and how you want it dealt with. If the response is not satisfactory, write to the shop head office with details of the problem and a copy of your receipt. If you have lost the receipt use other evidence, such as a credit card or bank statement.

Help in solving problems
Sometimes consumers need advice and guidance on everyday shopping problems. Local authority trading standards officers enforce and advise on a wide range of legislation relating to consumer protection and deal with problems and complaints. Details of local Trading Standards departments can be found on the Trading Standards website: www.tradingstandards.gov.uk.

Task
Making a complaint
Sometimes a product does not perform as expected and a complaint must be made. Imagine you have recently bought a portable DVD player that fails to work properly. Describe the actions you would take to complain about the product:

a) in person
b) by telephone
c) in writing.

3. Advertising and marketing

Advertising and the role of the design agency in communicating between manufacturers and consumers

Advertising is any type of media communication that informs and influences existing or potential customers. The cost of advertising is a major marketing expense and in spite of all the money that is spent on it no one really knows what really makes it work!

Many large companies employ advertising agencies to run their advertising campaigns. Successful campaigns sell products and they are often the ones that we as consumers remember. There are said to be two approaches to advertising – hard sell and soft sell.

Hard sell
A hard sell advertisement has a simple and direct message about the benefits of a product. It projects a product's **unique selling proposition (USP)**; its unique features and advantages over a competitor's product. Hard sell works best with functional products such as computer equipment, e.g. in a recent campaign a company grinds to a halt because it does not use IBM computers.

Soft sell
On the other hand, soft sell advertisements promote a product's image, with which consumers can identify. This approach takes for granted the benefits of the product and focuses on creating positive emotional associations with the product. Soft sell is often used to create brand loyalty and is successfully associated with frequently bought items like detergent.

Task
Find one hard sell and one soft sell advertisement in a newspaper or magazine.

a) Devise a list of changes that would make the hard sell softer and the soft sell harder.
b) Explain how your changes would make the advertisements still appeal to the same target market groups.

Successful advertising campaigns
Successful advertising campaigns are often associated with brands. For example Jamie Oliver, the celebrity chef, has represented Sainsbury's supermarket for the past three years. Oliver's association with the Sainsbury brand is

said to have boosted the supermarket's profits by £153 million a year. He is said to 'appeal to everybody', because he has a passion for quality food and wants to make it available and enjoyable to everyone every day. The effect of Oliver on sales is said to be so great that when he recommended spreading truffle butter under the skin of Christmas turkeys, more than 50 000 jars of truffle butter were sold.

Advertising standards

The Advertising Standards Authority (ASA) regulates all British advertising in non-broadcast media. The code has three basic points which state that advertisements should:

- be legal, decent, honest and truthful
- show responsibility to the consumer and to society
- follow business principles of 'fair' competition.

The role of the media in marketing products

The main aim of marketing through the media is to influence customers' buying decisions, to help sell more products and promote a good public image for a company. Paid-for media options include:

- the press – newspapers, magazines and direct mail
- broadcast media – television and radio
- cinema
- outdoor – posters and sports grounds
- electronic – direct e-mail and the Internet.

An important source of information when choosing an advertising medium is a large-scale regular survey called the Target Group Index (TGI). This researches the consumption patterns of a representative sample of consumers – asking questions about the products they buy and use, what they watch, read and listen to. Subscribing to this survey allows a marketing organisation to match its target market with the media it uses most.

Table 4.2.2 is a guide to the strengths and weaknesses of the major types of media.

Tasks

1 Collect a variety of advertisements from magazines and newspapers. For each advertisement explain:

- the image given about the product – is it luxury, bottom end of the market, or something in between?
- the target market, relating to demographics and lifestyle
- any 'hidden' messages about the product – the use of emotions, concern, fear, compassion, persuasion, politics
- any 'values' attached to the product
- if it is hard sell or soft sell
- if the product is advertised in other media or is part of wider campaign. If so, do any other adverts differ? How?

2 You need to promote your coursework product to the consumers in your target market group.

a) Describe the features and characteristics of your product.
b) Describe the buying behaviour and lifestyle of your target market group.
c) Choose two different media for an advertising campaign and explain why each is appropriate for your product.
d) Explain how an Internet marketing campaign could help you reach your target market.

Table 4.2.2 *Strengths and weaknesses of the major types of media*

	Strengths	Weaknesses
Television (around 33% of UK advertising expenditure)	• High audiences, but spread over channels • Excellent for showing product use	• Short time span of commercials is limiting • High wastage – viewers not in target market
Newspapers and magazines (around 60% of UK advertising expenditure)	• Can target the market with detailed info • Can get direct response (reply coupon)	• Can have a low impact on consumers • Timing may not match marketing campaign
Radio (around 2–3% of UK advertising expenditure)	• Accurate geographical targeting • Low cost and speedy	• Low numbers compared to other media • Listen to it in the background to other tasks
Posters (around 4% of UK advertising revenue)	• More than 100,000 billboards available • Relatively cheap	• Seen as low impact/complicated to buy • Subject to damage and defacement

Market research techniques

Questions such as 'Who are my customers?' and 'What are their needs?' must be answered before any decision is made about what to make, what to charge, how to promote it and how to distribute it. In order to find answers to such questions, it is necessary to undertake marketing research. This involves the use of market research techniques to identify:

- the nature, size and preferences of current and potential target market groups and sub-groups (called **market segments)**
- the buying behaviour of the target market group
- the competition – and its strengths and weaknesses (this includes pricing and marketing policies)
- the required characteristics of new products – matching these to target market needs (this information is also used to improve existing products and to identify 'gaps' in the market)
- the effect price changes might have on demand – how sales would be affected by a price increase, how the price compares to that of a competitor, the price to set for a new product
- trends in design, colour, demographics, employment, interest rates and inflation.

Conducting marketing research

Marketing research comes from two types of sources:

- Primary sources provide original research, from things like internal company data, questionnaires and surveys.
- Secondary sources provide published information from things like trade publications, commercial reports, government statistics, computer databases. Other sources of secondary research include the media and the Internet.

When conducting primary research to find out, for example, about the size and buying behaviour of a target market group, two types of data can be collected: quantitative and qualitative.

Quantitative research

This kind of research often uses a survey to collect data about *how many* people hold similar views or display particular characteristics. Normally, the information is collected from a small proportion (a 'sample') of a target market group. The views of the whole target market group are then based on the responses from the sample.

Qualitative research

This kind of research collects data about *how* people think and feel about issues or why they take certain decisions. Qualitative research explores consumer behaviour and is conducted among a few individuals. It is often used to plan further quantitative research to see if the views of a few are representative of the whole target market group.

Surveys

A survey is a way of collecting quantitative data, often about behaviour, attitudes and opinions of a sample in a target market group. The most effective way to conduct a survey is to follow a series of key activities.

Task

There are four main types of survey – interview, telephone, mail and self-completion surveys. They all have benefits and disadvantages. Some of the benefits are listed in the table below. For each of the four survey types, list and explain the disadvantages of using that particular method.

Table 4.2.3 Benefits of different survey types

Survey type	Benefits
Interview	The most reliable way to get data using a questionnaire. Questions can be clarified by the interviewer
Telephone	Quick and inexpensive if the questionnaire is short
Mail	Usually very inexpensive (no interviewer or phone costs). Can use longer questionnaires to reach a wide population
Self-completion surveys	The least expensive. Can distribute questionnaires to a captive audience on trains and planes

Questionnaires

The design of any questionnaire is critical in order to get the kind of answers that allow decisions to be made. Always test a questionnaire before using it. In general, shorter questionnaires are usually better than long ones. The wording of questions is important too. They should be:

- relevant – only include questions that target the information required
- clear – avoid long words, jargon and technical terms; questions must be easy to understand

> **What is your age? (tick one only)**
> Under 18 ☐   18–64 ☐   65 or over ☐
>
> **Which types of kettles do you prefer?**
> **(tick as many as apply)**
> Jug ☐   Stainless steel ☐   Patterned ☐
>
> **Where did you buy your kettle?**
> Supermarket ☐   DIY store ☐
> Department store ☐   Mail order ☐
> Don't know ☐
> Other (please specify) _____

Figure 4.2.9 *Multiple-choice questions from a questionnaire on kettles*

- inoffensive – take care with questions about age, social class, salary, ethnicity, so as not to cause offence (use these types of question at the end)
- brief – short questions of less than 20 words
- precise – each question should tackle one topic at a time
- impartial – avoid 'leading' questions that influence the answer.

Types of questions
- Open-ended questions, such as 'What is your opinion of the C5 car?', allow very wide and ambiguous answers that may not be useful.
- Closed questions provide a limited number of optional answers to choose from. They are easier to answer and are often used in surveys. Avoid offering two options such as yes/no because they are of limited use.

Multiple-choice questions offer a range of different types (see Figure 4.2.9).

Task
Imagine you are designing a folding stool for young people going camping. Draw up a questionnaire to find out their requirements.

Product analysis

SIGNPOST
'Developing product analysis skills' Unit 1 page 12

As we have seen in Unit 1, product analysis is an important tool for the designer. It enables the analysis of a competitor's product and the development of specifications for new products.

Task
Planning a shop report
A shop report can involve:

- window shopping
- going into shops to look for trends, themes, styling and colour ideas
- going to art galleries and museums to look for ideas and design information

Plan a shop report on interior domestic lighting. Use your research to match the design, style and price of desk lamps to the potential customer. Sketch the products and record the materials and components used.

Test marketing (test selling)
The purpose of **test marketing** is to find out potential problems with a new product before its full-scale marketing. It is done under real market conditions to find out customer and retailer reactions to the product – who are the buyers? how often do they buy? This enables forecasts to be made of future sales and profitability and enables production planning to take place.

Taking risks
The risk of not doing any research is high, because it can mean making the wrong decisions. Imagine that a company is thinking about adding a new product to its range, but is put off by the high cost of research. The company has the option of launching the product on the market and risking failure rather than funding research costs. A product failure could undermine the reputation of the company and its existing products. Should it decide to undertake the costly research or risk launching the product?

The two case studies below illustrate the importance of matching the product to the target market.

The C5 car
The image of the one-seat, three-wheeler C5 came somewhere between a car and a bicycle. The C5 designer, Sir Clive Sinclair, admitted that he did not believe in market research, because demand could not be assessed for an innovative product that was not yet produced. Sinclair predicted sales of 100,000 C5 cars a year and planned to start production in the autumn of 1984. Any research that was done was mainly connected with product development. Sixty-three families were asked to try out the prototype car.

There was no test marketing of the C5, which

was launched in January 1985. The Consumers' Association tested the C5 and produced the following criticisms:

- On the road, the C5 was the same height as the bumpers of other cars – this made it difficult to see and increased the chance of accidents.
- The C5's low height meant that exhaust fumes, spray from rain and dazzle from the headlights of other cars were in the direct line of vision of the C5 driver.
- The C5's headlight beam was not strong enough, the horn was too quiet and mirrors and indicators were optional extras.
- There was no reverse gear, so drivers had to get out of the C5 to move it backwards.
- The top speed of the car was 15mph – this caused problems even where the speed limit was 30mph.
- The C5 only went for 15 miles on a fully-charged battery, not the 20 miles as was planned.
- Drivers had to pedal on hills because the motor kept cutting out.

After only three months, production was cut back by 90 per cent. It was stopped within six months, after only 14,000 cars had been produced. Could it have been that the marketing was wrong? Many consumers thought that the C5 was a 'fun' car, rather than a serious form of transport to rival the moped.

Coca-Cola

The Coca-Cola Company has one of the most impressive marketing operations in the world and one of the most recognised trademarks. The success of Coca-Cola rests on the belief in the product and that consumer demand drives everything the company does. The Coca-Cola brand projects an emotional attachment that has been used to great effect, such as in a recent TV advertising campaign featuring the Diet Coke 'hunk' outside an office block.

Coca-Cola's marketing philosophy is that a global brand is a local brand many times over. The secret of being local is for the brand to connect with the individual. Coca-Cola achieves this by using the product to bond people to positive memories, so that Coke becomes more than a product, it becomes a way of life, 'the real thing'. Just recently Coca-Cola has been working on what the brand should stand for and it has found that Coke has a 'magic', which relates to the role it has in people's lives. Coca-Cola believes everyone has a 'Coke story' and it has used storytelling to create another advertising campaign called 'life tastes good'.

Figure 4.2.10 *The Coca-Cola trade mark is one of the most recognised in the world*

Task

Read the two case studies then answer the following questions:

a) Describe your reactions to the C5 case study. Do you think it is humorous or tragic? Justify your answer.
b) Describe what went wrong with the C5 design and development. Explain how market research could have resulted in a different outcome. Describe what this outcome might have been.
c) Describe how the Coca-Cola marketing strategy matches the product to the target market.

The marketing research process

There are three key stages in the marketing research process. These are planning, implementation and interpretation.

Planning

Identify a clear reason and purpose for marketing research. This reason usually relates to a problem, an issue or opportunity for design. Once the reason is found (often in the form of a design brief), it is used as a starting point for research – what data needs to be found out and how to collect it.

Implementation

There are many ways of collecting data, depending on the research plan and what needs to be found out. Surveys and questionnaires can be useful, but can sometimes be expensive. Always think about alternatives.

Interpretation

The information that is created when data is collected should be interpreted in relation to the design problem. All findings should be used to influence decision making.

Task

Planning market research
Read the marketing research process above.

a) Choose a target market group for the C5 car.
b) Write a design brief to include purpose of the car, what it should do and the range of potential users.
c) Draw up a market research plan that includes planning, implementation and interpretation.

The basic principles of marketing and associated concepts

SIGNPOST
'Design and marketing' Unit 3B1 page 123

Marketing involves anticipating and satisfying consumer needs while ensuring a company remains profitable. The main objectives of marketing include:

- generating profit
- developing sales
- increasing market share
- diversifying into new markets.

In order to do this companies must create opportunities for meeting customer needs, for managing change within design and production and for promoting a corporate image. The main method of achieving this is through a marketing plan.

Marketing plan

A good sales and marketing plan involves developing a competitive edge through providing reliable, high-quality products at a price customers can afford, combined with the image they want the product to give them. This is sometimes called lifestyle marketing.

The basic structure of a product marketing plan includes:

- background and situation analysis, including **SWOT** (product strengths, weaknesses, opportunities and threats from competition) and **PEST** (political, economic, social and technological issues)
- information on markets, customers and competitors
- a plan for action and advertising strategies
- planning marketing costs
- time planning – the best time to market the product to an achievable timetable
- a plan for monitoring the marketing.

Target market groups

Companies supply their goods to customers. A market consists of all the customers of all the companies and organisations supplying a specific product, for example the car market or the domestic lighting market. As a result of undertaking marketing research, some companies decide that they cannot possibly supply all the potential customers in their markets – maybe the market is very large, geographically scattered, maybe the competition is strong or customer needs too varied. In this case companies have to decide which types of customers to aim for and then to target their products at them – at a selected part of the market (known as the market segment). The process of identifying market segments and developing products for it is called target marketing.

Consumer demand and market pull

Customers in a market demand or 'pull' products and services to satisfy their needs. The job of the marketing department is to maximise the demand for its own company's products. Existing customers must be satisfied with these products and not be tempted to buy from other suppliers. At the same time the customers of other suppliers need to be persuaded to change their product brand loyalty. Further to this, potential new customers need to be attracted, in order to expand the company's market share.

Task

Identify three products that you think have been developed as a result of market demand. Identify the characteristics of these products.

Lifestyle marketing

Lifestyle is used as a basis for target marketing. Different lifestyles have been identified by investigating the geographic and demographic characteristics of the population. People with similar demographic characteristics often live in similar types of houses and have similar lifestyles; for example, young professional people living in towns and cities and who are either working or studying to improve their lifestyle form one group. They tend to be young, highly educated, live in high-status areas and have a high level of mobility. This group may then be broken down into further types, such as well-off town and city areas, singles and young working couples, furnished flats and bedsits, younger single people.

Lifestyle marketing is the targeting of these potential market groups and matching their needs with products. Market research identifies the buying behaviour, taste and lifestyle of these potential customers. This establishes the amount of money they have to spend, their age group and which products they like to buy. New products can then be developed to match their needs.

Brand loyalty

Branding is a key marketing tool for many manufacturers. A 'brand' is a marketing identity for a generic product that sets it apart from its competitors. The brand name protects and promotes the identity of the product, so that it cannot be copied by its competitors. Typical brand names include Coca-Cola, Nike and Microsoft. A branded product usually has additional features or added value over and above other generic products – something that makes the product 'special' in the eyes of consumers.

Figure 4.2.11 *The Nike 'swoosh' is such a recognised brand that the company does not even have to include the word Nike!*

Advantages of branding for the customer

For the customer, buying branded products provides an expected and reliable level of quality. Brand names also give consumers benchmarks when making their own purchasing decisions. For example, some people might use the AppleMac personal computer (see Figure 3.2.1 on page 108) as a reference brand against which other computers are compared. The status and image of a strong brand and brand loyalty can therefore be powerful influences over which products consumers buy. For consumers, the benefit of knowing which brands provide certain reliable or good quality products can also save time when deciding which product to buy.

> ### Task
> Investigate the design features of a range of branded hand tools sold in your local DIY store. Explain how this product range promotes a brand image and why it might be attractive to the target market.

Competitive edge and product proliferation

If there are several companies producing a wide range of similar products for the same customers with similar needs, the only criterion upon which customers can base their buying decision is price. In theory the most expensive products would not sell, but this is rarely the case – since most customers are not aware of every product and price on the market.

To ensure that their products have a successful market share, many manufacturers try to make their products different from a competitor's. This involves creating unique features for the product in order to give it a 'competitive edge'. For example, some manufacturers might offer 'special' features or different levels of quality. Different price levels can then be set; these are often based on the value that customers might put on these features or qualities.

> ### Task
> Compare the cost and design features of a branded product with the cost and design features of a similar 'own brand' product. Explain how the unique features of the branded product give it a competitive edge over the own brand.

Price range and pricing strategy

Price is one of the most important aspects of marketing because a product's price affects profit, the volume of products manufactured, the share of the market and the image of the product. We've probably all heard the term 'cheap and nasty'!

Tasks

Market share and promotional gifts

1 The aim of marketing departments and advertising companies is to increase the market share of brands and products. Research has shown that in advertising success breeds success. Increasing the market share of an already successful product or brand is easier than for a less successful or less well-known brand. Try to explain the reasons for this.

2 Direct marketing often includes mail shots with free 'gifts' or samples, often with a coupon that provides money off for the next purchase. Explain why you agree or disagree with this kind of marketing.

The key to successful pricing is the attractiveness of the price to the customers. The concept of price, value and quality are difficult to separate.

How much we are prepared to pay for a product depends on how much we value it. The justification for a higher price may depend on the following:

- a product's extra features, characteristics or innovative design
- the perceived quality of the product
- the increased reputation of the product brought about by advertising and promotion
- the possibility of paying by credit, which justifies a higher final price
- the guarantee of a specific delivery date and easy access to the product through mail order or the Internet.

Distribution

The distribution of products covers a variety of operations. They make the product available to the maximum numbers of target customers at the lowest cost. The way that technology and communications have changed in the past few years has had an enormous impact on the distribution of products and the way we shop.

Task

List three ways that consumers shop these days in comparison with shopping in the 1970s. Explain the main factors that have brought about these changes in shopping.

4. Conservation and resources

Environmental implications of the industrial age

One of the greatest problems relating to the industrial age is the consumption of non-renewable, finite resources which will eventually be exhausted (see Figure 4.2.12). Recognition of this problem leads to an understanding of the need for conservation and the better management of resources. It also needs to lead to a better understanding of the meaning of 'design' and the **'purchase-attraction'** culture.

The future – designing where less is more

There are a number of questions relating to product design that need to be answered. For example:

- What will be the aims of product design in the future?
- How can we manage the changes necessary to our technological-industrial society to make life on earth bearable in the future?

One of the greatest problems for product designers of the future will be how to design with the environment in mind. Previous sections of this unit discussed mass production and the marketing of products. We saw that the aim for

Figure 4.2.12 *Open cast mining is now used to obtain the bulk of most minerals, although the environmental damage is severe*

most manufacturers is to sell as many products as they can in order to make a profit. Designers now and in the future will need to decide if this is an ethical way forward.

Influencing the future

The key question is how design can influence the environmental safety of products and how it can contribute to a reduction in the number of products made.

At present we have what is called a 'purchase-attraction' culture, which results in the proliferation of products that we see in the market place today. In the future this may have to make way for a culture that supports long-term use and the conservation of resources. Changing the purchase attraction culture could be achieved by:

- changing our attitude to products from purchase-attraction to their long-term use and usefulness
- developing products that would not be bought, but that remain the property of the manufacturer
- paying for the use of the product and its maintenance
- returning products to the manufacturer to be serviced, repaired, recycled and reused.

It is the task of all of us, including designers and technologists, to find starting points for changes to our purchase-attraction culture in order to ensure the future of the planet. Good design has an ethical and moral value – something that you have been asked to take account of, for example, when drawing up a design specification. Industrial production is becoming an increasing problem, with the production of more and more products. This is placing an enormous burden on the environment.

Tasks

1 Explain what is meant by the term 'purchase-attraction' culture.
2 Put forward arguments for and against mass production and the marketing of products.

Conservation

Conservation is concerned with the protection of the natural and the manufactured world for future use. In urban areas, for example, buildings may be protected because of their historic interest. In rural areas, plant species, animal habitats and landscapes may all be protected.

Resource management

Conservation is also concerned with the sensible management of resources and a reduction in the rate of consumption of non-renewable resources, such as coal, oil, natural gas, ores and minerals. The aim of conservation is to achieve sustainable development. The 1987 Brundtland Report, *Our Common Future* (World Commission on Environment and Development), defined sustainable development as:

'development that meets the needs of the present, without compromising the ability of future generations to meet their own needs'.

Efficient management of resources includes:

- using less wasteful mining and quarrying methods
- making more efficient use of energy in manufacturing
- reducing fuel consumption in motor vehicles
- using cavity and roof insulation in buildings
- using low-energy light bulbs.

The use of non-renewable raw materials and fossil fuels during the manufacturing process

SIGNPOST
'Environmental costs' Unit 4A page 195

Many modern products are made from non-renewable resources such as metals and plastics. The electrical energy used in their manufacture comes from coal, gas, oil or nuclear power. The management of these finite resources will increasingly become the responsibility of us all. Existing British, European and international legislation already places demands upon companies to design and manufacture products with the environment in mind. In this respect, product designers need to consider:

- reducing the amount of materials used in a product
- using efficient manufacturing processes that save energy and prevent waste
- reusing waste materials within the same manufacturing process
- recycling waste in a different manufacturing process (see Figure 4.2.13)
- designing for easy product maintenance, so that parts can be replaced, without the need to dispose of the whole product at the end of its useful life
- designing the product so that the whole or parts of it can be reused or recycled.

Figure 4.2.13 *Glass is one of the most cost-effective materials to recycle*

Figure 4.2.14 *Several large wind farms have been built in Europe and the USA, principally at windy coastal sites*

Task

We are all responsible when it comes to the environment. Explain your views on the following issues:

a) The need for manufacturers to continually produce new products.
b) The use of lifestyle marketing to encourage consumerism.
c) Buying cheaper, short-lifetime products rather than more expensive but more durable ones.
d) Throwing away products because they are old fashioned or out of date.
e) Recycling or reusing products.

Tasks

1 Explain your views on the use of renewable energy when compared to using coal, oil or gas.
2 Describe the benefits and disadvantages of renewable energy in relation to set-up costs, accessibility, production processes and environmental impact.

Renewable sources of energy, energy conservation and the use of efficient manufacturing processes

Renewable resources are those that flow naturally in nature or that are living things which can be regrown and used again (see Table 4.2.4). These include the wind, tides, waves, water power, solar energy, geothermal, biomass, ocean thermal energy and forests (see Figure 4.2.14).

Forests are renewable as long as they are not used faster than they can be replaced. In recent years, the indiscriminate destruction of the world's rainforest has led to a severe shortage of tropical hardwoods, such as Jelutong from South-east Asia. In order to conserve valuable renewable resources such as these, manufacturers are encouraged to use only those woods grown on plantations or in managed forests.

The use of efficient manufacturing processes

Even the most industrially advanced economies still depend on a continual supply of basic manufactured goods. This production will continue to require large amounts of raw materials and energy. Product manufacturers can contribute to sustainable development and to reducing costs by using more efficient manufacturing processes. This can often be achieved through redesigning an existing product.

Reducing costs and environmental impact through the redesign of an existing product

'Good old *Yellow Pages*' characterised twenty years of advertising for the most commonly owned book in the country, used by almost 27 million households. Although *Yellow Pages* was the market leader, its design had not changed for

Table 4.2.4 Renewable energy sources

Renewable energy source	Process	Advantages	Disadvantages
Wind	Power of wind turns turbines	Developed commercially Produces low-cost power	High set-up cost Contributes small proportion of total energy needs Wind farms sometimes seen as unsightly
Tidal	Reversible turbine blades harness the tides in both directions	Occurs throughout the day on a regular basis Reliable and non-polluting Potential for large-scale energy production	Very high set-up cost Could restrict the passage of ships Could cause flooding of estuary borders, which might damage wildlife
Water	Running water turns turbines and generates hydro-electric power (HEP)	Clean and 80–90% efficient	High set-up cost Suitable sites are generally remote from markets Contributes small proportion of total energy needs of an industrial society
Solar	Hot water and electricity generated via solar cells	Huge amounts of energy available. Could generate 50% of hot water for a typical house Relatively inexpensive to set up	Cost of solar cells Biggest demand in winter when heat from Sun is at its lowest
Geothermal	Deep holes in Earth's crust produce steam to generate electricity	Provides domestic power and hot water	Only really cost-effective where Earth's crust is thin, e.g. New Zealand, Iceland
Biomass	Burning of wood, plant and waste generates heat	Produces low-cost power	Environmental pollution Potential for deforestation

15 years. The covers were unimaginative and the typeface was too dense for easy reading.

The new *Yellow Pages* design needed to keep its traditional qualities of reliability and accuracy, reduce costs by saving space *and* increase its advertising. *Yellow Pages* ran to 76 local editions and used up vast amounts of raw materials. The new design included:

* an emphasis on the *Yellow Pages* brand, necessary to confirm its number-one status.
* a new cover design, a smaller, easier-to-read 5.5 point font and a new layout with more information fitted on to each page. This reduced the size of the index, which was cut down from twenty-three pages to seven.
* an emphasis on the local character of each edition, using cartoon-style images to inject humour into the text.

A national survey found that 94 per cent of people thought the new look *Yellow Pages* was easier to use, 98 per cent liked the new covers and advertising sales grew. The reduction in paper use saved £1.5 million per edition, showing that good design can cut costs and boost profits.

Task

a) Identify a product that you think is badly designed. Give reasons for your choice.

b) Using annotated diagrams, explain how the product could be redesigned to improve efficiency of its manufacture.

New technology and environmentally friendly manufacturing processes

SIGNPOST
'New materials, processes and technology'
Unit 3B1 page 119

Redesigning a product is one way of achieving efficiency in manufacturing. For many companies this also improves their environmental performance and helps them to increase profits.

Since 1994 one organisation that has been helping UK companies to do just this is Envirowise (see Figure 4.2.15). This is a joint initiative of the Department of Trade and

Figure 4.2.15 *The Envirowise logo*

Industry (DTI) and the Department for Environment, Food and Rural Affairs (DEFRA). Envirowise aims to help manufacturing companies improve their environmental performance and increase their competitiveness.

The main themes of the Envirowise programme are waste minimisation and cost-effective **cleaner technology**:

• Waste minimisation often generates significant cost savings, through the use of simple no-cost or low-cost measures.

• Cleaner technology is the use of equipment or techniques that produce less waste or emissions than conventional methods. It reduces the consumption of raw materials, water and energy and lowers costs for waste treatment and disposal.

Reducing waste in the paper and board industry

The paper and board industry uses large amounts of non-fibrous materials in the paper-making process, to improve machine performance and paper quality. Non-fibrous materials include cleaning chemicals for papers and machines, fillers to improve paper opacity, sizing agents, dyes and optical brightening agents, and paper coating chemicals. The poor control of non-fibrous materials costs money and has been found to result from some of the following:

• a lack of operating manuals
• specifying the wrong type and dose of chemicals
• using incorrect sizes of dosing pipes
• using incorrect pump speeds
• incorrect connections between chemical storage systems and papermaking machinery
• poor labelling on storage tanks, pumps and pipe work.

It is estimated that a 1 per cent reduction in the use of non-fibrous materials would save the paper and board industry around £4 million per year. The improved management of non-fibrous materials can:

• reduce raw materials costs
• reduce the generation of waste
• reduce waste disposal costs
• increase the amount of saleable paper produced
• reduce machine downtime
• reduce production losses by up to 5 per cent.

To get the most out of HVLP spray guns, operators need to change the way in which they spray. They should:

• Spray a little closer (150–300 mm is a good distance).
• Keep the spray at right angles to the object being sprayed.
• Reduce the paint flow or spray a little faster, to save material.

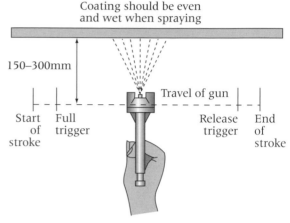

Figure 4.2.16 *Useful tips for using the new HVLP spray guns*

Cleaner technology using high volume, low pressure spray guns

A good example of cleaner technology is the use of high volume, low pressure (HVLP) spray guns which reduce solvent use and meet the requirements of the 1990 Environmental Protection Act. Compared with conventional spray guns, HVLP guns atomise paint using a higher volume of air at a lower pressure. Paint spray is less likely to bounce off the work and overspraying is reduced, making HVLP spray guns as effective as conventional guns in producing a good finish.

The new HVLP guns use less paint and solvents than conventional spray guns and use less compressed air, giving reduced running costs through energy savings. The benefits of the use of modern HVLP spray guns include:

• an initial reduction in paint use of up to 21 per cent
• a short payback period on the purchase price of the spray guns
• substantial savings in the cost of paint
• reduced environmental impact.

Cleaner design

Cleaner design is aimed at reducing the overall environmental impact of a product from 'cradle to grave'. It uses life cycle assessment (LCA) to evaluate each stage of the product life cycle from raw materials to end-of-life.

Table 4.2.5 Cleaner design using LCA

Raw materials	Reduce materials use Use materials with less environmental impact
Manufacture	Use fewer energy resources Produce less pollution and waste
Distribution	Reduce the impacts of distribution
Use	Use fewer energy resources Cause less pollution Optimise functionality and product life
End-of-life	Make reuse and recycling easier Reduce the environmental impact of disposal

Case study: Avad contemporary furniture

Avad is a small company that manufactures high-quality contemporary furniture. It makes effective use of resources by:

- using sustainably managed hardwood
- developing sustainable construction methods, such as joints that require no adhesives or additional fixing materials.

This has resulted in production savings, minimal use of non-sustainable materials, increased product life and the possibility of greater material reuse at end-of-life.

These cleaner design features have helped increase demand for Avad's contemporary, environmentally sustainable products.

The importance of using sustainable technology

In the previous sections you have been introduced to different aspects of manufacturing that relate to the environment: using cleaner

Figure 4.2.17 Avad contemporary furniture

design and cleaner technology. These are both aspects of what is known as 'sustainable development'. This concept puts forward the idea that the environment should be seen as an asset, a stock of available wealth. If each generation spends this wealth without investing in the future, then the world will one day run out of resources. The concept of sustainable development includes a number of key concepts:

- that priority should be given to the essential needs of the world's poor
- meeting essential needs for jobs, energy, water and sanitation
- ensuring a sustainable level of population
- conserving and enhancing the resource base
- bringing together the environment and economics in decision making
- making industrial development more inclusive.

Sustainable development is a problem for the whole world and many countries are involved in trying to develop policies which support it. In the UK, for example, many government programmes are involved, such as Envirowise. Bio-Wise is another government initiative that supports and advises companies and organisations on developing sustainable practices that make use of biotechnology.

For more information on both initiatives, visit the Envirowise website on www.envirowise.gov.uk and the Bio-Wise website on www.biowise.org.uk.

Information about biotechnology can be found in the next section.

Tasks

1. Discuss the importance of using sustainable practices when manufacturing products. Include references to raw materials and manufacturing processes.
2. Explain how you could adapt sustainable technology to your own manufacturing.
3. Research further examples of good practice in manufacturing, using the Internet, CD-ROMs or libraries.

Management of waste, the disposal of products and pollution control

SIGNPOST
'Life cycle assessment' Unit 4A page 192

As we saw from the Envirowise Programme, waste minimisation often generates important cost savings, through the use of simple changes to manufacturing processes. There are three key approaches to reducing waste:

- reduce the amount of materials used in manufacture
- reuse materials in the same manufacturing process where possible
- recycle materials in a different manufacturing process if possible.

Reducing materials use

The possibility of reducing the amount of materials used to manufacture products is often found in processes that involve cutting and stamping shapes from sheet materials. For example, in can manufacture careful calculations must be made to limit the amount of aluminium used for making the circular tops of the cans. There are two ways of arranging the can top on a rectangular sheet, as shown in Figure 4.2.18. In a) the tops are in a square formation, each sitting in its own square of aluminium. In b) the tops are in a triangular formation, which is the closest that they can be packed together. When the scrap for each is worked out a) is found to produce a staggering 21.4 per cent scrap, while b) works out at only 9.3 per cent scrap. Clearly the placement of the can tops on the aluminium sheet will have an enormous impact on the amount of aluminium required to manufacture the cans.

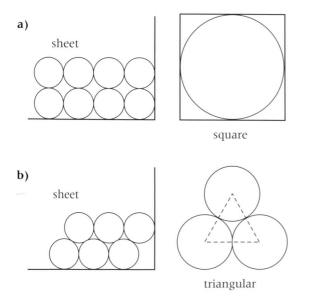

Figure 4.2.18 *Reducing the amount of aluminium required to make can tops depends on how they are arranged on the aluminium sheet*

The disposal of products and pollution control

The disposal of products when they have reached the end of their useful life is a major problem. Around 90 per cent of household rubbish in the UK is buried in landfill, five per cent is incinerated and only five per cent is recycled. One of the best options for products is, firstly, to create the need for fewer of them to be made and, secondly, to design for recycling.

For the disposal of industrial waste a simple solution is 'skip and tip' into either landfill or the sewers. Disposal by landfill is currently inexpensive and popular, but legislation is expected to enforce change. Pollution control is the responsibility of a variety of agencies which enforce the 1990 Environmental Protection Act (EPA). This Act introduced wide-ranging legislation with tight controls on the discharge of waste into air, water and land in the UK. It also reinforced the policy of 'polluter pays'. The aim of this policy is to restrict potentially harmful materials from entering the environment and to place greater responsibility on those generating, handling and treating wastes. Legislation is strictly enforced so that any company or organisation that causes pollution can be fined huge sums.

Task

Discuss the impact of the policy 'polluter pays' on product manufacture. Relate your discussion to one product type, such as PET bottles. Explain how product pricing might be affected.

Impact of biotechnology on manufacture

Biotechnology is the use of biological processes which make use of living proteins, called **enzymes**, to create industrial products and processes. These enzymes are the same kind that help us digest food, compost garden rubbish and clean clothes. The use of yeast in making bread and wine was one of the first examples of the use of biotechnology. Technologists also developed enzymes that could be added to detergents to improve their cleaning properties, resulting in 'biological' washing powders.

Biotechnology is being used in an increasing number of industries to provide efficient and ecologically sound solutions to environmental problems. Current applications of environmental biotechnology include the treatment of industrial air emissions; **aerobic** and **anaerobic treatment** and reed beds to treat industrial effluent; composting to treat domestic, industrial and agricultural organic waste; and **bioremediation** techniques to clean up contaminated land. Not only are many companies using biotechnology to increase their competitiveness, but they are using it to meet strict new environmental legislation and to improve their competitive edge.

Compared with more traditional methods, biotechnology can often produce better, faster, cleaner, cheaper and more efficient ways of doing things. One example, in the packaging industry, is the development of new biodregradable packaging material.

Biodegradable packaging material

Around 185,000 tonnes of polystyrene is used in the UK each year, most of which goes straight to landfill. The electronics industry is one of the largest users of polystyrene foam packaging, which is used to protect fragile products in transit.

A new type of biodegradable packaging material, called Biosystem™, has been developed as an alternative to polystyrene. The new packaging, made from renewable corn starch, can be composted in landfill with paper and cardboard. Biosystem™ is designed to compete on performance and cost with the polystyrene and polyethylene foams currently used for heavy white goods and hi-fi equipment. The new material, which will soon be on the market, has been featured on BBC TVs *Tomorrow's World* where it was tested for effectiveness at the Eden Project in Cornwall. The starch-based eco-friendly foam was able to protect a working television set in a fall of 1.5 metres and was shown to be biodegradable in water.

Figure 4.2.19 *Polystyrene packaging provides excellent insulation, but most goes straight to landfill*

Turning waste wood dust into garden compost

The UK furniture and timber industries generate around three million tonnes of waste wood dust each year. Wood dust is a serious health hazard and requires costly disposal to landfill.

A recent biotechnology project demonstrated how to turn waste wood dust into garden compost. The main difficulty in this process is the lacquers, sealers and solvents left in the dust after furniture manufacture. Adding microbes that feed on these substances overcomes this problem. Hardwood, softwood, wood chips and MDF dust can all be composted using a computer programmable system. This controls variables such as temperature, oxygen and airflow to produce uniform, clean compost.

This environmentally friendly composting system will not only cut the costs of disposal by landfill but will provide furniture manufacturers with a saleable product.

The advantages and disadvantages of recycling materials

> **SIGNPOST**
> 'Eco-design and environmentally friendly processes' Unit 3B1 page 120

The advantages for the environment of recycling are numerous. Recycling conserves non-renewable resources, reduces energy consumption and greenhouse gas emissions, reduces pollution and the dependency on raw materials.

For manufacturers the advantages of recycling are cost-related, with metal, glass and paper being the most cost-effective materials to recycle:

- Scrap non-ferrous metals are sorted into different grades for recycling, due to their relatively high commercial value.
- Steel and cast iron are graded by size, due to their lower commercial value.
- Items made of glass, rubber and plastics are more difficult to sort and have a much lower value than metals.
- Plastics, paper and glass recycling are growing industries.

Scrap materials from the manufacturing process are the most valuable because their materials content is known and they are easily available. The disadvantage of recycling scrap from old products is that the chemical or physical make-up of their materials is often very complex, so for some products recycling is too expensive or impossible. The result is that 'old' scrap from products has a lower commercial value.

Design for recycling

Under proposed European regulations, manufacturers may have to recycle electrical and electronic parts rather than disposing of them by landfill or incinerating them. By 2006 customers may be able to return electrical goods to stores for recycling. This will have a major impact on the way that products are designed.

With this in mind British engineers have developed a mobile phone that falls apart when heated, so that the component parts can be recycled. The phone is made from **shape memory polymers**, plastics that revert to their original form when heated. Different components of the phone will change shape at different temperatures, allowing them to fall off at different times, when they pass along a conveyor belt. The reusable parts can be recycled. The phones will only fall apart under extreme temperatures – so they will be safe if left in the car in the sun!

Recycling packaging

According to a recent government report the volume of plastic packaging in the UK is 1.7 million tonnes per year, of which less than

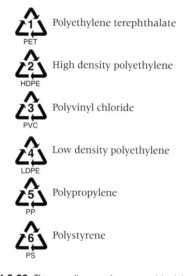

Figure 4.2.20 *The coding system used to identify different types of plastics*

15 per cent is recycled. The rest goes to landfill where it remains for at least a hundred years. An EC Directive on packaging waste, which is being adopted throughout Europe, sets a target of at least 50 per cent of plastic packaging to be recycled. Recycling is therefore a growing feature of the packaging industry and local government. Although the packaging industry helps by using a coding system to identify the type of plastic used in packages (see Figure 4.2.20), many local councils currently do not have the facilities to recycle more than the basic paper, aluminium and glass. For example, there are mountains of old refrigerators waiting to be disposed of because there are not enough recycling centres to deal with them.

As a result, although the potential for recycling plastic packaging is enormous, tonnes of plastic packaging still ends up in landfill. Different types of plastic can be identified by putting them in a water tank. Polyethylene (HDPE and LDPE) floats on the surface, so is easily identified. Although polyethylene terephthalate (PET) and polyvinyl chloride (PVC) sink, they can be identified chemically using sensors.

PET is the main type of plastic used in recycling. PET bottles can be chopped, melted, extruded and blow moulded into new products. PET can also be extruded into polyester fibres and yarns for knitting into jumpers. It takes around 27 PVC bottles to make one polyester fleece sweater!

Another solution to the plastics problem is the degradable plastic bag. The new d2w™, a degradable polythene packaging, is already used by Somerfield and Kwik Save supermarkets. The manufacturer, Symphony Plastic Technologies, is in talks with another eight large UK retailers and expects to supply more than 100 councils with degradable green waste bags. The manufacturing technology for degradable polythene uses a special additive that helps break down the polythene leaving only water, carbon dioxide and environmentally safe biomass. The added bonus is that this new polythene can be recycled in the same way as non-degradable plastics.

Exam preparation

You will need to revise all the topics in this unit, so that you can apply your knowledge and understanding to the exam questions. Some questions may ask you to give answers related to a product example. If you cut out and save newspaper and magazine articles about products, it will help keep you up to date with the latest information.

In preparation for your exam it is a good idea to make brief notes about different topics, such as 'Relevant legislation on the rights of the consumer when purchasing products'. Use sub-headings or bullet point lists and diagrams where appropriate. A single side of A4 should be used for each heading from the Specification.

It is very important to learn exam skills. You should have weekly practice in learning technical terms and in answering exam-style questions. When you answer any question you should:

- read the question carefully and pick out the key points that need answering
- match your answer to the marks available, e.g. for two marks, you should give two good points that address the question
- always give examples and justify statements with reasons, saying how or why the statement is true.

Practice exam questions

1 a) Economic factors need to be taken into account when manufacturing products.
    Explain three of the following terms:
    - production chain
    - primary sector
    - secondary sector
    - tertiary sector.   (6)
    b) Describe how productivity is measured and its impact on profitability.   (4)
    c) Explain the issues a company needs to consider when developing a new product range.   (5)

2 a) Explain why hardwoods are more costly than softwoods.   (4)
    b) Outline the factors that cause costs to be lower in high-volume production.   (5)
    c) Explain the cost of making a product 'right first time'.   (6)

3 a) Outline what is meant by a 'quality management system'.   (5)
    b) Describe the following terms:
    - risk assessment
    - hazard.   (4)
    c) Explain what is meant by the term 'statutory rights'.   (6)

4 a) Explain **two** of the following terms:
    - quantitative research
    - qualitative research
    - surveys.   (4)
    b) Product development starts with the identification of a need. Discuss the factors that
       must be identified before product development can begin.   (6)
    c) Advertising is used to inform and influence potential consumers. Describe the
       characteristics of soft sell advertising used to promote everyday products.   (5)

5 a) Describe the basic structure of a marketing plan.   (5)
    b) Discuss **two** of the following terms:
    - target market group
    - market pull
    - brand loyalty.   (6)
    c) Explain how and why a manufacturer might attempt to give a product a 'competitive edge'   (4)

6 a) Describe **three** ways in which efficient management of resources could be achieved. (6)

b) Renewable sources of energy include tidal, geothermal and biomass. Explain which of these sources of energy would be most suitable for an environmentally conscious company to use. (3)

c) Explain, using an example for each, the difference between cleaner design and cleaner technology. (6)

Total: 90 marks

UNIT 4 B2 CAD/CAM (G403)

Summary of expectations

1. What to expect

CAD/CAM is one of the two options offered in Section B of Unit 4. You must study the same option that you studied at AS level.

2. How will it be assessed?

The work that you do in this option will be externally assessed through Section B of the Unit 4 examination paper. The questions will be common across all three materials areas in product Design. You can therefore answer questions with reference to **either**:

- a Graphics with Materials Technology product such as a toy robot
- a Resistant Materials Technology product such as a robotic arm
- **or** a Textiles Technology product such as a 3D knitted jumper.

You are advised to spend **45 minutes** on this section of the paper.

3. What will be assessed?

The following list summarises the topics covered in this option and what will be examined:

- Computer-aided design, manufacture and testing (CADMAT)
  - computer-integrated manufacture (CIM)
  - flexible manufacturing systems (FMS).
- Robotics
  - the industrial application of robotics/ control technology and the development of automated processes
  - complex automated systems using artificial intelligence (AI) and new technology
  - the use of block flow diagrams and flow process diagrams for representing simple and complex production systems

  - the advantages and disadvantages of automation.
- Uses of Information and Communications Technology (ICT) in the manufacture of products
  - the impact and advantages/disadvantages of ICT within the total manufacturing process.

4. How to be successful in this unit

To be successful in this unit you will need to:

- have a clear understanding of the topics covered in this option
- apply your knowledge and understanding to a given situation or context
- organise your answers clearly and coherently, using specialist technical terms where appropriate
- use clear sketches where appropriate to illustrate your answers
- write clear and logical answers to the examination questions, using correct spelling, grammar and punctuation.

There may be an opportunity to demonstrate your knowledge and understanding of CAD/CAM in your Unit 5 coursework. However, simply because you are studying this option, you do not have to integrate this type of technology into your coursework project.

5. How much is it worth?

This option, together with Section A, is worth 15 per cent of the full Advanced GCE.

Unit 4A + option	Weighting
A2 level (full GCE)	15%

1. Computer-aided design, manufacture and testing (CADMAT)

CADMAT

Systems that fully integrate the use of computers at every level and stage in the manufacturing process can be described as CADMAT systems. Unit 3 looked at the benefits of CAD and the advantages that CAM brings and you were introduced to the wider role that computers now play in all sectors of graphic design and manufacturing. In addition to their application in CAD/CAM, computers are also used extensively for decision making within CADMAT in a variety of ways because of the operational flexibility they provide. The graphic design and product manufacturing industries use computers for activities like:

- information control through the gathering, storage, retrieval, and organisation of data, information and knowledge
- simulations in which computer 'models' are used to help to provide answers to 'What if …?' questions; for example, 'What would be the consequences if we changed to a flexible manufacturing system compared to our present sequential system of manufacturing?'
- narrowing the field of design choices available using number-based and other analytical methods; for example, graphs can be used for analytical and mathematical purposes rather than simply as visual aids to explain data

- communicating and discussing product designs with clients
- managing the full range of manufacturing data including the tracking of components and stock
- controlling equipment and production processes to include the routine scheduling of maintenance or minimising the effects of machine **downtime** when production tools have to be replaced
- monitoring quality and safety to the appropriate national standards.

As with all systems, the efficiency of a CADMAT system is determined by the effectiveness of many interrelated sub-systems. A failure in any of the input, processing or output sub-systems leads to production lines that malfunction or operate below capacity. Work in progress can be delayed in many ways, especially when the required materials and components are unavailable, which means that suppliers need to be part of the production planning process. The complex relationships in the design and manufacturing process can be described in a systems diagram (see Figure 4.3.1). Maximising profit and reducing costs is important for all manufacturers but the global economy that has been made possible by the use

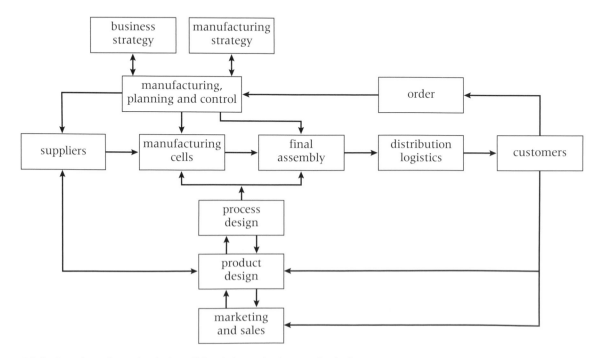

Figure 4.3.1 *A system flow chart describing integrated manufacturing*

of computers has created many new pressures. Increasingly, manufacturers have to compete with low-cost imports, satisfy customers who require shorter production runs and improved delivery times. In order for companies to compete for business more effectively they need to manage data in 'real time' by using customised computer programmes to manage orders and control production times.

Product data management (PDM)

Product data management (PDM) is a complete (holistic) data management system that aims to integrate all aspects of manufacturing from product modelling to the management of the business. All the design and production processing data is generated once and stored electronically on a secure database that can be accessed at different levels according to a specific need. PDM enables instant and rapid communications between departments, manufacturers and retailers across the world so that a product that has been designed in one country can be manufactured to the design specification in another.

Online order tracking is another example of the potential power of PDM because of the advances in information and communications technology (ICT) and the growth of the Internet as a way of selling goods and services across the world. **Telematics** is an ICT-based technology that allows a product to be managed from the receipt of a customer's order. The product will be tracked from development into manufacturing, on through delivery and finally to after-sales support via real-time feedback from telematic reporting systems. An everyday example of real-time feedback is the customer service section on an Internet website that sells CDs and DVDs. On this type of site it is possible for the customer, who has an unique access code, to track the progress of his or her order, when it was or will be dispatched and what the final costs of the transaction will be. If the customer requires further information, then an email facility is provided that allows an instant and low-cost dialogue with the company, especially if it is based in another country.

The advantages of using PDM include:

- increased productivity because of fewer bottlenecks
- improved quality control
- more accurate product costing
- more effective production control
- a reduction of work in progress and a smaller inventory of components and materials
- increased flexibility
- higher profits.

Production data from a manufacturing plant or a volume production line such as a printing press can be simultaneously merged and analysed along with other data, including product demand and financial figures. This means that raw materials and components for the production line are bought in as required. As a result, financial efficiency improves as the size of the stock inventory and its storage space reduces. Reduced business overheads equal increased profitability. This type of approach is known as 'just-in-time'.

Just-in-time (JIT)

Just-in-time (JIT) is a management philosophy which is applied in businesses around the globe. It was first developed in Japan in the 1960s by Taiichi Ohno within Toyota's car manufacturing plants as a 'quick response' method of meeting consumer orders with the minimum of delay, to the required level of quality and in the right quantity. It has also come to mean producing with minimum waste. Waste here is taken in a general sense to include such things as time and resources as well as raw materials (see JIT manufacturing below). JIT is achieved by a strategy of computer-aided inventory management in which raw materials, components or **sub-assemblies** are delivered from the 'supplier' just before they are needed in each stage of the manufacturing process. The advantages of JIT include reduced stocks of raw materials leading to a reduction in the area, volume and cost of storage space required. It also reduces the levels of finished goods kept in stock, waiting to be sold.

JIT 'suppliers' and 'customers'

A 'supplier' can either be the organisation that provides the original raw materials, components or stock needed for the product or it can be the previous stage of the manufacturing process (downstream). A 'customer' can either be the purchaser of the final product or it can be another stage of the manufacturing process further along the production line (upstream).

JIT manufacturing

This is a systems approach to developing and operating all the processes in a manufacturing system. JIT has been found to be so effective that it increases productivity, work performance and product quality, while still saving costs. Information flow through the system is vital to its success, and advances in computer-based applications, such as PDM, have further enhanced its effectiveness. Before discussing the specific elements of JIT, it is useful to identify the underlying principles:

- JIT is a continuous operation, not a one-off event, and workers are responsible for the quality of their individual output.
- There has to be synchronisation (matching) or balance between operations so that all production occurs at a common rate that reduces 'bottlenecks'.
- Simple is considered to be better, so that a continuous effort is made to improve (**Kaizen**) by operating with fewer resources and less waste in terms of time, personnel, materials and equipment.
- Work is carried out in less-complicated ways through employing foolproof (**poka-yoke**) tools, jigs and fixtures to prevent mistakes.
- It is necessary to concentrate on fundamentals and remove anything that does not add value to the product.
- Factory layouts should be product-led so that less time is spent moving materials, components and parts reducing 'waiting time' to a minimum.
- It looks for opportunities to use machines that are self-regulating (**Jidoka**-autonomation) and that make production decisions based on the data they are processing rather than relying on human intervention.

The key features of JIT workplace organisation
The control systems for JIT are the most visible manifestation of the JIT approach but there are other important features:

- Operational set-up times are reduced, increasing flexibility and the capacity to produce smaller batches more cost effectively.
- There is a multi-skilled workforce that is capable of operating multiple processes leading to greater productivity, flexibility and increased job satisfaction.
- Production rates are either levelled or varied, as appropriate, to smooth the flow of products through the factory.
- **Kanban** is a control technique that is used to 'pull' products and components through the manufacturing process. The Kanban controls and schedules production rates including the flow of materials by using computer applications.

Task
Research what role computers played in manufacturing operations in the 1960s compared to now. Write a short report explaining what it was like then and how advances in ICT since then have supported the further development of modern manufacturing systems.

Computer-integrated manufacture
In Unit 3 we saw that to achieve full **computer-integrated manufacture (CIM)**, all aspects of a company's operations must be integrated so that they can share the same information and communicate with one another. A CIM system uses computers to integrate the processing of production and business information with manufacturing operations in order to create cooperative and smoothly running production lines. CIM ensures that all existing systems can talk to one another within both local and wide area networks. The tasks performed within CIM will include:

- the design of the graphic product to be produced
- planning the most cost-effective workflow
- controlling the operations of the machines or equipment needed to make the product
- performing the business functions such as ordering of stock, materials and customer invoicing.

Digital print solutions
The print industry has been revolutionised by the application of CIM. Digitalisation means change, and change brings new business. The demand for short, personalised print runs and faster job turnaround has quickly created a new, dynamic market for printers. Producing short colour runs – with extremely fast delivery times, last-minute changes and uncompromising print quality – is now a must for one-to-one marketing.

On-demand printing – quickly supplying exactly the right number of copies to meet each customer's needs – provides a competitive edge by providing greater flexibility and less need for expensive stockpiling. State-of-the-art digital technology means that images are scanned or transmitted electronically into memory, to be stored for further use. Data only needs to be input once, reducing much of the set-up time involved in traditional printing and copying methods. In addition, less set-up time means lower costs for the customer. Printing with variable data enables print products to be easily personalised as required.

Some printing companies allow Internet access with online ordering and they also provide a fast delivery service, so that customers can get the job printed and delivered back to them without ever leaving their office. Digital ICT can now be employed effectively in all three stages of the printing process – pre-press, press and post-press.

Pre-press solutions

The software a company has used to produce its document or artwork defines the speed of production and the printing hardware has to keep up. Digital technology and media grow more important with the increasing data volumes involved. The goal of large volume printers is to minimise production times with completely digitised production workflows and database-supported networks. Digital solutions provide a full range of image-handling capabilities, which enables greater flexibility in DTP, colour management, proofing and workflow systems. Image-handling capabilities of digital technology include: image clean-up, reducing, enlarging, photo screening, cropping, masking, rotating, inline collating, moving of text and graphics to enable printing on unusual paper sizes and other media.

High-volume printing

Printing costs reduce with **digital printing** machines that can operate at up to 14,400 pages per hour. Digital printing is also well suited to the production of images in low-volume or 'short runs', as it does not require the making of **plates**, as does offset printing. Digital technology means that images are scanned or transmitted electronically into memory, to be stored for further use. When higher print volumes are required, other digital methods are used.

Computer to plate

In conventional print making, images are converted into film, which is used to create the printing plate, which is then used to print the final image. The introduction of computer to plate (CTP) has increased workflow times as images can now be sent direct from a computer to a digital plate-making machine without the need to create a film first.

Adding value after printing

Post-press is about the finishing of print or graphic products which are increasingly complex. It adds value and leads to increases in profit. Customers' wishes are rapidly driving technological innovations so their ideas can be produced as originally envisioned by the designer. Whether folding, gluing, cutting, perforating, die cutting, wire-stitching or perfect binding is involved, maximum precision is necessary, and so is a standard of productivity that lets finishing machines keep up with the output of the latest digital printing presses. Digital finishing equipment can be more easily programmed and provides greater accuracy when compared to traditional mechanical labour-intensive equipment.

The development of CIM in the packaging industry

CAD/CAM technologies were first developed for the aerospace and automotive industries. For the companies that invested in these technologies they provided a competitive advantage by allowing the faster development of new ideas. Once the financial advantages they provide were realised they quickly spread to many other manufacturing sectors, especially those where there is an increased emphasis on quality, accuracy and consistency. One of the fastest growing applications for computers is in the graphics area and particularly in packaging.

Designers in all manufacturing disciplines are under increasing pressure to develop distinctive design concepts in order to create a strong corporate or brand image. Recognisable product identities can establish a secure and viable market share for their products. Graphic designers, particularly in the packaging industry, must ensure that their clients' products stand out clearly on the shelf. In many cases, this means that products are becoming more difficult to design and manufacture by conventional techniques.

CAD systems, no matter how sophisticated, are not a substitute for the creative abilities and visual skills of a graphic designer, especially during concept development when ideas are first being generated. The computer provides real time savings in all the subsequent operations that need to take place to convert an idea into reality. As an example, photo realistic visualisations rather than costly one-off models can be presented to the client electronically allowing unsuitable designs to be eliminated more quickly. On the functional side of packaging the computer's mathematical capability can be used to calculate the 'volumes' that can be contained by different designs more quickly and accurately. In the case of manufacturing technologies, such as injection or blow moulding used to produce plastic containers, reliable manufacturing data can be generated from the design drawings to ensure accurate and effective dies and moulds.

For graphic product designers the power of modern CAD systems is their ability to allow designs to be visualised quickly in different material and colour combinations within different settings. For example, the designer can combine the digital images of an actual supermarket display with the images for proposed product design that he or she has created. In effect, a shampoo or cosmetic manufacturer could be given a visualisation of what the pack would look like on a supermarket display when compared with its various competitors.

Many printing operations including those in the packaging sector are now employing CNC machines in the key production areas of pre-press (artwork and printing plate preparation), press and binding (assembly of the finished article). Traditionally, films are used to transfer images from original artwork to the printing plates. Computer to plate (CTP) processing equipment means that digital artwork can be used to output a set of plates directly from the desktop. The quality of reproduction is greatly improved because printing plates can be imaged directly by laser and workflow for plate production is greatly improved.

Other CIM developments include computer-controlled digital presses that mean four-colour printing in relatively small quantities is more efficient and cost-effective. These types of press can produce images on a wide variety of materials such as plastic sheets and laminates. This means that personalised and short-run printing such as promotional mouse mats or printed products with an individual customer's name on can now be produced competitively.

Task

Use a digital camera to take pictures of a product display of your choice. Transfer the images into a CAD system. Create a new package design for a product of your choice and produce a visualisation of what your product might look like on the digital images of the real display that you captured.

Flexible manufacturing systems (FMS)

In the late 1960s, all the leading Japanese manufacturers moved towards the flexible factory as a new way of gaining a competitive advantage. Since that time there has been increasing global competition and the creation of more open markets. Additionally, consumers are more sophisticated and are demanding an increasingly diverse range of products with regular updating beyond simple cosmetic changes. This, in turn, is influencing manufacturing to invest in more flexible plant and equipment. A piece of flexible equipment is one that has the ability to perform multiple processing tasks on a wide range of products, for example CNC machining centres for machining a variety of parts, or robots for material handling.

The introduction of machinery that can operate flexibly enables manufacturing to explore various processing sequences that might be available through alternative configurations of plant and equipment. The following are characteristics of most flexible manufacturing solutions:

- A system that responds quickly to changes from whatever source on the 'supply' or the 'demand' side.
- A range of techniques to increase operational flexibility are used and their effectiveness is constantly monitored and evaluated.
- The lead times from design to manufacture are significantly reduced and future changes in design can be made quickly.
- Levels of stock are kept to a minimum.
- Vertical partnerships, with effective two-way, ICT-based communication, exist between suppliers, product manufacturers and retailers.
- Increased sales and stock turnover for all partners in the enterprise.
- Computer-based management tools such as manufacturing resource planning (MRP) are used to collect real-time processing data as well as to track work accurately through the production cycle and to re-plan production in response to changing demand.

Creative and technical design: the role of computer-aided engineering

CAD is at the core of the graphics industry. **Computer-aided engineering (CAE)** can be part of a CAD application or it can form a 'stand-alone' system that can analyse the effectiveness of a design by creating simulations in a variety of conditions to see if it actually achieves what the designer intended. The success of modern CAD/CAM and CAE systems has been the dramatic reduction in design and development time. This allows a company to develop products in quick response to the needs of the market.

Modelling and testing ideas

With the rapid developments in mobile phone technology over the last few years the outward appearance of the phone as a visual or graphic product has changed dramatically from the early models that were the size and appearance of a small brick. The changes in technology have allowed product designers more freedom to change the visual appearance of the phone without compromising its technical efficiency. Consumers can also customise their phones by downloading different ring tones or adding covers with different graphic designs on them.

CAD modelling techniques allow the graphic

designer to try out different shapes and styles on screen. They can experiment with button layouts and the placing of graphic images or logos. CAE enables the production designers to simulate the manufacturing process, such as plastic injection moulding, to see if the design can be manufactured. Software is used that converts the solid graphic representation of the design to be converted into the thin shelled case that is needed to protect the operational circuitry and control devices on the phone. The 'engineered model' can be looked at from any angle and all the strengthening ribs and flanges required in the moulding process can be put in place. Once the design dependencies are determined the software can be used to generate the data needed for the computer-aided manufacture of the product.

Task

A more recent development in computer-aided modelling is **Rapid Prototyping (RPT)**. This process was first developed in the 1990s as a means of using a computer to generate a 3D model of an object such as a perfume bottle or similar 3D objects that had been drawn by CAD software. Use the Internet to investigate the emerging technology of RPT and compile a short illustrated report showing its application in CAD/CAM.

Virtual Reality Modelling Language (VRML)

3D virtual product modelling is an emerging technique, looked at in Unit 3. **Virtual Reality Modelling Language (VRML)** further extends the power of product modelling as a design tool. It is a specification for displaying 3D objects on the Internet. It is the 3D equivalent of HTML. Files written in VRML have a *wrl* extension (short for world); HTML files have the extension *html*. The VRML script produces a virtual world or 'hyperspace' on the computer display screen. The viewer 'moves' through the world or around an object by pressing computer keys or using another input device to turn left, right, up or down, or go forwards or backwards. This technology is still in the early stages of development; the first VRML standards were only set in 1995. It is set to become a powerful tool for the computer-based modelling of products (see Figure 4.3.2).

Task

Use the Internet and other sources of information such as printed media to collect information on how **virtual reality** is being used increasingly in the 'e-marketing' of products on the world wide web. Present your findings in a suitable graphic format.

Production planning and control

Production planning and control are areas where computers are used to good effect. Earlier in this unit we learned how modern manufacturers regard 'time' as a weapon that gives them an edge over their competitors. Scheduling is part of the planning process and its key features are to specify:

- the scope and detail of the work to be done
- the date when production has to start
- the latest date that production can be completed by
- any specialist machinery or manufacturing processes that are required
- what labour capacity is available.

The range of scheduling software applications that is available is evidence of the many different planning approaches used across different manufacturing sectors. The term finite capacity scheduling describes a processing schedule that is based on the overall manufacturing capacity available. By contrast, infinite capacity schedules use the customer's order due date as an end stop. The aim of these schedulers is to complete the order by working back from the due date using the available capacity.

Figure 4.3.2 *3D virtual reality model of a CIM system*

Modern manufacturers regard time and responsiveness as strategic weapons; to them time is the equivalent of money or increased productivity. They focus on reducing 'non-value-adding' time rather than trying to make people or machines work harder or faster on value-adding activities. The ways that leading companies use computers to manage time – in production, in new product development, in sales and in distribution – represent the most powerful new sources of competitive advantage. Today's new-generation companies compete with flexible manufacturing and rapid response systems, expanding variety and increasing innovation.

A company that has a time-based strategy is a more powerful competitor than one with a traditional strategy based on low wages, lower scales of production or a narrow product focus. These older, cost-based strategies require managers to do whatever is necessary to drive down costs. This could involve moving production centres to a low-wage country, building new facilities by consolidating, in effect closing down, old plants to gain economies of scale, or focusing operations down to the most economic activities. These tactics reduce costs but at the expense of responsiveness.

In contrast, strategies based on the principles of flexible manufacturing and rapid response systems need factories that are close to the customers. They serve to provide fast responses rather than low costs and control. The whole process is coordinated via computer databases and software applications that provide a full range of scheduling functions.

Master production schedule

The Master Production Schedule (MPS) is a top-down scheduling system that sets the quantity of each product to be completed in each week or **time bucket** over a short-range **planning horizon**. It is derived from known demand, forecasts and the amount of product to be made for stock. The planning assumption is that there is always sufficient manufacturing capacity available. For this reason, this method is sometimes called infinite capacity scheduling.

Types of computer-aided scheduling

- *Resource scheduling* is a finite scheduling function that concentrates on the resources that are required for converting raw materials into finished goods. Other scheduling functions may include the entering and invoicing of sales orders (sales order processing), stock recording and cost accounting. These functions combine to provide a powerful integrated database for the company. The only real problem with these applications is related to the accuracy of the data that is input into the database – inaccurate data causes cumulative problems as products move through the processing stages.
- *Electronic scheduling board*. The simplest scheduler is the electronic scheduling board, which imitates the old-fashioned card-based loading boards that were used to sequence machining operations. The advantage of this computer-based system is that it calculates processing times automatically and warns the production staff of any attempt to load two jobs on to the same machine.
- *Order-based scheduling*. In order-based scheduling the sequencing and delivery of individual component parts and resources is determined by the overall priority of the order for which the parts are destined. It is a distinct improvement on infinite capacity schedulers but its biggest drawback is that it allows gaps to appear in resources. Some schedulers allow the process to be repeated to try to reduce gaps before it is put into practice in order to reduce the time through the system but this can be a very time consuming process.
- *Constraint-based schedulers*. With these the aim is to locate potential bottlenecks in a production line and ensure that they are always well provided by synchronising the flow through the MPS.
- *Discrete event simulation*. This computer simulation loads all the required resources at a chosen point on the production line. When all processing problems and queues are resolved at that point, it moves on to the next point on the line. Because the simulation moves from one set of processing events to the next, there are far fewer gaps in the schedules and consequently production lines are far more stable.

Task

Describe the costs and benefits of using computer systems to aid production planning.

Control of equipment, processes, quality and safety

Different types of systems are used in modern manufacturing centres to control production tools, to monitor and evaluate quality and ensure safe working environments. These systems can be electrical, electronic, mechanical pneumatic, computer or microprocessor controlled. On modern machines such as printing presses or plastic moulding equipment these systems are often used in combination and their effective operation is dependent on the feedback of information on the state of each part of the system. Feedback makes control systems more efficient. Control systems improve efficiency, accuracy, reliability, safety and reduce waste. They can be found in:

- materials handling, such as moving resources so they are in the right place at the right time
- processing of materials, for example the automated manufacture of plastic-based packaging, using temperature control
- joining materials, such as the bonding and heat-sealing of plastics in a point-of-sale display using electronic or computer control
- monitoring quality by using feedback from colour sensors to regulate the feed rate on a digital printing press to ensure more even inking of the paper or card
- ensuring safety using feedback from electronic sensors to stop machines such as printing presses if hands or other obstructions are in the way
- graphical displays in a central control room that often show the floor-plan of the factory, the manufacturing cells or machines in use, the position of materials and components and a visual and audible alert system for when there are problems.

Total Quality (TQ)

One of the characteristics of a 'world class business' is that a 'quality culture' exists throughout all levels of the organisation. There is a strong and clear commitment to getting things right first time, every time. Total Quality links together quality assurance and quality control into a coherent improvement strategy. **Total Quality Control (TQC)** is the system that Japan has developed to implement Kaizen or continuous improvement for the complete life cycle of a product. It operates both within the function of the product and within the system to develop, support and retire the product. Total Quality Management (TQM) is the US equivalent. What both these systems have in common is that they are 'purpose driven', they involve comprehensive change and both are long-term processes rather than short-term fixes for problems that arise in design and manufacturing operations.

Implementation of TQ systems

Computer-based technologies are used to ensure TQ throughout design and manufacturing organisations. The use of computers ranges from supporting employee training by providing interactive learning methods to computer-aided statistical tools and methods for checking quality. For example, a product can be assigned a unique

bar code that can be read by laser readers at different points in the manufacturing process. The data that is collected is fed into the central PDM system and that information can be used in a variety of ways. The quality control team can use the data collected to trace a faulty product back through the manufacturing process to pinpoint where the particular problem occurred. They can then take the necessary remedial action to remove the cause of the problem so the manufacturing process is further improved.

There are three stages in the development of TQ in an organisation:

1 Awareness raising – to recognise the need for TQ and learning its basic principles.
2 Empowerment – learning the methods of TQ and developing skills in practising them.
3 Alignment – harmonising the business and TQ goals with the manufacturing practices of the company.

TQC and TQM tools

Thinking tools are equally as important as the physical tools we use to make things. Thinking or intellectual tools enhance the way we plan to do things at a strategic, tactical or operational level. At a strategic level, such as business planning, thinking can be highly symbolic or abstract and this relies on accurate qualitative information. At a tactical level, such as production scheduling, the information required is a mixture of qualitative and quantitative data. The importance of qualitative information decreases as you move from the strategic management level down to the operational level of factory coordination where achievement is judged against the planned outcomes (see Figure 4.3.3).

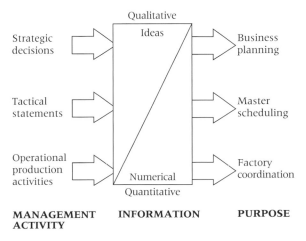

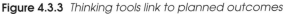

Figure 4.3.3 *Thinking tools link to planned outcomes*

In the context of TQC or TQM, computer-aided systems contribute to the collection, dissemination and analysis of the qualitative and quantitative data that is needed to bring about improvement. In general, computer-based thinking tools can be classed under the following functions:

- management
- product planning
- quality functions
- statistical process control (SPC).

Using computers in SPC

There are several established graphically based methods for representing statistics and analysing processes. Computers now play an important role in representing and analysing the causes and effects of different actions. For example, Pareto charts are used as a decision-making tool. The Pareto Principle suggests that most effects come from relatively few causes. In quality control this is known as the 80–20 rule – 80 per cent of faults come from only 20 per cent of the causes. If it is possible to identify the 20 per cent accurately, you can eliminate 80 per cent of your faults (see Figure 4.3.4). Pareto charts can be used to compare before and after situations. This allows managers to decide where to apply the minimum of time for maximum effect. Other areas where computers can be applied effectively to improve quality include the following:

- Flow charts are useful for modelling processes, feedback and **critical control points (CCPs)**. They use symbols, text and arrows to show direction of information or data flow (see Figure 4.3.5).

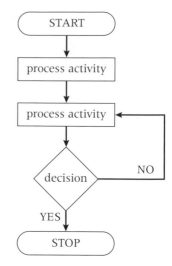

Figure 4.3.5 *A flow chart – used to show the movement of people, materials, paperwork or information and processes*

- Cause and effect, fishbone or Ishikawa diagrams describe a process or operation as a sequence. They provide a method for analysing processes to establish a cause and its effects (see Figure 4.3.6).
- Bar graphs or histograms are used to analyse variations in data in graphic format. This makes it easier to 'see' graphical variations than it would be to read a table of numbers (see Figure 4.3.7).
- Check sheets can be used to collect quantitative or qualitative data during production. They are used in high-volume manufacturing where there is greater repetition than you would find in shorter production runs.
- Checklists are an important management tool used in specific operational situations to ensure quality by establishing consistency and reliability. They list all the important steps or actions that must be correctly sequenced to achieve the best possible quality outcome.

Monitoring and inspecting quality

Despite the progress in designing for quality and incorporating quality into manufacturing processes, inspection still remains a necessary component of many quality assurance systems. In all areas of manufacturing, whatever the product, the cost of defects escalates as materials and components move through the supply chain to the customer. The manual or 'human' monitoring of any continuous process, such as in the printing industry, becomes less efficient as

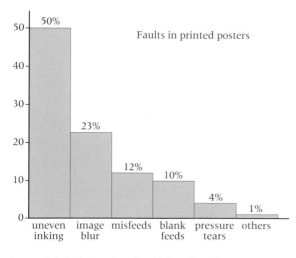

Figure 4.3.4 *A Pareto chart identifies the production areas where there are faults*

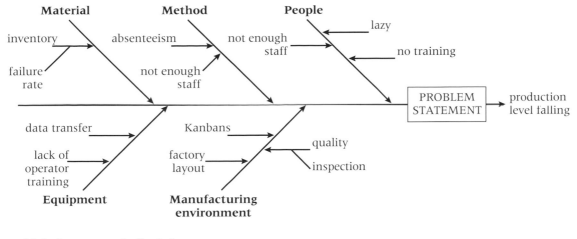

Figure 4.3.6 *A cause and effect diagram*

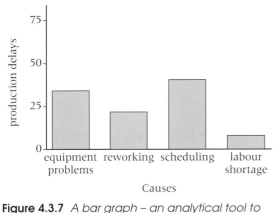

Figure 4.3.7 *A bar graph – an analytical tool to compare outputs, percentages or activities*

the time spent on it increases. In many industries, this effect is minimised by the setting up of teams that work together in a manufacturing 'cell'. Each cell is responsible for the product from when it is received by the cell until it moves on to the next stage of production. As a team the cell has to meet specified **quality indicators**. Each member of the team is responsible for maintaining quality level so that the burden of monitoring and inspection is shared. This leads to the improved detection of faults.

In high-volume, continuous production industries such as packaging and printing 'artificial vision' is now used to good effect in the monitoring and inspection of the processing operations. In a print room, for instance, the optical quality of a colour print can be more accurately judged over a longer period of time by using digital image-processing technology rather than relying on an experienced operator who might be subject to many physiological defects such as tiredness. The benefits of optical quality control are:

- high-speed, real-time fault detection, recording and reporting throughout a processing operation from receipt of raw materials and components through to the delivery of the product
- reduced labour costs
- low running costs after initial installation
- specific quality specifications can be programmed into a computer and checked using the digital data collected during the full range of processing operations.

Manufacturing to tolerances

Most mass-produced 3D products are assembled from components that are manufactured or machined to specific or 'nominal' measurements so that they can all be fitted together. However, it would be too expensive and wasteful to produce each part to the exact dimension every time so dimensions on a drawing often have a '**tolerance**' that indicates what the acceptable maximum and minimum dimensions are. Providing the dimensions of the component or part lie within this range it is acceptable, as the product will still function effectively. There are two types of dimensional measurement that can be applied:

- On-process measurement where a part is measured while under production. This is used where manufacturing cycles for a component or part are long or high-cost materials are being processed. The early detection of error removes the high cost of failure.

- Post-process measurement involves making measurements of critical operational dimensions after the component or part has been produced. It is a method that is often used in high-volume, low-cost manufacture or in specialist one-off manufacturing such as the production of promotional gifts using a CNC machine. The measuring methods used are mostly mechanical in nature such as vernier **calipers**, height gauges and profile projectors. Increasingly, computers and lasers are being used because of their increased accuracy, as they do not rely on the visual interpretation of a scale or rule.

Controlling complex manufacturing processes

A modern integrated system of manufacturing is heavily reliant on the use of computers throughout the process of design and make. Computers are also used to determine the optimum factory layout of plant and equipment, the effective deployment of labour and the scheduling of processing operations. Computers can also be used to monitor and control 'workflow' and the dissemination of information about work in progress, especially in high-volume manufacturing situations such as printing and packaging.

Managing workflow

Production plans and control systems manage the input of materials and components to the different processing departments within a manufacturing operation. They also monitor output and work in progress. This information informs and controls the workflow through the factory to ensure that orders are met on time and to cost. It is particularly important in batch and short-run manufacturing operations such as those involved in the production of promotional gifts or sign making.

Monitoring workflow

The role of artificial vision and the use of lasers were described earlier in this unit and they are a vital component in computer-controlled parts recognition systems. These visual recognition systems rely on **data communication tags** (bar codes) as a way of monitoring products through a high-volume manufacturing system. They can be attached to pallets of stock that are to be processed or to individual assemblies. A sensor or bar code reader attached to an individual workstation can read these tags. Laser scanners are used to read bar codes because they are well suited to applications requiring high reading performance, small size and low cost. A

digital pre-processor receives and decodes the signal into data that can be read and analysed by a computer. This data is then transmitted to the 'supervising' computer controlling the production line. The development of these systems has improved the operational effectiveness of manufacturing systems based on the concurrent model.

Controlling workflow

The term workflow describes the tasks required to produce a final product. Project management software allows a manufacturer to define different workflows for different types of jobs and coordinate production cells in an overall MPS. For example, in a design environment a CAD file might be automatically routed from the designer to a production engineer to purchasing for comment or action. At each stage in the workflow, one individual or group is responsible for a specific task such as ensuring the right materials are in the right place when required. Once the task is complete the workflow software ensures that the individuals responsible for the next task are notified and receive the information they need to execute their stage of the process.

Sequential or concurrent manufacturing?

The design of an effective and marketable product that will be sold at a profit is dependent on the input from a range of specialists and the efficiency of the manufacturing or processing system that is chosen. The scale of the manufacturing operation is a key factor in deciding on a system of design and make.

Sequential manufacturing

In a design studio, where a small group of graphic designers is working together, each person may take responsibility for a particular project or graphic outcome. The designer works in a similar way to you when you are producing your practical coursework for this examination. In this linear or sequential approach to design and making an idea or a product will pass through a series of discrete, self-contained stages and at each stage outcomes are evaluated before moving on to the next. If there is a fault, it is passed back through the stages until the fault is found and corrected when it then begins its journey through the process again. This linear or 'function-based' approach is slow to respond to change or demand. It has longer lead times and is often characterised by low product quality because of the separation of the design and manufacturing functions and the costly design and redesign loops.

Concurrent manufacturing

Concurrent manufacture is widely regarded as an effective system for scales of production ranging from batch to mass and high volume. The key feature of concurrent or simultaneous manufacturing is a team-based approach to project management, so that the right people get together at the right time to identify and resolve design problems. The underpinning philosophy of this approach is that quality decisions have to be made at every stage to ensure the intended outcome of a quality product that makes a profit. This means setting appropriate specifications and 'quality indicators' to evaluate both the design and intended manufacturing processes. The concurrent approach also forces manufacturers to consider all elements of the product life cycle from conception to disposal.

Many companies also involve their suppliers and retailers at an early stage in the product development cycle so that they create a product development team (see Figure 4.3.8). They work closely in a vertical partnership known as a 'value chain'. For this partnership to work effectively, information has to flow quickly between the partners. Increasingly, this takes place electronically through **electronic data interchange (EDI)** systems, which we shall look at in more detail later in this unit.

Computers are becoming increasingly useful when companies decide to adopt a concurrent approach to design for manufacture (DFM). The importance of this is clear when you consider that 70 per cent of the manufacturing costs of a product, materials, processing and assembly, are determined by design decisions. Production decisions such as process planning or machine selection are only responsible for 20 per cent.

Companies can now create their own 'expert systems'. These are databases of established good design practice and specialist knowledge that are available on a network of computers within the company known as an **intranet**. This internal computer network enables fast, efficient communication between all the members of the product development team.

The development of Internet technologies means that this information can also be accessed via the world wide web from anywhere in the world. Different access codes or passwords can control access to commercially sensitive areas of development. Concurrent systems enable the use of just-in-time (JIT) and Quick Response Manufacturing, which reduces the time from idea to market. The times to market for sequential and concurrent manufacturing are compared in Figure 4.3.9.

In concurrent systems, to ensure that the time to market is further reduced, different members of a team will take responsibility for ensuring specific production deadlines or 'milestones' are met. The team agrees these milestones and they are recorded in a **Gantt chart**, a simplified form of which is shown in Figure 4.3.10. This chart shows

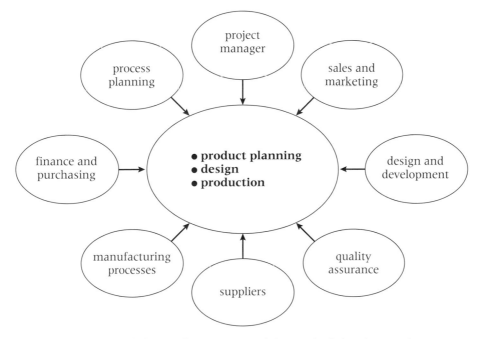

Figure 4.3.8 *Concurrent manufacturing – a team approach to product development*

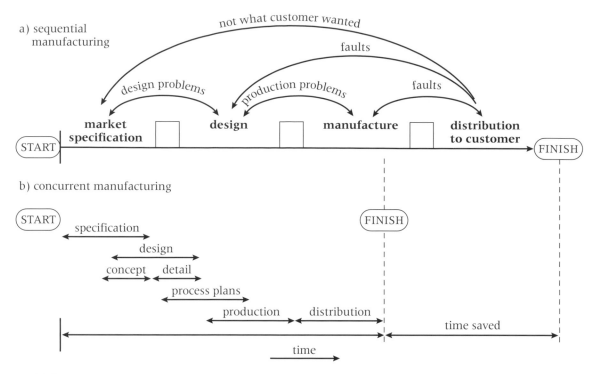

Figure 4.3.9 *A comparison of sequential and concurrent manufacturing. The latter reduces time to market*

where operations are to start and finish and also those activities that can be run concurrently because they are not dependent on other operations being completed first. You will see, for instance, that the design and production of advertising, promotional and sales literature that would include printed and electronic media starts before the first prototype has been produced. You will also see that the final list of suppliers is only completed once the milestone of a successful prototype has been reached.

Tasks

1 Explain why the use of ICT is such an important feature in a concurrent manufacturing system.
2 For a product of your choice, explain the benefits of using a concurrent manufacturing system in comparison with a sequential manufacturing system.
3 Explain how **milestone planning** reduces the 'lead in' time for a new product.

Week 1	Time ⟶	End date	Responsibility
▶ specification agreed			sales and marketing
▶ ▶ ▶ concept agreed			design
▶ ▶ ▶ ▶ suppliers finalised			purchasing
▶ ▶ ▶ ▶ ▶ prototype			design and production
▶ ▶ ▶ materials components in stock			purchasing
▶ ▶ ▶ ▶ ▶ ▶ ▶ sales literature			sales and marketing
▶ ▶ ▶ product test			quality
▶ ▶ ▶ ▶ test batch			production
▶ ▶ ▶ ▶ quality checks			quality

Figure 4.3.10 *A milestone plan for a typical manufactured product*

2. Robotics

Automation is a term describing the automatic operation and *self-correcting* control of machinery or production processes by devices that make decisions and take action without the interference of a human operator. Robotics is a specific field of automation concerned with the design and construction of self-controlling machines or robots. More recently, mechatronics, a Japanese term, is an exciting development in the field of automation. Mechatronic devices integrate mechanical, electronic, optical and computer engineering to provide mechanical devices and control systems that have greater precision and flexibility.

The first operator-guided robotic manipulating devices were developed in the 1940s for use in radioactive conditions encountered in the development of the atomic and hydrogen bombs during World War II. The first industrial robots developed in the late 1950s and early 1960s included sensors and other simple feedback devices. The use of robots is growing with the development and wider application of CIM and FMS. Figure 4.3.11 shows a bank of robots working in a manufacturing cell in a car plant. The next generation of manufacturing robots will have enhanced sensory feedback systems and possess a degree of **artificial intelligence** (AI).

The robots we see in the movies, such as C-3PO in *Star Wars* and in the *Terminator* series, are portrayed as fantastic, intelligent, even dangerous forms of artificial life. However, the robots of today are not exactly like these walking and talking intelligent machines. Most of the world's robots are working for people in factories,

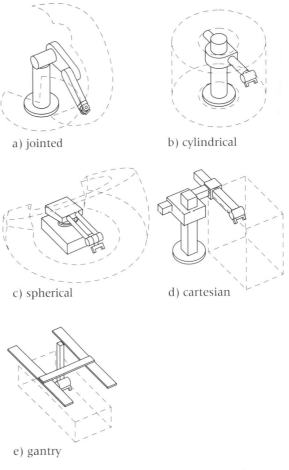

a) jointed   b) cylindrical

c) spherical   d) cartesian

e) gantry

Figure 4.3.12 *Robot configurations and work envelope*

warehouses, and laboratories. Specialist suppliers now offer a large range of robot sizes/payloads/reach (work envelope) and joint capabilities with differing degrees of precision and repeatability. Figure 4.3.12 shows different robot configurations and their work envelope.

Robots have been applied to assembly, painting, palletising, packing, welding, dispensing, cutting, laser processing, material handling and other emerging applications in the advertising, promotional, leisure and entertainment industries such as animatronics. Animatronics, which has a lot in common with robotics and mechatronics, encompasses automata and is concerned with bringing inanimate sculptures to life. These devices known as animatrons (see Figure 4.3.13) require many design and manufacturing skills drawn from mechanical engineering, pneumatics, sculpture and animation.

Figure 4.3.11 *A manufacturing robot*

Figure 4.3.13 *An animatric iguana puppet hams it up during the taping of a children's show for National Geographic television*

Basics of robot design

Most robots are designed to be a helping hand. They are in effect robotic arms. Figure 4.3.14 shows a schematic view of a typical jointed-arm robot. They help people with tasks that would be difficult, unsafe or boring for a real person to do alone. At its simplest, a robot is a machine that can be programmed to perform a variety of jobs, which usually involve moving or handling objects.

For a machine to be classified as a robot, it usually has five parts:

• *Controller*. Every robot is connected to a computer, which keeps the pieces of the arm working together. This computer is known as the controller. The controller functions as the 'brain' of the robot. The controller also allows the robot to be networked to other systems, so that it may work together with other machines, processes or robots.

Robots today have controllers that are run by programs – sets of instructions written in code. Almost all modern robots are entirely pre-programmed by people; they can do only what they are programmed to do at the time, and nothing else. In the future, controllers with artificial intelligence (AI) could allow robots to think on their own, even program themselves. This could make robots more self-reliant and independent.

• *Arm*. Robot arms come in all shapes and sizes. The arm is the part of the robot that positions the end-effector and sensors to do their pre-programmed business. Many (but not all) resemble human arms, and have shoulders,

elbows, wrists, even fingers. This gives the robot a lot of ways to position itself in its environment. Each joint is said to give the robot 'one degree of freedom'. So, a simple robot arm with three degrees of freedom could move in three ways: up and down, left and right, forward and backward. Most working robots today have six degrees of freedom (see Figure 4.3.15).

• *Drive*. The drive is the 'engine' that drives the links, the sections between the joints, into their desired position. Air, water pressure or electricity powers most drives.

• *End-effector* (end of arm tooling). The end-effector is the 'hand' connected to the robot's arm. It is often different from a human hand – it could be a tool such as a gripper, a vacuum pump, tweezers, scalpel, blowtorch or heat

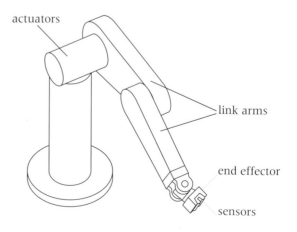

Figure 4.3.14 *The parts of a robot arm*

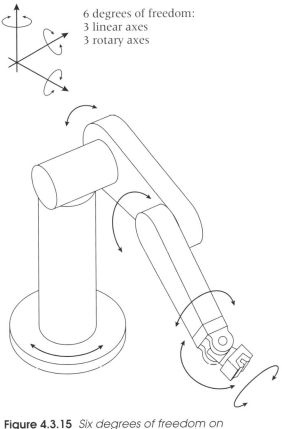

6 degrees of freedom:
3 linear axes
3 rotary axes

Figure 4.3.15 *Six degrees of freedom on a robotic arm*

sealing gun, just about anything that helps it do its job. Some robots can change end-effectors and be reprogrammed for a different set of tasks. If the robot has more than one arm, there can be more than one end-effector on the same robot each suited for a specific task.

- *Sensor.* Most robots have limited awareness of the world around them. Sensors can provide some feedback to the robot so it can do its job. Compared to the senses and abilities of even the simplest living things, robots have a very long way to go. The sensor sends information in the form of electronic signals back to the controller. Sensors also give the robot controller information about its surroundings and let it know the exact position of the arm or the state of the world around it. Sight, sound, touch, taste and smell are the kinds of information we get from our world. Robots can be designed and programmed to get specific information that is beyond what our five senses can tell us. For instance, a robot sensor might 'see' in the dark, detect tiny amounts of invisible radiation or measure movement that is too small or fast for the human eye to see.

The industrial application of robotics/control technology and the development of automated processes

Ninety per cent of all robots used today are found in factories. These robots are referred to as industrial robots. Although many kinds of robots can be found in manufacturing, jointed arm robots are particularly useful and common. Ten years ago, car manufacturers were buying nine out of ten robots – now, car manufacturers buy only 50 per cent of robots made. Robots are slowly finding their way into warehouses for automatic stock control, laboratories, research and exploration sites, energy plants, hospitals, even hostile environments like outer space (see Figure 4.3.16).

Japan is a world leader in robotics. It has 400,000 robots working in factories – ten times as many as the United States. Robots are used in dozens of other countries. As robots become more common, and less expensive to make, they will continue to increase their numbers in the workforce.

The manufacturing industry, including those sectors involving graphics, is in a non-stop state of change and is evolving rapidly because of the availability of a whole range of improved computer-based control technologies and the recent emergence of the global marketing of

Figure 4.3.16 *The loading bay on the Space Shuttle showing the Canadian Robot Arm in use by astronauts*

products on the Internet. These new digital technologies are supporting rapid changes in working practice from the shop floor through to business processes such as the web-based advertising and selling of products. The application of robotic technologies has brought about significant changes in those sectors of industry concerned with batch and mass production.

Robots in batch production

Robotic devices have supported the batch and short-run production of items such as CD-ROMs that contain entertainment or business software. Figure 4.3.17 shows a CD printing machine that is loaded using a robotic application. In mass-production they are used in large, highly automated operations such as printing and packaging.

The advantage of these 'robotic' systems is that they support continuous production and they allow many processes to run concurrently, sequentially or in combination. The numbers of human operators that work are minimal and those that do are working in safe and secure operating conditions. This is because electronic, computer or microprocessor controlled systems have many sensing devices that monitor events in real time and are designed to shut down to a 'fail-safe' state in the event of a failure or accident. These robotic technologies are **enabling technologies** that allow innovative solutions to complex processing operations.

Figure 4.3.17 *CD printing machines use a robotic application*

Control systems

The common factor in all automated processes, including the use of robots, is that they all require a computer or microprocessor-based control system. Control systems regulate, check, verify or restrain actions. Automatic or robotic manufacturing systems are able to sense and control how processes are operating by combining sub-systems of electrical, electronic, mechanical and pneumatic devices (see Figure 4.3.18). All these devices produce performance data that may need to be monitored by instrumentation systems.

Instrumentation systems provide data in the form of visual displays or electronic signals that indicate how effectively the manufacturing sub-systems are operating, both individually and in relation to one another. Because of their ability to store, select, record and present data, computers and microprocessors such as programmable **logic** controllers (PLCs) are widely used to direct and control actions and process operations.

In many manufacturing and assembly applications robots are internally programmed to perform a programmed cycle of operations when given a simple start signal. However, there are many robots which combine an internal computer program with compliance to commands from outside the program. These can be classified as sensor or remotely controlled robots. Many robot tasks cannot be programmed exactly in advance because real-world operating conditions cannot be predicted exactly and they rely on sensors to provide the information required by the robot. Below are some examples.

Automatic storage and retrieval systems (ASRS)

Automatic storage and retrieval robots that are used in automated warehouses are commanded from outside either to transfer an object from a designated pickup point to a designated position in storage or vice versa. The required motions are internally programmed into the device. Such robots are made in sizes to handle 'objects' from tape cassettes, DVDs and other data storage devices to pallets carrying cardboard in a carton-making factory. The commands may come from a computer, which controls a larger operation, in which case the robot computer and the computer that commands it are said to form a control 'hierarchy'.

Mobile robots or automated guided vehicles (AGVs) are typically used in component or pallet transfer. The AGV is an unmanned vehicle that carries its 'load' along a pre-programmed path. AGVs use different navigational systems and travel around under some combination of automatic control and remote control. Other

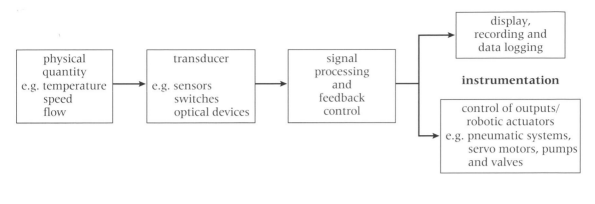

| physical quantity e.g. temperature speed flow | → | transducer e.g. sensors switches optical devices | → | signal processing and feedback control | → | display, recording and data logging |

Figure 4.3.18 *A simple systems model can represent a manufacturing control system*

uses of AGVs are in mail delivery, surveillance and police tasks, material transportation in factories, military tasks such as bomb disposal, and underwater tasks like inspecting pipelines and ship hulls and recovering torpedoes.

Robots for advertising and promotional purposes
There are other uses for automated or mobile robots. 'Fun' robots are innovative graphic products that are used widely in the United States, and their use is increasing in the UK. Fun robots are extremely effective in situations where there is a need to grab people's attention or where a company wants to 'stop traffic' by engaging potential customers. When robots talk, people listen!

These robots make a lasting visual impression at trade shows, parades, theme parks and museums such as NASA Houston. They are also used to publicise special educational programmes such as anti-drugs campaigns and fire safety. The types of robots found most effective for advertising promotions are mobile robots, talking signs and animated figures, especially musicians.

Mobile robots, which are powered by rechargeable batteries, can seize anyone's attention by driving up to or chasing him or her and starting an interactive 'conversation'. The robots can 'talk' via wireless microphone systems using a transmitter and microphone that allows the operator to speak through the character. In some products such as futuristic or 'alien' robots, a voice modifier is included to electronically modify the operator's voice to sound like an alien, robot, monster or normal, simply by changing a four-position switch. The character's mouth movement is automatically synchronised with the operator's voice and there are other movements such as mouth lights that are automatically synchronised with the 'voice'.

Robots can be 'customised' by painting and the application of vinyl decals designed to convey corporate images in the form of logos and specific colour combinations. Some robots are equipped with a digital messaging system that allows a customer to record messages up to three minutes long on a digital chip and activate message playback either automatically by infra-red sensor or manually by push buttons. In noisy situations such as a trade show or shopping centre the robots can be equipped with a remotely controlled siren or back-up alarm, or a combination of these attention-getting noisemakers.

A remotely controlled futuristic mobile robot known as 'Z-Tron' features a remote camera that allows moving graphic images to be relayed between the operator and the person that is engaged. It has full 360-degree mobility and a full 'talk-back' system that simulates hearing and allows the operator to respond to questions the robot is asked.

Figure 4.3.19 *Robots, like this one entertaining children at a fair, are often used to promote companies and their products*

Talking signs are stationary robots that are used for promotional and advertising campaigns (see Figure 4.3.19). They 'talk' by a wireless microphone system or by a recorded message on a cassette tape or digital chip. The message can be activated automatically by infra-red sensor that senses when a person is nearby or manually by push buttons or by a combination of both. Their mouth movement is automatic, in synchronisation with the sound, and they can also turn their heads automatically. They are excellent as 'greeters' in restaurants, retail stores or anywhere you want to capture customers' attention. Talking signs can be strategically placed in a store to talk about sale items, special promotions, directions, instructions and forthcoming events.

New uses for robots in manufacturing

The majority of the plastics industry is based on a limited number of production processes. Technical advances have been made in recent years by applying six-axis robot work cells to the primary and secondary processes of injection moulding, blow moulding, extrusion and sheet fabrication. For example, within a robotic work cell for the trimming of plastic components, a six-axis robot would manipulate a trimming device around a plastic part to accomplish drilling, routering, blow moulded bubble removal, flash removal, de-gating or any other type plastic removal application. It has applications in the packaging and display areas as well in the production of other plastic components such as packaging and point-of-sale displays. The key benefits of this robotic cell are:

- consistent cutting quality
- safer work environment
- reduction of labour costs.

Other applications of robots in the plastics industry include:

- insert loading systems in which the robot will load metal threaded inserts, other plastic components and appliqués into the moulds to keep a continuous cycle of production
- part removal systems in which the robot will remove parts from the moulding machines to maintain consistent production
- infra-red plastic welding systems in which the robot presents two halves of an assembly to an infra-red heat source to melt the weld areas and then assembles the two halves together
- laser cutting systems in which the robot will manipulate a laser around the plastic part or manipulate the part around a stationary laser to remove unwanted plastic.

Complex automated systems using artificial intelligence and new technology

Artificial intelligence (AI) was a term defined by John McCarthy at the Massachusetts Institute of Technology (MIT) in 1956. It describes the branch of computer science concerned with developing computers that think and act like humans. AI is a broad area of development covering a wide range of different fields, including engineering, ranging from 'machine vision' to 'expert' systems. The common thread that links research in this area is the creation of devices that can 'think'. Computers are not yet able to fully simulate human behaviour in manufacturing environments but this is a major research area in manufacturing and it is only a matter of time because there have already been significant developments in the use of AI in other areas. In 1997, an IBM super-computer called 'Deep Blue' defeated the world chess champion.

What is a thinking machine?

In order to 'think', it is necessary to possess intelligence and knowledge. Intelligent

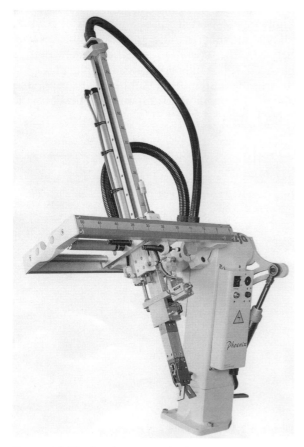

Figure 4.3.20 *A swing arm robot*

behaviours may consist of solving complex problems or making generalisations and constructing relationships. To do these things requires perception and understanding of what has been perceived. Intelligent systems should be able to consider large amounts of information simultaneously and process them faster in order to make rational, logical or expert judgements. Perhaps the best way to gauge the intelligence of a machine is British computer scientist Alan Turing's test. He stated that a computer would deserve to be called intelligent if it could deceive a human into believing that it was human.

Knowledge-based or expert systems

A knowledge base stores the knowledge related to a particular area or domain. Expert systems in which computers are programmed to make decisions in real-life situations already exist to help human experts in several domains including engineering, but they are very expensive to produce and are helpful only in special situations or in hostile working environments like working in space. Expert systems are designed by knowledge engineers who study how expert designers and others make decisions. They identify the 'rules' that the expert has used and translate them into terms that a computer can understand.

Application of AI in design and manufacture

In Unit 3 we saw that CAD systems already provide a quick and efficient means of representing the technical form of a product. Present systems currently provide limited design information or advice to inform the decision making part of the design process. CAD/CAM software readily generates sets of manufacturing instructions but a designer receives limited information from the software to decide whether the designed part is capable of being economically manufactured. At present, these types of decision are reached by combining and applying the experience and expertise of the whole product development team.

Applying design or production rules

It is possible to represent some knowledge in the form of a set of linguistic facts or logical rules. In electronics, for instance, there are the IF and THEN statements which can be combined with **logic gate** truth tables. When designing an electronic system, IF a set of conditions is true, THEN a conclusion can be made or an action can be taken. In a warning system on a printer, for example, if two inputs are true – the machine is on but there is no paper – then an audible alarm will sound or an on-screen warning is displayed to alert the operator. If two inputs are both true and an audible warning is required, then logic dictates that you would use an AND gate.

Future uses for new technologies

Enabling technologies such as vision recognition and AI are being used to improve production planning such as the optical analysis of visual images on the production line for quality, safety and process control. Vision systems are used in product distribution and bar coding systems. These computerised warehouses can make electronic links (see below for a detailed explanation of electronic data interchange) between suppliers and customers. These intelligent 'vision' systems can 'see', 'make decisions', then 'communicate' those findings to other 'smart' factory devices, all in a fraction of a second. Modern digital vision technology means that little to no extra external requirements such as special lighting are required. With a central computer hooked up, it can update the pictures it takes very quickly with high resolution making the system more responsive and easier to program.

Developing artificial intelligence

Because of the pressure to develop more responsive manufacturing systems considerable research and development time is focusing on developing manufacturing systems that fully integrate the use of **CAD modelling**, artificial intelligence, ICT and knowledge-based databases. These areas include:

- neural networks
- voice recognition systems
- natural language processing (NLP).

Neural networks

A neural network is a computer system modelled on how the human brain and nervous system operate. Whereas a computer manipulates data in zeros and ones, a neural network reproduces the types of processing connections (neurones) that occur in the human brain. Neural networks are particularly effective for predicting events, when they have a large database of examples to draw on. They are proving successful in systems used for voice recognition and natural language processing (NLP).

Voice recognition systems

These are computer systems that can recognise the spoken word and currently they can take dictation but cannot understand what is being said to them. Since such systems are high cost and have operating limitations, they have only been used as an alternative to a computer keyboard. They are used when working in hostile environments such as space, or when the use of a keyboard is impracticable because the operator is disabled. In the future, an operator will be able to talk directly to an expert system for guidance or instruction.

Natural language processing (NLP)

If successfully developed, it is hoped NLP will enable computers to understand human languages. This would allow people to interact with computers without the need for any specialised knowledge. You could simply walk up to a computer and talk to it. Unfortunately, programming computers to understand natural languages has proved to be more difficult than originally thought. Some rudimentary translation systems that translate from one human language to another are in existence, but they are not nearly as good as human translators.

Tasks

1 Explain and give examples of what is meant by an enabling technology.
2 Choose three enabling technologies and describe how they support the development of flexible manufacturing systems.

The use of block flow diagrams and flow process diagrams for representing simple and complex production systems

Graphical system diagrams

Systems thinking and design are evident all around us in the manufactured world. For example, an audio system can contain many different components and devices that can be configured to operate in a variety of ways. We can make sense of the natural world by applying systems thinking, as when we talk about a weather system. Block flow diagrams have been developed to explain how the parts or activities in a system are organised and related to one another. They explain the processes that change an input into an output (see Figure 4.3.21). There is a set of drawing or graphic conventions for representing what the blocks that make up a block flow diagram do (see Figure 4.3.22).

Complex processing within a manufacturing system can be modelled by breaking down the system into a collection of sub-systems each with their own input and output. Together they describe the flow of information or actions through a process. These are sometimes called flow diagrams. Figure 4.3.23 shows a simplified flow diagram of a plastic injection moulding process.

Open- and closed-loop control systems

A system operating **open-loop control** has no feedback information on the state of the output. It will continue without interference from the system even when the output changes. This is a major disadvantage in an automated process, as we shall see later when we consider 'lag'. A system operating **closed-loop control** can have either positive or negative feedback. In these systems, information about the state of an output is fed back into the processor where it is combined with the input signal in order to control or change the size of the output.

Positive and negative feedback

In an automated process positive feedback would result in the system becoming unstable because an increase in the output leads to an increase in the input which creates an increase in the output and so the loop starts again. Negative feedback is

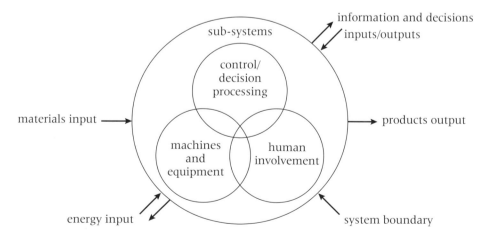

Figure 4.3.21 *A system has a boundary that defines the limits of the system*

often used in automated control processes because it works to change the input in such a way that the output is decreased to provide stable operating conditions. Kanbans are used in some manufacturing systems to control the flow of work on a production line. Without negative feedback they would not operate effectively.

Error signals

In any system with feedback, the difference between the input signal and the feedback signal is called the error signal. The size of the error signal determines how much the system output will need to be changed. When the system is in a near stable state, the error signal will be nearly zero. In a high-volume industry such as printing, for instance, an error signal will be generated when there is a difference between the projected printing production and the actual production. If production is higher than required, this will generate a positive error and signal the need to decrease production levels. If production falls, this generates a negative error and production is increased.

Lag

In any large-scale system or automated process, it will take time for the system as a whole to respond to the feedback signals it is getting. If a faulty batch of products has ended up in the distribution chain, it will not be noticed until customers or the outlets selling the product notify the company of the problem. This time delay before the system is able to respond is known as lag and it is a common feature in closed-loop control systems. The ways in which manufacturers and others are using ICT to improve the speed of communications at all levels of business processes will be discussed later. The measures being taken to reduce lag include electronic data interchange (EDI) and improved 'real-time' sales data from electronic point of sale (EPOS) information systems. These inform the manufacturer of the need to adjust production to correct the 'fault'.

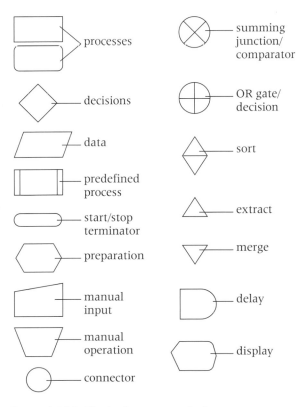

Figure 4.3.22 *Block diagram symbols*

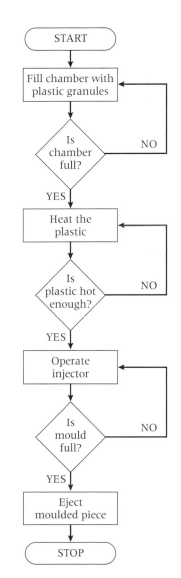

Figure 4.3.23 *A simplified material processing system and its sub-systems of injection moulding*

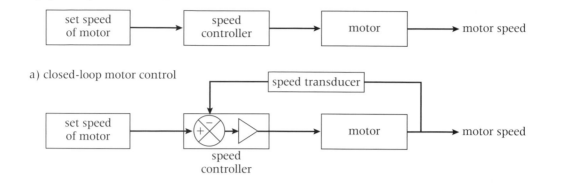

Figure 4.3.24 *Open- and closed-loop control of a motor*

Automated systems using closed-loop control systems

Open-loop control systems have a major disadvantage when it comes to controlling devices such as the motors on a conveyor belt. If the motor is put under too great a load its speed decreases and it may stop completely with disastrous effect such as overheating. The alternative is a closed-loop system in which negative feedback is used to provide proportional control of the motor speed. Any difference between the required motor speed and its actual speed produces an error signal that is fed back to the speed controller in order to stabilise the conveyor system by adjusting the speed of the motor. Figure 4.3.24 shows open- and closed- loop control of an electric motor.

Sequential control

Robotic and automated processes often use sequential control programs in which a series of actions take place one after another. For instance, on the automated CD loading system discussed on page 246 each action depends on the previous one having been carried out. If the system has not 'sensed' that the CD to be loaded is in place, the screen printing process will not operate. It will sit and wait until the sensor sends it the required information signal that the CD is in place.

Logical control of automated and robotic systems

Combinational logic is used in situations where a series of conditions have to be met before an operation can take place. This is also known as 'multiple variable' control. The conventional logic gates that you may be familiar with are the AND/OR/NAND gates. **PLCs** can also perform similar operations. Figure 4.3.25 is a simplified diagram of how multiple inputs are controlled in the operation of a CNC machine.

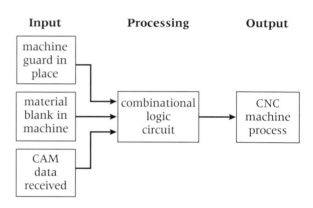

Figure 4.3.25 *All three inputs have to be 'on' for the CNC machine to operate*

Fuzzy logic

Fuzzy logic was developed in the 1960s and it recognises more than the simple true and false values such as those used in digital electronic systems that are based on zeros and ones. An input or output is either on or off. With fuzzy logic, propositions can be represented with degrees of truthfulness and falsehood. For example, the statement 'Today is sunny' might be 100 per cent true if there are no clouds, 80 per cent true if there are a few clouds, and 50 per cent true if it is hazy and 0 per cent true if it rains all day. Fuzzy logic allows conditions such as rather warm or pretty cold to be formulated mathematically and processed by computers in an attempt to apply a more human-like way of thinking. Fuzzy logic has proved to be particularly useful in expert systems, other artificial intelligence applications, database retrieval and engineering.

Fuzzy control

Fuzzy logic controllers (FLC) are the most important applications of the recently developed fuzzy logic theory. They work rather differently compared to conventional logic controllers employed in process control operations involving electrical, hydraulic and pneumatic devices. Expert knowledge is used instead of algebraic equations or linguistic conventions to describe a system. This knowledge can be expressed as fuzzy sets in a very natural way using linguistic variables similar to the way that humans express their ideas. A fuzzy set is a collection of objects or entities without clear boundaries.

Many people think that until computers can think like humans their wider application will be limited. Fuzzy control is useful for very complex non-linear processes when there is no simple mathematical model or if the processing of (linguistically formulated) expert knowledge is to be performed. Fuzzy control is less useful if the conventional single or multiple variable control theory based on mathematical models described earlier provides an effective result. This is a rapidly developing area and application is found in several areas ranging from electronics design automation (PCB design and manufacture) and product analysis, sampling or testing to e-commerce applications such as document management systems, data warehousing and marketing.

The advantages and disadvantages of automation and its impact on employment, both local and global

The pressure for automated manufacturing

In many large manufacturing organisations the main focus of research and development is on improving the ability of all their production operations to respond to unpredictable disturbances and increasing change on a global scale. Global competition forces companies to compete on all fronts in terms of cost, quality, delivery, flexibility, innovation and service.

Traditionally, the performance of all production systems, including those that are automated, has been assessed in terms of their output under steady-state operating conditions. However, greater product variety, smaller batch sizes and frequent new product introductions, coupled with tighter delivery requirements are becoming the norm. This means that manufacturers have to devise and develop automated operations that are capable of performing consistently under continually changing or disturbed conditions. Disturbances may be external to a production process (for example sudden changes in demand for the product, variations in raw material supply) or internal (for example machine breakdowns).

Automated production will play an increasing part in production responsive strategies. We saw in Unit 3 that a characteristic of these approaches is the negative impact it has on local and global employment patterns.

Advantages of automated manufacturing systems

- They increase 'value-added' by reduced labour costs including compensation costs for physical problems associated with repetitive tasks and back injuries.
- Precision and high speed improves cycle time, reliability and reduces downtime.
- They offer a faster time to market .
- Multi-axes robots are capable of servicing multiple machines/stations/operations in addition to having the ability to re-orient parts between operations without expensive options or use of complicated jigs and fixtures.
- They enable increased production rates not only with the accuracy and speed of a robot but also through the elimination of load-out inaccuracies and downtime required to change from product to product.
- They improve machine tool uptime productivity as much as 30 per cent by eliminating 'door open' time.
- Several sensing, motion, process and system options allow for greater control, consistency and quality output in less time with less chance of scrap parts.
- Typically, production rates are improved and do not vary by more than 3 per cent.
- There is no indirect labour training of potentially large numbers of operators.
- There is a shorter pay-back time on the capital invested in the machinery and equipment for an automated production line compared to machines with human operators.

Disadvantages of automated manufacturing systems

As we have seen, most types of robotic or automated devices are designed for manufacturing assembly or for materials handling, retrieval and storage. They cannot be used in all manufacturing situations – the greater the manufacturing complexity, the greater the complexity of calculations that need to be done before instructions can be given by the computer control system to the robot. Too complex a manufacturing process slows down a robot's speed of action and therefore increases manufacturing time. This could be a no-win situation, resulting in no manufacturing advantage – a human workforce could be more cost-effective. There are other significant cost factors:

- the cost of buying and installing and commissioning new technology so that it is effective
- the cost of recruiting and training operators with the necessary skills to enable them to use the new technology
- the cost of keeping up with new technological advances that enables the company to maintain its competitive edge.

Sensing, control and AI technologies are continually evolving and there are expected to be significant increases in the processing power and operational capability of computers in the future. The range and complexity of tasks that can be performed by manufacturing robots will therefore increase.

Tasks

1 Explain the difference between automation and robotics, giving examples of their application in your chosen materials area.
2 What are the advantages and disadvantages of flexible automation using robots in comparison with employing direct labour?

The impact of automation on employment

The most obvious impact of automation is on the numbers of people that are required to service the manufacturing process. Those people that are working in an automated manufacturing environment have to be able to operate and think in a more flexible way than ever before. This requires education, training and the development of ICT skills such as the collection and analysis of production data in order to measure their effectiveness. In the UK there is a chronic shortage of people with technical skills who are willing to take up a career in manufacturing. The people who want to work in this sector of industry need a wider range of basic skills including literacy and numeracy than previously and they must be able to transfer their skills and knowledge into new situations because the speed of change is so great. 'A job for life' is no longer the option for millions of workers across the globe. Workers in an automated working environment have to be capable of multi-tasking so that they are capable of responding to different machines within a production cell.

3. Uses of ICT in the manufacture of products

Since the 1990s some of the market forces driving manufacturing improvements have included an increased emphasis on design, innovation, operational flexibility, quick response to changing demand and the need for improved product quality. ICT is the key to enabling companies, large and small, to respond to new challenges effectively. Computers talk to one another across networks such as the Internet, which is a global network of computers. The recent development of **Integrated Service Data Networks (ISDN)** and broadband technology now means that huge amounts of information and data can be transferred across computer networks at far greater speeds than ever before. This is having a profound effect on the range and scope of electronic communications that are now possible in modern manufacturing.

Electronic communications
E-manufacturing

The term electronic manufacturing or e-manufacturing reflects the impact that ICT has made on the way that manufacturing is organised and managed. ICT is used from the boardroom, throughout the business and out into the supply and distribution (**logistics**) chains. Figure 4.3.26 shows the complex organisation of an e-business that is only cost-effective because of readily available electronic information and communication.

Beyond electronic mail (e-mail)

ICT has improved the range and reach of electronic communication. Electronic mail or email is the simplest form of electronic communication and it has a comparatively low level of reach and range when it is used for messaging or sending files to an individual or a work group. Reach refers to the level of communication that is possible with other users across a communications network and range refers to the types of data transfer that can take place. When business or manufacturing data can be shared by anyone, anywhere, irrespective of a computer's operating system, then the building blocks for an integrated design and manufacturing system with extended range and reach is in place. Electronic data exchange (EDI) makes the system work.

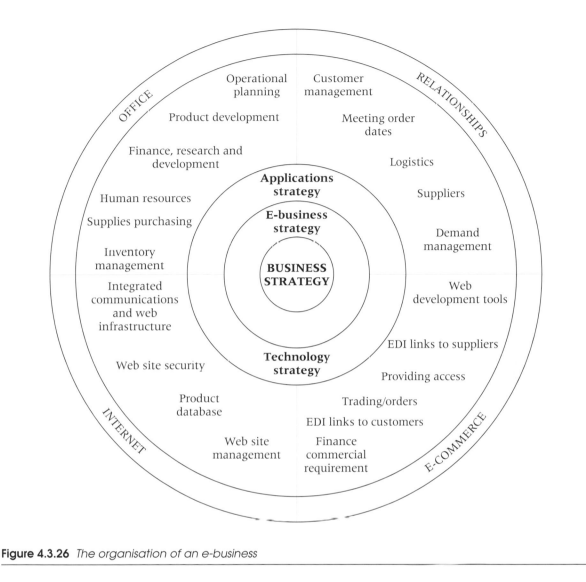

Figure 4.3.26 *The organisation of an e-business*

Electronic data interchange

ICT is the tool that integrates the use of computer systems and electronic links to create paperless trading. Figure 4.3.27 shows how this can happen in the manufacturing business so that all the participants or trading partners can communicate with each other electronically. This cost-effective and efficient technology known as electronic data interchange (EDI) is gradually being adopted by many industries as an essential tool to increase competitiveness. EDI has been a core element for many companies who have adopted the quick response and just in time as manufacturing strategies discussed earlier.

Electronic data exchange

The development of EDI into a means of exchanging technological data about all aspects of a product is well under way. CAD/CAM data interchange (CDI) is the process of exchanging design and manufacturing data. The system by which EDI and CDI are combined to provide automated transfer of data over a computer network is called electronic data exchange (EDE). There are various networks available for implementing EDE systems. The key to their usefulness in the field of graphics is their connection speed and the rate at which data can be transferred (throughput).

The modem was a big breakthrough in computer-based electronic communications. It allowed computers to communicate by converting their digital information into an **analogue signal** to travel through the public telephone network. The amount of information an analogue telephone can transfer is limited to about 56 kilobytes per second (kb/s). Commonly available internal or external modems have a maximum speed of 56 kb/s, but the actual speed

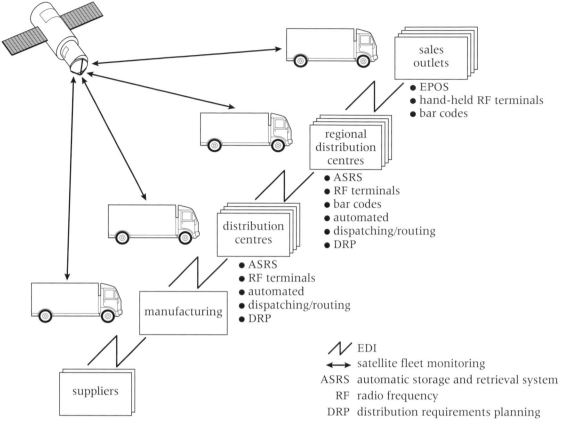

Figure 4.3.27 *The future uses of EDI in the total manufacturing process*

of data transfer is limited by the quality of the analogue connection. Computer modems routinely transfer data at about 35–40 kb/s. CAD files and other large files such as workflow data take a vast amount of time to be transferred from computer to another. This is ineffective and costly for all business, especially those involve in the graphics sector.

Integrated Services Data Network (ISDN)

Integrated Services Data Network (ISDN) allows multiple digital channels to be operated simultaneously through the same regular telephone wiring used for analogue lines. This permits a much higher data transfer rate than analogue lines. In addition, the amount of time it takes for a communication to begin on an ISDN line is typically about half that of an analogue line. This improves response for interactive applications with high graphic content and decision making such a games or conferencing.

Multiple devices

Previously, it was necessary to have a separate telephone line for each device you wished to use simultaneously. For example, one line each was

required for a telephone, fax, computer, or live videoconference. To transfer a CAD file to someone while talking on the telephone or seeing his or her live picture on a video screen requires several potentially expensive telephone lines. ISDN allows multiple devices to share a single line. It is possible to combine many different digital data sources and have the information routed to the proper destination. Since the line is digital, it is easier to keep the electronic 'noise' or interference out while combining these signals.

ISDN interfaces

The ISDN Basic Rate **Interface** (or BRI) is ideal for home, small business and remote working, also known as teleworking, since it gives fast access to most of the services available in an office including telephone, fax, email, Internet. ISDN BRI offers two simultaneous connections (any mix of fax, voice and data). When used as a data connection, ISDN BRI can offer two independent data channels of 64 kb/s each or 128 kb/s when combined into one connection compared to 30–40 kb/s for a modem. The fast dial-up speed and high throughput (when compared to a conventional modem) offers a more effective remote working environment for individual graphic designers or computer artists contracted to provide graphic services such as desktop publishing or website design to larger companies or end users

The ISDN Primary Rate Interface (PRI) is intended for high-volume telephone users such as large businesses as it is well suited for central-site tasks where many concurrent connections or calls will be handled in one place. ISDN PRI provides 30 channels of 64 kb/s each, giving 1920 kb/s. As with BRI, each channel can be connected to a different destination, or they can be combined to give a larger **bandwidth**. These channels, known as bearer or B channels, are at the heart of the flexibility of ISDN. A server-based connection to ISDN can act as a gateway offering telephone services to users on the **local area network (LAN)** in the office. For example, a server connected to ISDN can accept incoming data such as CAD files, and route them to individual users on the LAN.

Broadband

Most recently, ISDN services have largely been displaced by broadband Internet service, such as cable modem services. These services are faster, less expensive and easier to set up and maintain than ISDN. Still, ISDN has its place, as backup to dedicated lines, and in locations where broadband service is not yet available. Broadband technologies offer the possibility of extending a product visualisation right into the customer's home through the electronic equivalents of mail order or catalogue shopping. This kind of selling is often referred to as e-marketing and it is commonly found on the Internet and digital television channels.

Local area networks

Local area networks (LAN) are closed networks that are used in a variety of ways to meet individual needs. They are used for e-mail and the exchange of other electronic data within an organisation. For example, a LAN can be used to communicate manufacturing data from the design office directly to CNC manufacturing equipment within the same building.

Wide area networks

A wide area network (WAN) allows data to be transferred to processing or business centres around the world using existing digital telephone systems. To ensure compatibility between different systems WANs follow agreed standards and **protocols**. This can be so costly and ineffective, as they require dedicated equipment and software, that many companies

are now using the global network provided by the Internet as it is easier to access through standard **Internet service providers (ISPs)**. There are other types of network in use.

Intranets and extranets

An intranet is an Internet-based system used for communications and data exchange with web-type pages but there is a '**firewall**' that prevents unauthorised access to the site. Companies use intranets in the same way as the Internet is used, to share information and expertise, but they are much less expensive to set up compared with a WAN.

An extranet is an increasingly popular way for a company to share its data with its business partners though usernames and secure password-controlled access. The 'identity' of individual users will determine which parts of the network they can enter. These networks are also used for subscription services on the Internet to provide access to the 'expert' knowledge systems that were discussed earlier. Some companies now combine the ease and flexibility of access provided by an Internet website with web pages where access is password controlled. For instance, a company providing graphics equipment might have a series of pages that describe its range of products and retail pricing. The secure section accessed by retailers will include details of local suppliers, wholesale costs and conditions of resale to a customer.

Global networks

The world wide web is a global network of Internet **servers** that process and communicate data via cable, radio and satellite. **Web browsers** and search engines with which you will be familiar, such as Netscape Navigator, Microsoft's Internet Explorer and Ask Jeeves, make it easy to access information sources. Each website has a unique address called a **URL (universal resource locator)**. Web pages are specially formatted documents that are written in a language called **HTML (Hypertext Mark-up Language)** that supports links to other web pages, graphics, audio and video files. **Hyperlinks**, the specially marked 'hot spots' sometimes underlined, sometimes in a different colour or shown as a graphic 'button', allow the user to jump quickly from one website to another.

Advantages of using the Internet and the web

The Internet is becoming an invaluable tool for designers, manufacturers and suppliers. It provides:

- an easily accessible means of sharing ideas at a relatively low cost

- a vast and ever-growing body of knowledge and information (but remember it might not always be correct as there is no control on information that can be put into the web)
- a medium for communicating with current and potential customers and for seeing what the other designers or manufacturers are producing
- a readily accessible online source of reference product information and design data ranging from product or material specifications and catalogues of parts and components, to marketing trends and other commercial data.

Disadvantages of using the Internet

Industrial espionage is a real problem in a paper-rich environment but there is no guarantee that the Internet is any more secure without sophisticated data encryption software. The problem is that the more secure a system, the more difficult it becomes to use and this is said to be putting off potential customers of commercial websites. The growth of e-business on the Internet can be further restricted by scare stories of computer viruses destroying computers after accessing e-mails via the Internet. Again, virus software that can be updated via the Internet will eventually ease the fears of users.

Tasks

1 What are the differences between a LAN, WAN and a global network? Explain where they are used.
2 Describe the benefits to designers and manufacturers of using the Internet.
3 It is easy to waste a lot of time when trying to find information on the world wide web, so it is important to be clear about what you want to find out. Using a web browser or search engine of your choice, produce a short list of websites that provide information about CAD, robotics or any other area studied on this course. To share the information you have found present your links as a reference source for others to access. This might be in the form of a web page produced using web-publishing software or a presentation put together on a software package like PowerPoint.

Videoconferencing

Computers, electronic communications and video technologies have revolutionised the way people live and work. **Videoconferencing (VC)** is a rapidly growing segment of the ICT sector that integrates these three technologies to enhance communications and speeds up decision making by eliminating the need for time-consuming travel to meetings, which may be across the other side of the world. The use of ISDN and broadband technologies with improved rates of data transfer means that problems such as video pictures that are jerky or 'choppy', sound and picture out of synchronisation and poor quality or fuzzy images are a thing of the past. Two main types of VC organisation are used today:

- Desktop videoconferencing (DTVC) works like a video telephone between two people. Each person has a video camera, microphone and speakers mounted on to a desktop computer, equipped with a sound card. This means that each person can hear and see the other talk in a small window on his or her computer. Lower hardware costs have meant that this a readily available system that has even reached into the home computing market.
- Multi-point VC enables three or more people to sit in a 'virtual' conference room and communicate as if they were sitting next to each other.

Benefits of videoconferencing for manufacturers

We have seen how ISDN lines allow simultaneous file transfers. After a design meeting ends, the design information which has just been discussed needs to be sent to a regional office. This electronic file can be transferred either after the VC meeting has ended or, if a two-channel ISDN line is in use, it can be sent in the background on one channel while sight and sound information is still being exchanged on the other channel. VC can be used in other applications:

- Marketing presentations. In an era of global companies, marketers and manufacturers may not operate in the same country. When a new product is to be launched at a trade show or directly to the public, it is important to know how the product will be presented to persuade people to buy it. VC provides the opportunity to see the presentations as they develop and enables instant opinions on their effectiveness.
- Corporate training. If there is something employees all over the world need to know, it may be much faster and cheaper to train them using videoconferencing.
- Remote diagnostics. Experts in a particular process may work in one office but the problem requiring their immediate expertise

might occur halfway around the world. VC offers the experts the opportunity to solve the problem without travelling from the office. Production downtime is reduced and the expert and the other employees can get back to whatever it was they were originally doing without wasting time.

Remote manufacturing of components using VC
ICT provides the opportunity for expertise and expensive equipment to be made available to a number of schools, colleges and universities from one central location. Denford Ltd has pioneered the educational use of CAD/CAM via video conferencing in the UK, enabling staff and students to manufacture parts at a distance in real time using industry standard equipment and software (see Figures 4.3.28 and 4.3.29).

Task
Describe the benefits that videoconferencing might bring to the design, manufacture and sale of a graphic product.

Figure 4.3.28 *Denford has identified the potential of videoconferencing as an educational tool by incorporating the vital dimension of data sharing to allow staff and students to open notebooks, documents, drawing files and work together discussing problems and jointly solving them*

The steps involved in remote manufacturing through video conferencing
(see Figure 4.3.29)

Step 1. The student creates a design on CAD/CAM software on a PC (stand-alone or networked).

Step 2. The student and teacher participate in a live VC with the Remote Manufacturing Centre or the Denford on-demand facility.

Step 3. The student's design is downloaded to the Remote Manufacturing Centre where it is discussed and amendments made, where necessary.

Step 4. The Remote Manufacturing Centre creates the CNC file containing the student's design.

Step 5. The student's design is manufactured on a CNC machine while he or she views via the remote video camera.

Step 6. The finished component is evaluated, shown to the student on the VC and then posted back to the school/college.

Source: Denford Ltd

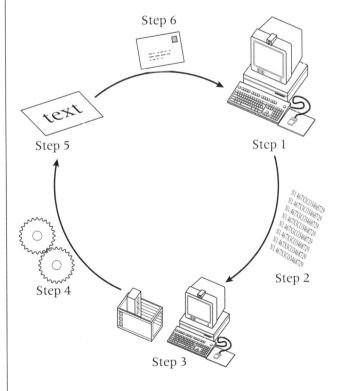

Figure 4.3.29 *The steps involved in remote manufacturing through videoconferencing*

The use of ICT for graphic communication

Electronic whiteboards

Sometimes called a 'smart board', this recently developed communication technology integrates the simplicity of a whiteboard with the power of a computer to make a tool for graphic communication. It is an interactive, flexible visual device for use in presentations, videoconferences, training sessions and for recording data (see Figure 4.3.30). Simply touching the board allows access to:

- a sensitive writing surface with a scanner and a thermal printer for producing hard copy
- computer software such as Microsoft Word, PowerPoint, spreadsheets and databases, for file manipulation and storage, printing or e-mailing
- data and video images, either stored on a computer or in the form of a live stream from a camera, video recorder or the Internet
- an automatic way of recording a video conference, which is available for playback and reference at a later date.

Information centres or PC kiosks

These interactive kiosks process, communicate and display specific graphic information and data stored on a computer (see Figure 4.3.31). Designed to automate information distribution, they are easily accessed by hotel guests, council tenants, tourists, etc. The kiosks can operate 24 hours a day and often include maps, display forthcoming events, employ graphic symbols and important contact information.

Figure 4.3.31 *A modern PC information kiosk*

Automated information appears in various forms:

- multimedia presentations of local information such as the location of restaurants, shopping centres, movies that are showing, hospitals, schools, addresses, businesses or local attractions, all accessed from a touch-screen, keyboard or mouse-driven interface
- information broadcasts via a customised mini-Internet broadcasting system using web-cams for video teleconferencing
- interactive services at a range of trade and product shows that allow potential customers immediate access to critical marketing information and product reviews. This type of interactive system can be used to create newsgroups, chat-rooms and portals to a vast range of community information in organisations such as local and regional authorities.

Kiosks have also been designed for museum displays, public Internet access and information booths in public spaces. Changes in the size and power of personal computers have revolutionised the design of these information centres. The large and bulky kiosks of yesterday have been transformed into the compact, slim-line designs of today. The PC kiosk with back-lit advertising signage shown in Figure 4.3.31 has a state-of-the-art Pentium processor with 128 Mb RAM at its heart, combined with a 15" LCD 'touch-screen', touch-pad and standard keypad. Extra functionality and security is provided by the

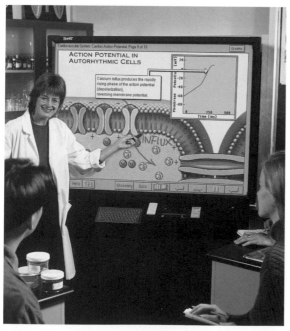

Figure 4.3.30 *Electronic whiteboard*

Figure 4.3.32 *An information kiosk for a public place*

inclusion of a credit card reader, camera and fully linked 'windows' type information screens. Figure 4.3.32 shows a PC kiosk of the future.

Task
Discuss how electronic communications have impacted on manufacturing and business practice worldwide.

Electronic information handling
Agile manufacturing
An agile manufacturer recognises the uncertainty that change brings and puts flexible manufacturing systems (FMS) and mechanisms in place such as **quick response manufacturing (QRM)** to deal with it. The organisation moves from being production-driven to customer-driven and it realises that customers will not pay a premium for product quality: quality is always assumed. Agile manufacturers work in partnerships with customers and suppliers, and understand that the so-called 'soft' business information processes are important to the entire manufacturing process. In an agile manufacturing environment, information is the primary enabling resource and ICT is the enabling technology. In order to find out what the customer needs and wants manufacturers set up a range of computer-based systems for collecting market information.

Computer-aided market analysis (CAMA)
Market research is a term that describes the collection and analysis of data about consumers, **niche markets** and the effectiveness of marketing programmes. Market analysis focuses on the collection, analysis and the application of research data to predict the future of a particular market (trends).

The analytical process includes the examination and evaluation of relevant information (data) in order to select the best course of action from the possible business options. The analysis can be undertaken in-house by the manufacturers themselves, through specialist market research agencies, or consultancy firms who will provide tailor-made or customised data sets related to specific companies, customers or markets.

The analysis of business data and market trends generates a vast flow of information and data that is now most efficiently managed in a computerised relational database. Such a database can be interrogated in various ways depending on the type of analysis required:

- A qualitative analysis will tell a company who is buying the products and why, or what customers like or dislike about a product and its after-sales support.
- A quantitative analysis will provide facts and figures such as where the products are being bought and when, or a comparative analysis of the company's financial ratios over time.
- A trend analysis will tell a company what is happening in a particular sector of a market and put the company's performance into a local, regional, pan-European or global context.
- **Market timing** is about attempting to predict future market directions, usually by examining recent product volume or economic data, and investing resources based on those predictions.
- In situations of high product volumes, existing customer information can be profiled against lifestyle surveys and demographic data to give a detailed picture of the company's ideal customer.

Benefits of CAMA
In a modern e-business environment, a manufacturer needs access to up-to-date research, in-depth product and market analysis and industry-specific expertise to make the best ICT-led decisions in relation to the core business goals. Using advanced computer-based marketing tools such as CAMA will help a manufacturer to:

- convert data into actionable information for sound marketing and planning decisions
- calculate demand for products and services more accurately and set the right sales targets
- identify markets and find out where potential customers are shopping
- employ a marketing technique – **market segmentation** – that targets a group of customers with specific characteristics
- launch new products with focused strategies such as regional or mini product launches rather than whole country product 'roll outs'.

Computer-aided specification development

An effective product design that satisfies functional requirements and can be manufactured easily requires vast amounts of 'expert' knowledge. Integrated ICT-based systems already exist where design features can be generated by CAD software and then checked for ease of manufacture (design for manufacture – DFM) and for ease of assembly (design for assembly – DFA) by a knowledge-based expert system. With the increased volume of design information generated by computer-based systems, designers now have to analyse and evaluate large data sets of design constraints to find a design specification that satisfies those constraints.

A design specification is a document which explores and explains the internal design of a given component or sub-system. It should be sufficiently detailed to permit manufacturing to proceed without significant reworking resulting from flawed or missing design details. It will often draw on previous specifications or other design information. The computer-aided specification of products saves time and costs through basing new product specifications on those already held in a product data management (PDM) system which is, in effect, an intelligent design system.

There are three classifications of information or knowledge held within an intelligent design system: CAD data, a design catalogue and a knowledge database. The CAD data contains specific information about the physical characteristics of each component part being designed. The design catalogue is a reference for data such as the cost, weight and strength characteristics of the standard materials, or parts and fasteners that are available to the manufacturer. The knowledge database contains 'rules' about design and manufacturing methods. These automated and intelligent design environments when fully developed will enable a designer to perform and manage complex automated design tasks including decision making.

> ### *Task*
> During this course you have developed different products to meet different demands. In other words, you have created your own 'expert' database. Using this knowledge, list six design and six manufacturing 'rules' that you have applied. Write your rules as an if/then statement. For example 'if the logo has been designed on a CAD package, then I can use a CNC plotter to manufacture the vinyl product.'

Automated stock control

Earlier in the unit, we looked at the benefits that arise out of applying computers to implement the just-in-time (JIT) philosophy where materials and components are ordered just in time for production. A characteristic of JIT is that product variety increases as the range of processing systems decreases and waste is reduced by a variety of means including automated stock control systems. These ensure that the size of the **inventory** is optimised but available on demand when and where it is needed. These automated stock control systems support the move from batch to continuous flow production. They are also an advantage in the process of line balancing, which is a scheduling technique that reduces waiting times caused by unbalanced production times.

Production scheduling and production logistics

Computer-based scheduling and logistics systems ensure that a production plan is implemented and that production is 'smoothed' so that small variations in supply and demand are managed without causing problems. This is achieved by spreading the product mix and the quantities of each produced evenly over each day in a month. Different manufacturers have different **planning horizons** that determine when the detailed production plan is produced; typically, it is one month in advance. The advantage of these computer-based scheduling and logistics applications is that:

- they are flexible and easily adapted if the product mix or quantity required is changed at short notice
- they minimise work in progress and reduce the inventory
- they maintain balance between the stations on a production line
- they raise productivity levels.

Flexible Manufacturing Systems

Quick Response Manufacturing (QRM)

QRM focuses on reducing product lead times and the production of small batches such as advertising or promotional gifts. QRM provides a better 'time to market'. Real-time reprogramming of manufacturing is a production management tool that offers high-volume, rapid turnover manufacturers the potential for enormous savings in time to market and increased business efficiency.

The ability to re-programme automatically both manufacturing and business processes in response to market pressures is on the horizon. Stock levels will be constantly re-evaluated as demand patterns change. The demand patterns also affect capacity planning and so the relevant individual within the business would be automatically alerted. The impact that a change to the business or a manufacturing process may have will be immediately available for review from anywhere in the world. The change to real-time re-programming also presents the possibility for disastrous levels of confusion unless the information flow is carefully managed by an effective PDM system using highly integrated knowledge bases.

Production control

Quality monitoring

Mechanical methods of inspection involve direct physical contact between a probe and the component or product being evaluated. The probe sends data back to a computer or microprocessor for checking against the product specification. These systems are concerned with checking the dimensional accuracy of a product such as the injection mould for producing a plastic package such as a shampoo bottle.

The computer-aided optical monitoring of quality involves the use of electronic sensors that use scanning and optical devices, digital cameras and vision systems. They are used in a range of situations such as checking the inking pattern on a poster or other printed product. They can be used for ensuring that colour mixes and inks are consistent throughout a high-volume printing process. They can be used for comparing the quality of a visual product against the original artwork or design.

Using digital cameras for monitoring quality

Inspection has always been part of the manufacturing process. However, this often involved the inspection of a random sample of finished components; if any errors are detected at this late stage, the main processing operations have already been carried out and putting the fault right is costly and may not be possible. This results in high scrap rates and wasted processing time. To help alleviate this problem, 100 per cent piece-by-piece inspection of work in progress can be carried out using low-cost digital cameras at critical points on the production line. The cameras are connected to a computer or to a dedicated microprocessor containing the original specification. Differences between what has been tested and the design specification are automatically evaluated and an audible or visual signal identifies when there is a problem. This allows rapid real-time quality control. The other main advantages of optical methods of monitoring quality are that:

- there is no direct mechanical contact with the product so the quality of the product, especially if it is a graphic product, is not compromised
- the distance between the optical sensor and the piece being measured can be large
- the response time is fast because of the electronics built into the system.

Product marketing, distribution and retailing

Using ICT for the business processes in manufacturing

Throughout this unit we have considered how ICT is being employed to improve profitability and product quality on the production side of manufacturing. CAD/CAM technologies and the use of robotics lead to new ways of organising manufacture. As we have seen, ICT can also be used to speed up business processes such as

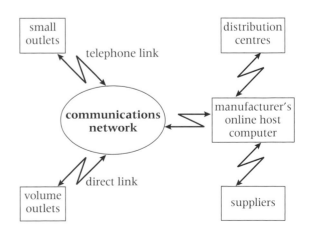

Figure 4.3.33 Manufacturer's communication links with distribution centres, suppliers and retail outlets

invoicing and stock control but this on its own will not necessarily increase productivity. One of the key benefits that increased use of ICT brings is in better business integration. This means that manufacturers not only have to change their manufacturing processes but also the way that they do business outside the company. Refer back to Figure 4.3.26 on page 255 to review the complex organisation of a large e-business.

Electronic point of sale (EPOS)

Information from many sources, including EDI, is at the centre of any business organisation. EPOS is a form of EDI that allows companies to monitor product sales in order to respond by supplying and delivering their products and services to their retailers or customers faster (see Figure 4.3.34). For retailers it allows them to keep their costs down by reducing the stock that has to be held in reserve. Almost all products sold in retail outlets have a unique bar code that is scanned by fixed or hand-held laser-operated bar code readers. The bar code contains a range of electronic information about the product which is collected automatically or at scheduled times during the day. In addition, the two-way information flow can include financial data, emails and price updates from the supplier to the product retailer. EPOS and the associated management software provides manufacturers with:

- a full and immediate account of the financial transactions involving the company's products
- the data needed to undertake a sales/profit margin analysis, as it can be exported into financial management software such as spreadsheets
- the means to monitor the performance and popularity of all product lines, which is particularly important in high-volume situations as it allows the company to react quickly to demand fluctuations and be able to deal with surges in demand by being notified of them immediately
- accurate information for identifying buying trends when making marketing decisions rather than relying on guesswork
- a full and responsive stock control system by providing real-time stock updates for product sales, transfers between retail outlets, goods received, etc.
- a system to ensure that sufficient stock is available to meet customer needs without overstocking and tying up capital. This is known as a continuous product replenishment system (CPR) which is a feature of high-volume merchandising.

Internet marketing

Entering the Internet market is a major strategic decision for any manufacturer. It generally means an almost total restructuring of the internal and external business processes as well as supply chains and relationships with customers. For many manufacturers this redesign of the business process presents a bigger challenge than the change to an automated production system or other ICT-based manufacturing process.

For many companies the rush to the Internet and the development of websites was ill conceived and badly thought out. Many businesses invested heavily in terms of time and money merely to put their existing product catalogues out on the world wide web to little commercial effect. There was no underpinning business strategy and customers were able to do little more than browse the catalogue electronically. If they wanted to contact the company, it had to be by conventional means such as phone or fax. Some companies added email addresses in recognition of a need to communicate interactively. Using the Internet to market a business is more than that. Some large companies are looking to small and medium-sized graphic design consultancies companies who specialise in the design of websites or the development and use of 3D images.

The move towards using interactive 3D virtual products in e-marketing on the web is a natural extension of modern marketing and sales strategies. Manufacturers are faced with ever-shortening product life cycles, increasing product complexity and greater global competition. They must ensure that existing and potential customers understand their products better. They might even have to provide a means by which individual customers are able to configure a product to meet their own specific needs as we saw when we looked at digital print solutions.

The benefits of Internet marketing include:

- instant global reach, access to new markets and an increased customer base
- faster processing of orders and transactions, resulting in efficiency savings and reduced overhead costs
- detailed knowledge of user preferences, leading to improved customer service through better product and after-sales support
- reduced time to market
- an increased company profile on a worldwide basis
- for small and medium-sized companies, especially those offering bureau or other specialist graphic services, it creates new

markets because it is often easier and cheaper to create an Internet presence than it is to promote their products by conventional printed or visual means.

The future use of ICT in the graphics field

We have seen that advances in ICT are already providing manufacturing with benefits throughout its business. The impact of ICT is also powerful in the field of graphic art.

Interactive graphic design

Games playing and simulations are examples of interactive graphic design but there is a growing and exciting range of interactive and visual experiences to be found. Digital media is influencing the arts and is giving form to new artistic languages that make bridges between the real world and imaginary virtual environments. Electronic visualisation is the art and science of creating images on electronic screens and on virtual reality display devices. Electronic visualisation uses the tools of advanced computer graphics, computer animation, interactive graphics, video and virtual reality (VR) to create a more intensely aesthetic interpretation of imagery and visual design in a virtual reality environment.

The role of the printed book as the prime information source has also been transformed by the arrival of digital inputs that create new information pathways that are interactive, versatile and aesthetically challenging in addition to engaging the intellect. Interdisciplinary groups are being set up around the world to develop the creative potential that evolves from the integration of technology, communication, art and design to expand the boundaries of traditional artistic media and of visitor participation in artwork. For example, one group has created an emotional, virtual reality experience for the public of being in different centuries and in different worlds all in one unique container, the Multi-Mega Book in a virtual environment. The range and scope of this exciting digital fusion is beyond the scope of this book but you can gain an insight into the future of digital media by visiting the following web sites www.evl.uic.edu in the USA and www.fabricator.com in Italy. These will lead you to other areas where science technology and art come together to create informative and educational interactive graphic applications.

Exam preparation

You will need to revise all the topics in this unit, so that you can apply your knowledge and understanding to the exam question.

In preparation for your exam it is a good idea to make brief notes about different topics, such as 'Production scheduling and production logistics'. Use sub-headings or bullet point lists and diagrams where appropriate. A single side of A4 should be used for each heading from the Specification.

It is very important to learn exam skills. You should also have weekly practice in learning technical terms and in answering exam-type questions. When you answer any question you should:

- read the question carefully and pick out the key points that need answering
- match your answer to the marks available, e.g. for two marks, you should give two good points that address the question
- always give examples and justify statements with reasons, saying how or why the statement is true.

Practice exam questions

1 a) Explain the following terms related to the use of computers in manufacturing:
    i) CADMAT
    ii) product data management (PDM). (6)
   b) Describe what is meant by a manufacturing cell. (3)
   c) Explain why computer-aided engineering is important when developing a product. (6)

2 a) Describe the meaning of **three** of the following terms related to production planning:
    - master production schedule
    - resource scheduling
    - order-based scheduling
    - materials requirements planning. (9)
   b) Describe the importance of inspecting quality when manufacturing products. (6)

3 a) Explain the difference between automation and robotics. (8)
   b) Outline the limitations on the use of robots in product assembly. (3)
   c) Describe **two** of the following terms:
    - automatic storage and retrieval system
    - artificial intelligence
    - fuzzy logic system. (4)

4 a) Explain **three** of the following terms related to electronic communications: (6)
    - electronic data interchange
    - local area network
    - global network
    - integrated services data network. (6)
   b) Discuss the benefits of using video-conferencing for manufacturers. (4)
   c) Explain how desktop video conferencing works. (5)

5 a) Explain the meaning of quick response manufacturing. (4)
   b) Describe the advantages of using computer-aided inspection methods for production control. (5)
   c) Discuss the issues related to Internet marketing. (6)

Total: 81 marks

UNIT 5 Product development II (G5)

Summary of expectations

1. What to expect

You are required to submit one coursework project at A2. This project enables you to build on the knowledge, understanding and skills you gained during your AS coursework. The A2 project should, therefore, demonstrate a broader use of varying materials, a wider variety of skills and increasing knowledge of the technology associated with the A2 Specification.

At A2 level you will need to take a commercial approach to designing and manufacturing, working in a similar way to an industrial designer. This means working more independently, designing to meet the needs of others and taking responsibility for planning, organising and managing your own project work. This may involve the use of a wider range of people to support you in your work. For example you could use the expertise of a visitor from business, or make use of work-related resources. Don't forget to reference any help you get in your bibliography.

The A2 project should comprise a product and a coursework project folder. It is important to undertake a project that is appropriate and of a manageable size, so that you are able to finish it in the time available.

2. What is a Graphics with Materials Technology project?

An A2 Graphics with Materials Technology project should:

- reflect a study of the 'technologies' involved in the A2 Graphics with Materials Technology Specification
- include a 2D element developed from traditional and modern graphics media
- include a 3D model using at least one resistant material listed in Unit 3A (Classification of materials)
- include a coursework project folder that demonstrates good quality graphic communication
- include an increased emphasis on industrial applications and commercial working practices.

For example an A2 Graphics with Materials Technology project could be developed in collaboration with potential users, or with a client (such as a local business or organisation). This would enable the development of a designer–client relationship. Collaboration would need to include consultations with the client or users to develop a design brief and specification, together with input into the analysis and research. Discussions with the client or users would be expected to take place at critical stages throughout the project, enabling the use of feedback when making decisions. Remember to reference in your folder any changes made to the product design and/or manufacture as a result of feedback from the users or client.

3. How will it be assessed?

The AS project covers the skills related to designing and manufacturing. It is assessed using the same assessment criteria as the AS project (see Table 5.1).

Table 5.1 AS coursework project assessment criteria

Assessment criteria	Marks
A Exploring problems and clarifying tasks	10
B Generating ideas	15
C Developing and communicating design proposals	15
D Planning manufacture	10
E Product manufacture	40
F Testing and evaluating	10
G Appropriate project	10
Total marks	110

You must attempt to cover all the Assessment Criteria A–G. The G criterion has been included to reflect that your project meets all the requirements of the Specification (see 'What is a Graphics with Materials Technology project?'). You will need to undertake a project that has a level of complexity suitable for A2. It is very important, therefore, to check the appropriateness of your project with your teacher or tutor *at the start of the project*.

At A2 the assessment criteria demand a different level of response, requiring you to demonstrate a higher level of design 'thinking' and more in-depth knowledge and understanding. You are also expected to take more responsibility for your own project management.

Your A2 project will be marked by your teacher and the coursework project folder will be sent to Edexcel for the Moderator to assess the level at which you are working. It may be that after moderation your marks will go up or down.

4. Choosing a suitable project

At A2 level you will need to design and manufacture a product that meets needs that are wider than your own. This will enable you to include a range of designing and manufacturing activities that are similar to those used in industry. You will need to design and manufacture one of the following:

- a one-off product for a specified user or client
- a product that could be batch produced or made in high volume for users in a target market group.

The key to success is to identify a realistic user, client or target market need and solve this need through the design and manufacture of an appropriate product. Remember to evidence your understanding of industrial practices through the use of industrial type terminology and technical terms. You should make use of feedback from your user, client or target market group in order to access the full range of marks.

Planning considerations for the manufacture of your one-off, batch, mass or continuously produced product should include details of how one single product will be manufactured in your school or college workshop. You are not required to manufacture the product in quantity, although, depending on the type of product you develop, you may need to produce identical components for use in the product. Together with your production plan for one single product, you will need to include an explanation of how the product could be batch, mass or continuously produced. This will require you to highlight the changes that would be necessary to the manufacture of your one single product if it were to be made in quantity.

5. The coursework project folder

The coursework project folder should be concise and include only the information that is relevant to your project. It is essential to plan and analyse your research and be very selective about what to include in your folder. The ability to be selective is a high order skill which will enable you to access the full range of marks at A2.

Your coursework folder should demonstrate that you have achieved:

- a greater understanding of the design process (higher level design 'thinking')
- a higher ability to select and use relevant information (higher level research and evaluation skills)
- closer connections between relevant research and the development of ideas
- better understanding of the use of appropriate materials, processes and manufacturing techniques

- greater understanding of relevant technical terminology
- higher level communication and presentation skills
- appropriate use of ICT, including finding a balance between computer-generated images and those that are hand drawn.

Your coursework folder should include a contents page and a numbering system to help its organisation. The folder should comprise around 20–26 pages of A3/A2 paper. A title page, the contents page and a bibliography should be included as extra pages.

Table 5.2 gives an approximate guideline for the page breakdown of your coursework project folder. In the section on 'Product manufacture' it is essential to include clear photographs of the actual manufacture of your product. This should provide photographic evidence of modelling and prototyping, any specialist processes you have used including the use of CAD/CAM and show the stages of manufacture.

Please note, however, that the guideline for the page breakdown of your coursework folder is only a suggestion. You may find that your folder contents vary slightly from the guideline because of the type of project that you have chosen.

Table 5.2 Coursework project folder contents

Suggested contents	Suggested page breakdown
Title page with Specification name and number, candidate name, and number, centre name and number, title of project and date	extra page
Contents page	extra page
Exploring problems and clarifying tasks	4–5
Generating ideas	3–4
Developing and communicating design proposals	5–6
Planning manufacture	3–4
Product manufacture	2–3
Testing and evaluating	3–4
Bibliography	extra page
Total	20–26

6. How much is it worth?

The coursework project is worth 20 per cent of the full Advanced GCE.

Unit 5	Weighting
A2 level (full GCE)	20%

A Exploring problems and clarifying tasks (10 marks)

1. Identify, explore and analyse a wide range of problems and user needs

Look at Figure 5.1 and consider the question in the caption.

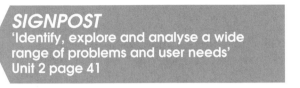

SIGNPOST
'Identify, explore and analyse a wide range of problems and user needs'
Unit 2 page 41

Developing a project at A2

The A2 project enables you to draw together and apply all the knowledge, understanding and skills related to designing and making, that you gained at AS level. This experience should provide you with a clear understanding of the assessment requirements and give you a solid basis for developing a new project at A2.

It may be helpful, at this stage, to read the AS and A2 exemplar projects in the Edexcel Coursework Guide, which demonstrate the standard at which you are required to work during the A2 project. You will be required to work more independently, which may involve using a wider range of people to support you in your work. Your most difficult decision now is to decide what product to design and make that will enable you to demonstrate a higher level of 'design thinking'. Your choice of A2 project must enable you to produce:

- a 3D model or prototype product, incorporating at least one resistant material, such as wood, MDF, plastic or metal. The 3D outcome should be semi-functioning.
- a 2D element developed from traditional and modern graphics media. The 2D element should be linked to and support the 3D outcome.
- a coursework project folder that summarises the development of the 2D/3D elements. The folder should demonstrate good quality graphic communication.

New ideas or problems may have come to mind during your work on Unit 2, or when you discussed your project with others. In order to develop your A2 project, you will need to decide on a design context and a target market group.

Developing a commercial approach

At A2 you are expected to take a more commercial approach. This means designing and making 2D and 3D elements that may have the potential for batch or high-volume production. In industry a prototype product is made prior to manufacture to test every aspect of the design before putting it into production. This commercial prototype product is as close as possible to the 'real' end product. Your product also needs to be as close as possible to the 'real' thing, so you should make it to the highest possible quality.

One way of developing a commercial approach to your project is to collaborate with a potential user(s) or with a 'client', which could be a local business or organisation. This will enable you to work in a similar way to a professional designer, who works to a client brief, meets the needs of others, makes use of client feedback and works to a budget and deadline. You will also need to make use of feedback from your user(s) or client, in order to access the full range of available marks. It is essential, therefore, to work out at the start of your project when and how you will consult with your user(s) or client, so you can obtain their feedback about your product.

There are a number of ways to develop a commercial approach to project work. You could consider some of the following:

Figure 5.1 *The Nike logo has been used on sports shoes since the 1970s and is one of the most successful logo of all time. Explain why you think that this is so*

- Use work-related materials produced by a business, e.g. using a company information pack to help develop a product that could be marketed by that company or retailer. In this case you would need to research the needs and use feedback from an appropriate target market group.
- Use the expertise of a visitor from business, e.g. understanding a company's marketing strategies and the type of products they make. In this case, you could use the visitor as a client, consult with them throughout the project and use their feedback when making decisions.
- Make an off-site visit to identify a specific problem in the local community, e.g. visiting a community centre, primary school or workplace, where different needs can be investigated.
- Use work experience as a context for designing and making a product, e.g. using a part-time job or work experience to spark off ideas about developing a product to meet specific needs.

Identifying and analysing a realistic need or problem and exploring the needs of users

> **SIGNPOST**
> 'Identifying and analysing a realistic need or problem' Unit 2 page 42

Once you have decided your approach for your A2 project, you can start the actual development process. It is always difficult at the start of a project to know exactly where and how to start, and what the end product will be – this is part of the excitement of product design. The best way forward is to undertake two tasks that are interrelated:

- identifying a realistic problem for a specified user(s) or a client that will lead to product development
- identifying the needs of potential users and developing a product that will fulfil these needs.

Whichever approach you take, the key to success is to develop a designer/client relationship so that you can make use of feedback when making decisions.

You can explore the needs of potential users and look for product ideas by undertaking market research. In many industries, market research is carried out to identify the taste, lifestyle and buying behaviour of potential customers. This establishes the profile of the target market group.

> ## Task
> ### Using market research techniques
> Collect information from your user, client or target market group and investigate products using market research techniques:
>
> a) Produce a product report through window shopping, going into stores, or visiting galleries or museums. Identify product types, price ranges, market trends and new ideas.
> b) Use questionnaires/surveys to identify user needs and values. Identify age groups, available spending money, favourite product types and brand loyalty.
> c) Use product analysis to find a 'gap in the market'. Identify different styles, manufacturing processes, quality of design and manufacture, and value for money.

For manufacturers the 'customer' plays a key role in the product development process. Without customers there is no need to make products, so their views are vital if a product is to sell in the market place. Manufacturers need to ensure that their products meet customer requirements, at the right quality, at the right price and at the right time. For example, it is no good trying to sell a product that is made from purple plastic if the customers' concept is that purple is last year's colour! Feedback from customers is, therefore, crucially important if their requirements are to be met. Figure 5.2 shows how one student sought the views of potential customers.

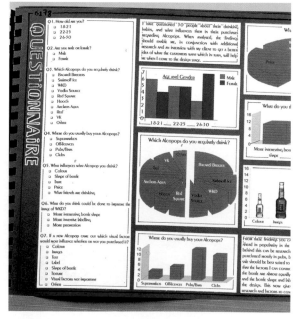

Figure 5.2 *Part of a student questionnaire*

Clarifying the task

SIGNPOST
'Clarifying the task' Unit 2 page 43

Clarifying the task means deciding on the purpose for your product and how it will benefit users, and the image it may need to project in order to promote its market potential. Market research about buying behaviour may help here. You should also consider the factors that affect customer choice when buying products. These may include considerations of:

- price and value for money
- aesthetic factors, like the product's appearance and the image it provides the users
- marketing factors, such as brand image
- functional requirements of packaging materials or point-of-sale display, such as ease and safety in use.

> **To be successful you will:**
> - Clearly identify a realistic need or problem.
> - Focus the problem through analysis that covers relevant factors in depth.

2. Develop a design brief

SIGNPOST
'Develop a design brief' Unit 2 page 43

Whichever approach you take in developing your A2 project, you should find that your early exploration of problems and user needs should help you develop a design brief. Take into account:

- the views of the potential users, client or target market group
- information about similar products
- the market potential of the product
- its benefit to the users.

Your brief should enable you to plan what, how and where you need to research in order to develop a product that meets the needs of your potential users.

> **To be successful you will:**
> - Write a clear design brief.

3. Carry out imaginative research and demonstrate a high degree of selectivity of information

SIGNPOST
'Carry out imaginative research and demonstrate a high degree of selectivity of information' Unit 2 page 43

At A2 level your coursework should demonstrate a greater understanding of the design process, higher-level research and evaluation skills, and increasing knowledge of the technology associated with the A2 Specification. Your coursework folder should demonstrate your ability to:

- Undertake research that targets the design brief more closely. Don't waste time doing unnecessary research.
- Select and use relevant research information. Only include in your coursework folder information that is directly connected to your project. For example, do not include everything you know about hardwood if you are not using wood to make your product. Even if you are using wood, you still do not need to include everything you know about it, only information relating to its appropriateness for making your product.
- Make closer connections between your research, your product design specification and your design ideas. Making connections is what this unit is all about (see Figure 5.3).

Collaborating with users or a client on your project will make it easier to target your research, because you will have specific requirements and user needs in mind. Include a range of primary and secondary research. This can include using questionnaires, analysing existing products, using statistical data and identifying market trends.

At Advanced Level you should be thinking about a range of issues related to your 3D model or 3D prototype product *and* the 2D elements, such as the ease and cost of manufacture; the ease of maintenance; how to ensure the performance, quality and safety of your product; if there are any environmental issues related to the materials or components you may use.

In industry, product design teams do not generally undertake market research themselves, but rely on market information produced by the marketing department. Many manufacturers use product data management (PDM) software to organise, manage and communicate accurate,

Figure 5.3 *Student research sheet*

up-to-date product information in a database. Everyone in the product design team has access to this information, making it easier and faster to get the product to market. Since you are not in this happy position, you have to do your own research, so bear in mind that you have to meet deadlines. If you have access to the Internet, you may be able to find useful information about existing products, for example about materials properties, product prices and styling. Many companies also provide information through company newsletters and case studies about their products. However, before you start any Internet research you need to know exactly what you are looking for and how to find it. Find out web addresses before you start and try to avoid following interesting but useless leads.

All the research that you undertake should enable you to write a product design specification. Review the criteria included in the AS specification because these basic criteria should be part of the A2 specification too. You should also refer to the further product design specification criteria required at A2 level to ensure that your research will enable you to address these criteria.

Stop and think!

Don't forget to reference all secondary sources of information in a bibliography at the end of your coursework folder.

To be successful you will:

- Carry out a wide range of imaginative research, with a high degree of selectivity of information.

4. Develop a design specification, taking into account designing for manufacture

SIGNPOST
'Develop a design specification, taking into account designing for manufacture' Unit 2 page 45

At A2 you may find that the development of your product design specification runs concurrently with writing your design brief, especially when you are working in collaboration with a client or users. Early exploration of the design context and possible discussions with the user(s) or client may provide enough information for you to develop the design brief and an *outline* product design specification at the same time. This can be very beneficial because it can help you to target more closely the research you need to do. As your research progresses, you can amend and develop your design specification until you reach your final product design specification. You may find that even this will need to be amended later on, after you have worked on and evaluated ideas for the product, or even after modelling and prototyping it.

Specification criteria

The A2 product design specification should take into account all the criteria that you worked with at AS level. At A2 you are expected to demonstrate a greater understanding of the design process. You can do this by developing a product design specification that takes into account more sophisticated specification criteria. The questions below should help you develop your A2 design specification:

* What influence will values issues, such as cultural, social, moral or environmental issues, have on the purpose, function and aesthetics of your product?
* What influence will market trends and user requirements have on the market potential of your product? The market potential of high volume products relies on developing a competitive edge – providing products at a price people can afford, combined with the image they want the product to give them.
* What will be the life expectancy of your product? What maintenance will it require?

What are the implications for the cost and quality of your product?
* What type of materials and components might be suitable? What are their properties and characteristics? How can you ensure availability at the right time and the right cost?
* What kind of processes, technology and scale of production might be suitable? Is the use of CAD/CAM to be specified in design and manufacture? What influence could the constraints of designing for high volume have on your product? Successful mass-produced products rely on manufacturing reliable, high-quality products at a price people can afford.
* What kind of quality assurance system can you put in place to ensure your product's quality and reliability? Will this system make use of quality control, inspection and testing?
* What legal requirements and/or external standards (British Standards) related to performance, quality and safety do you need to take into account when developing your product?

Take into account scale of production

At A2 you should be developing 2D and 3D elements for an individual user, client or target market group. Your identified scale of production will be one-off, batch or high volume production, but you will be designing and manufacturing *one* product in your school or college workshop. You are not required to manufacture the product in quantity, although you may need to produce identical components for use in the product. When you plan the manufacture of your one single product, you will need to explain how it could be adapted for batch, mass, or continuous production.

Your product design specification is the connection between research and ideas. It should guide all your design thinking and provide you with a basis for generating, testing and evaluating your design ideas, your final design proposal and the end product. The design specification is therefore a control mechanism that sets up the criteria for the design and development of your product.

> ### To be successful you will:
> * Write a clear design specification.

B Generating ideas (15 marks)

1. Use a range of design strategies to generate a wide range of imaginative ideas that show evidence of ingenuity and flair

SIGNPOST
'Generating ideas' Unit 2 page 46

You are expected to demonstrate a more mature approach to design when generating ideas at A2, which means making closer connections between your research, your product design specification and your ideas.

In industry the product design specification forms the basis for developing ideas. Although your specification should provide inspiration both for aesthetic and functional considerations for design, it will probably be aesthetic considerations that provide the most freedom when generating exciting ideas.

In your research, you will have investigated:

- user, client or target market needs relating to their preferences about products

Figure 5.4 This 1927 poster by A M Cassandre uses powerful geometric forms based on the most famous avante-garde movement of the day – cubism. It very cleverly integrates the company name with the image

- market trends relating to styling, colour, images or typography
- existing products that are similar to the one you intend to design and make.

This kind of research information is vital if you are to generate imaginative ideas. Industrial and commercial designers rarely start from scratch, mainly because they have to meet deadlines and because they have to design products that people will buy. They have to take shortcuts, which often involve using other product designs, the work of artists or other designers as inspiration or as starting points for new ideas (see Figure 5.4).

You can work in a similar way, using information about market trends and other aesthetic considerations to inspire ideas. You could:

- produce a '**moodboard**' of visual images to inspire ideas, e.g. make a collection of quirky images and products, a collection of colours, textures, typography and 'swipes' from magazines that suggest a mood or theme – these can be used to give your product an identity or image
- make connections between old and new technology, e.g. adapt the style of 1930s or 1960s advertising materials to produce a new look for packaging, point-of-sale or promotional materials
- use the work of a design movement to inspire the imagery for your 2D and 3D elements, e.g. reflect the 'Memphis' style
- use an art movement or the work of an architect to inspire ideas, e.g. use the art movement 'Futurism' or the work of the Spanish architect Gaudi to develop form or styling
- use the influence of cultural or traditional art or design to inspire ideas, e.g. use themes such as 'American Indian' or 'East meets West'
- use themes built around 'values issues' as a starting point for design, e.g. 'eco design' or 'recycling'
- use the natural or built environment to inspire ideas, e.g. shells, fruit, wrought-iron work, street furniture.

It is often a good idea to keep a notebook for jotting down ideas as thumbnails or quick sketches to help you develop design ideas. Thumbnails are small rough sketches showing the main parts of designs in the form of simple diagrams.

Sometimes it is helpful to try to put down as many initial ideas as possible in a set time, which is a bit like producing a brainstorm in image form. These initial ideas can be pasted, or scanned and pasted, into your coursework project folder, rather than be redrawn – your project folder should show evidence of creative thinking rather than stilted copied-out work. At this stage, the examiner is looking for evidence of your design thinking. Quick sketches need to be produced fast, using pencils, pens, markers or the like. Use arrows and brief notes to explain your thinking and do not include too much detail at this stage.

Think about this!

Many products in an industrial context are modifications rather than original ideas. Inspiration for this type of designing can come through product analysis. In industry it may involve:

- adapting existing products to compete with 'branded' products
- developing existing products to appeal to different target market groups
- developing existing products through following new legal guidelines, i.e. related to environmental or safety issues
- adapting existing products through the use of new or different materials or processes.

To be successful you will:
- Use a broad range of design strategies to generate and refine a wide range of imaginative ideas.

2. Use knowledge and understanding gained through research to develop and refine alternative designs design detail

As you become more involved with your ideas and start to think about them in greater depth, you may need to produce larger, slightly more detailed sketches. These should still be produced quickly, but may start to show alternative ideas or some parts in more detail. Always add brief notes to explain your design thinking so you provide evidence of how your research influences your ideas. An example of how one student tackled this is shown in Figure 5.5.

SIGNPOST
'Use knowledge and understanding gained through research' Unit 2 page 47

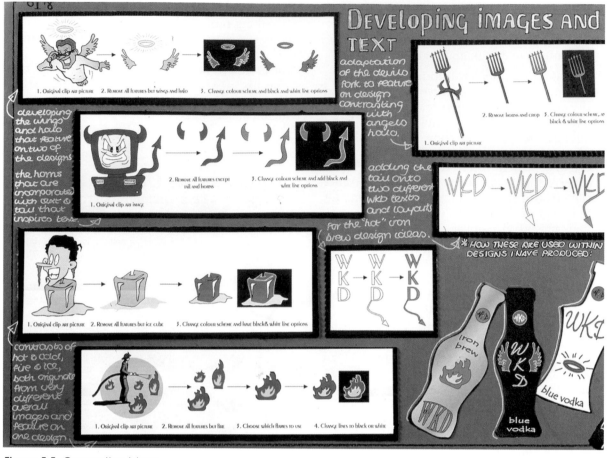

Figure 5.5 *Generating ideas*

At this stage, it may be helpful to use cut paper or simple 3D images to explore ideas. **One-point** or **two-point perspective** or the technique called '**crating**' are extensively used by designers to produce initial ideas. You may wish to add shading or texture, using pencil crayons or pale-coloured markers to convey information about possible materials, for example to show if they are smooth and polished or matt. Knowledge and understanding about the materials or components you may use and about suitable techniques, processes or finishes can be evidenced by annotating your drawings.

Sometimes it is helpful to use 2D or 3D modelling in paper or card for developing initial ideas, especially if complicated shapes or nets are involved. Simple modelling can give a real sense of the size and feel of a product, but if you work in this way, be sure to provide evidence in your coursework folder. It is fairly easy to include 2D modelling, but you may have to photograph any 3D modelling work that you do.

Think about this!

Solving design problems is a complex activity because there are many conflicting constraints and possible solutions. For example you must satisfy the brief, user or client needs, the constraints of manufacture, the limitations of materials and equipment, and the demands of selling in the market place. You have to respond to all these constraints and take on the many roles that in industry would be filled by design, marketing, planning and production teams. In industry the simultaneous design of a product and its manufacturing process is called 'concurrent manufacturing'. An individual taking on a range of design and manufacturing roles also works 'concurrently'.

To be successful you will:
- Demonstrate effective use of appropriate research.

3. Evaluate and test the feasibility of ideas against specification criteria

SIGNPOST
'Evaluate and test the feasibility of ideas against the specification criteria' Unit 2 page 48

As your ideas develop you should evaluate them against your design specification. You should also evaluate the feasibility and market potential of your ideas by getting the views of your client or user(s). It is always helpful to explain your ideas to others – always listen to their views as they can provide you with unexpected insights into your work. If you are using the expertise of a visitor from business, or are working with people in the local community, they may be able to offer constructive criticism, which will help you develop your design work further. Getting feedback is vitally important, not only because it will help you make decisions about your product, but it will also enable you to access the full range of available marks. Remember to reference in your folder any changes you make to your design ideas as a result of feedback from your user(s) or client.

Testing the feasibility of ideas

In industry ideas evaluation is sometimes done by constructing an evaluation matrix, which compares each idea against the specification criteria. Each idea is assessed against the criteria using the following technique:

- + (plus), meaning better than, cheaper than, easier than the specification criterion
- – (minus), meaning worse than, more expensive than, more difficult to develop than, more complex than, harder than the specification criterion
- S (same), meaning the same as the specification criterion.

Each idea in turn is evaluated against the specification criteria. Each idea is given a score, either +, – or S. Scores for each idea are added up, to show the strengths and weaknesses of each one.

In industry a design team would look at the weaknesses of all ideas to see what could be done to improve them. Very weak ideas are then eliminated, leaving the strong ideas which can be developed individually or combined in some way. This kind of exercise gives the design team a greater understanding of design problems and potential solutions and is a natural stimulus to produce design solutions.

Try using an evaluation matrix to evaluate your initial ideas. This should give you a clearer view about what is worth developing. Use written notes to explain this thinking.

You should also make a note of any further research you may need to do. After consulting with your user(s) or client, you may need to modify your design specification, to take into account any decisions made as a result of feedback from them. At A2 level you should be using a more refined approach to focus your ideas, so that they meet more closely the requirements of the specification.

> **To be successful you will:**
> - Objectively evaluate and test ideas against the specification criteria.

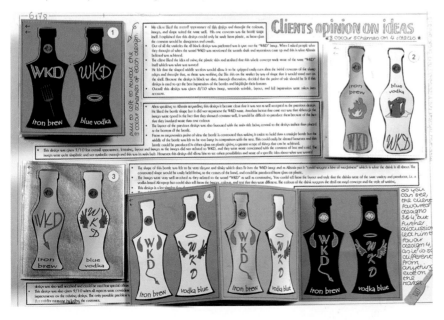

Figure 5.6 Evaluating/testing ideas

C Developing and communicating design proposals (15 marks)

1. Develop, model and refine design proposals, using feedback to help make decisions

SIGNPOST
'Develop, model and refine design proposals, using feedback to help make decisions' Unit 2 page 49

Your aim is to develop and refine your chosen ideas until you find the optimum solution – the best possible solution to your design problem. Be aware that you are developing 2D and 3D elements that could be made in batch or high volume so take into account how easy they will be to manufacture. Even if you are developing a one-off product for an individual client, you will need to consider how easy your product will be to manufacture. In industry, planning the manufacturing processes is a normal part of the design process. Modelling and prototyping should play a key role in your production planning. You can use them to trial your ideas, to explore materials and components, to work out manufacturing processes and for materials planning.

Modelling and prototyping will enable you to consider every aspect of your design. They will enable you to make judgements about the visual elements of your design, such as shape, form, proportion, styling, images, colour, texture, typography or layout, as well as the size and appropriateness of any components you may use. Many everyday products, such as toothpaste tubes or toys, are modelled to ensure they are the right size and easy for people to use. Figure 5.7 shows an example of 2D modelling.

> **To be successful you will:**
> • Develop, model and refine the design proposal, with effective use of feedback.

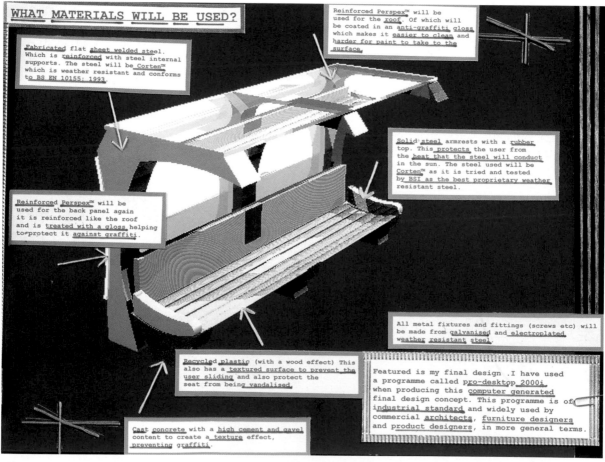

Figure 5.7 2D modelling

2. Demonstrate a wide variety of communication skills, including ICT for designing, modelling and communicating

SIGNPOST

'Demonstrate a wide variety of communication skills, including ICT for designing, modelling and communicating'
Unit 2 page 49

At A2 you are expected to demonstrate a higher level of communication and presentation skills to develop, model and refine your design proposals. This should include the use of relevant technical terminology when you are explaining technological or scientific concepts. You should use ICT if it is available to you, but you will not be penalised for non-use. ICT is useful for recording design decisions, data handling, identifying the properties of materials and for modelling ideas in 2D and 3D.

When using 2D modelling you can develop all the techniques you learned during your AS course. This may include modelling your product using:

- **exploded views** to show individual parts and how they fit together or to show hidden detail
- sections or cutaways to show the inside of the product
- CAD modelling – make sure that you find a balance between computer-generated and hand-drawn images.

Task
Using ICT in industry
The increasing use of ICT by industry has had an enormous impact on design, through the use of CAD systems for computer modelling.

Every aspect of a product's development can be modelled using **electronic product definition (EPD)**, in which all the data required to develop and manufacture a product are stored in a database. This means that even complex products, such as cars, can be modelled electronically as 'virtual products' and developed directly on the computer screen.

List the advantages to manufacturers of such a system.

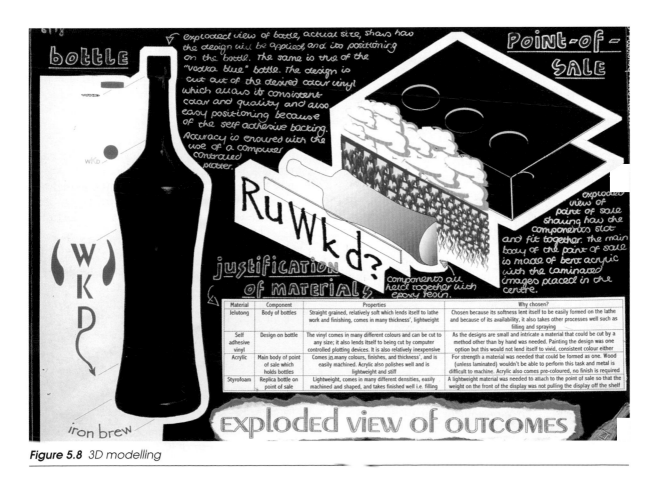

Figure 5.8 3D modelling

When using 3D modelling, you should also develop the techniques you learned in the AS course. Your modelling should ensure that your product will meet the performance and ergonomic requirements of your design specification.

> ## SIGNPOST
> 'Anthropometrics and ergonomics' Unit 3B1 page 130

Testing ergonomic requirements

Often when testing ergonomic requirements, designers work with 3D models to give a real impression of size and the relationship of the product with the user. Since there is no such thing as an 'average-sized' person, designers often use anthropometric data about measurements of the human body, such as height, weight, reach, leg and arm lengths and strength. If necessary, check out anthropometric data from standards organisations such as the British Standards Institute.

> ## Factfile
> **Prototyping in industry**
> A prototype product is a detailed 3D model made to test a product before manufacture. This is especially important when using new materials or manufacturing processes. CAD software can reduce the need to make expensive manufacturing prototypes because all the component parts can be modelled and assembled 'virtually' on screen.

Modelling techniques

You can use a variety of modelling techniques and a range of materials to make 3D working mock-ups or '**lash-ups**'. For example:

- Sheet materials such as paper, card and thin plastics can be used for prototyping ideas as they are quick, easy to use and relatively inexpensive. Look out for useful modelling materials such as recycled packaging, or use flat sheets cut from drinks cans and odd bits of string or wire.
- Simple techniques like cutting, scoring and folding will enable you to explore curved and rectilinear forms. Product designers often prototype products in card and add detail by drawing details like switches or dials on the surface. You could photocopy design details from magazines, cut them out and stick on to give a realistic idea of your product.
- Clay or plasticine may be used for moulding solid models, or polystyrene foam used for making block models.

- Frame models can be built to scale to explore and test structures, using anything that cuts easily, like drinking straws, strips of wood, uncooked spaghetti or wire.

> ## Think about this!
> A photocopier is a useful piece of equipment for developing design details. Copy mesh textures for detailing grills on speakers. Copy your outline drawings so you can experiment with different colours and textures to create different surface finishes and materials.

Choosing a colour scheme

Colour is an essential characteristic of many products and most advertising materials. It is a powerful marketing tool because when we look at any product or packaging, it is usually the first thing we notice. Colour can convey strong messages about the product. For example, colour is often associated with specific products such as pastel blue or pink for baby products. Would these colours look 'right' used in the design of breakfast cereals or for point-of-sale advertising?

Some of our responses to colour are through association; for example, red for danger, blues are cool, whereas browns and greens may suggest a natural quality. When choosing colours for your product, you need to be sure of the message that you wish to communicate. The creation of moodboards can often be helpful in selecting colours that are appropriate to your product.

It is important to try out your product colour scheme. You could use photocopies or CAD software to try out different colourways. Experiment with different tones and try unusual combinations. Before colouring your final design it may be helpful to test your colours as near to full size as you can. Large areas of colour can look entirely different to a small colour swatch.

> ## Think about this!
> In industry 3D prototypes are often made for products such as torches or telephones. Prototypes are made as accurately as possible to represent the appearance and function of the finished product.
>
> When James Dyson was developing his cyclone action vacuum cleaner, he made 5127 prototypes.

> ### To be successful you will:
> - Use high-level communication skills with appropriate use of ICT.

3. Demonstrate understanding of a range of materials/components/systems, equipment, processes and commercial manufacturing requirements

SIGNPOST
'Selecting materials' Unit 2 page 51
'Testing materials' Unit 3A page 102
'Selection of materials' Unit 4A page 174

At A2 you are expected to draw on and use a higher level of understanding about materials, components and processes. You should make relevant and real connections between that understanding and your own design and manufacture. Modelling and prototyping should enable you to trial materials and components. It may also give you the opportunity to explore a range of interesting materials and look for new ways of using them. Testing materials will enable you to select those that are the most suitable for your product. Testing may include researching known properties or the use of **comparative testing**.

Do not forget to annotate your final design proposal to explain how and why it meets the requirements of your product design specification. You should also consider the commercial manufacture of your product.

> **To be successful you will:**
> • Demonstrate a clear understanding of a wide range of resources, equipment, processes and commercial manufacturing requirements.

4. Evaluate design proposals against specification criteria, testing for accuracy, quality, ease of manufacture and market potential

SIGNPOST
'Evaluate design proposals' Unit 2 page 52

During the refinement of your final design proposal you will need to check its feasibility against your product design specification. You should also consult with your client or user(s) on the development of your product, as their views should provide you with feedback about its viability.

Figure 5.9 *Demonstrating understanding of materials and commercial manufacturing processes*

You will need to explain:

- how your design proposal will meet the specification – will the product 'work'?
- how its aesthetic characteristics will meet the needs of your client or user(s)
- how the proposal will meet market trends and ergonomic requirements
- how the performance characteristics of your proposed materials, components and manufacturing processes will meet the quality

and cost expectations of your client or user(s)
- how easy it will be to manufacture in the time available
- how you could manufacture the product in batch or high volume.

> **To be successful you will:**
> - Objectively evaluate and test your design proposals against the specification criteria.

D Planning manufacture (10 marks)

1. Produce a clear production plan that details the manufacturing specification, quality and safety guidelines and realistic deadlines

> **SIGNPOST**
> 'Planning manufacture' Unit 2 page 53

You should produce and use a clear **production plan** that explains how to manufacture your product.

The work that you have already done during the research, design and development stages of your project should enable you to do this. During those developmental stages you needed to take into account how easy your product will be to manufacture. In many industrial situations the planning of the production processes is a normal part of the design process. Your modelling and prototyping will have played a key role in your production planning. You used them to trial your ideas, to explore materials and components, to work out manufacturing and assembly processes and for materials planning.

Think about how long each different production process will take. Will you have enough time to make your product? Will you need to simplify anything? Do you need any special materials or tools? How soon do you need to order any materials or components so they are ready when you need them? At this stage, you need to estimate your production costs and a possible **selling price (SP)**.

Producing a production plan

Your production plan should include clear and detailed instructions for making your product. An example of production planning is shown in Figure 5.10. Although you will be making one product in your workshop, you will need to detail how identical products could be manufactures by batch or high volume production. Planning

quality into your work is therefore vitally important. Base your production plan on the one you used at AS level. You should also take into consideration the following:

- Your manufacturing specification should include a dimensioned working drawing (see Figure 5.11). You could use orthographic 3rd angle drawings or possibly CAD software.
- You should use quality control methods to ensure that identical products could be made. Plan your quality control at critical control points (CCPs) in the product manufacture. Use quality indicators to explain how to check for quality. Think about balancing the quality of your product against the cost of making it and the time available. What are the quality requirements of the client or user(s)?
- Use risk assessment procedures to check the safety of your manufacture. Are there any potential hazards? Does the design specification set out any safety standards that will help you assess risks? Are there any risks attached to the materials, processes and equipment you will use? Do you need to follow any safety regulations? Are there any risks in the use of your product? What about its disposal?

> **Task**
> **Risk assessment**
> a) Draw up a flow chart to show the key stages of manufacture of your product.
> b) For each stage list the risk assessment procedures required to make your product as safe as possible.

> **To be successful you will:**
> - Produce a clear and detailed production plan with achievable deadlines.

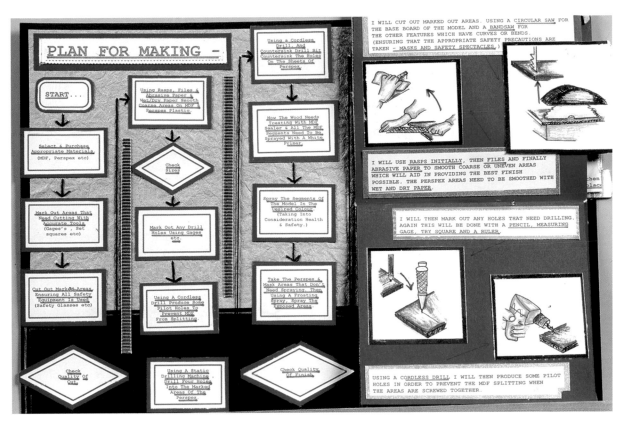

Figure 5.10 *Production planning*

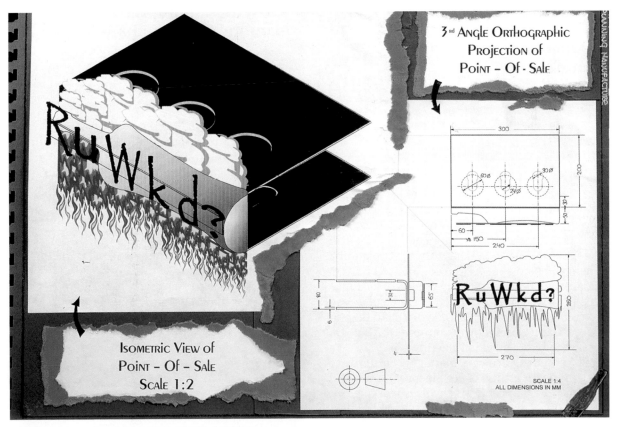

Figure 5.11 *A dimensioned working drawing*

Factfile

Quality assurance (QA) and quality control (QC)

1 The aim of QA is to make identical products with zero faults. A QA system monitors every stage of design and manufacture. QA makes use of written documents such as:

- production plans
- detailed specifications
- production schedules
- costing sheets
- quality control and inspection sheets.

2 QC is used to test and monitor the production of products.

- QC means checking for accuracy at critical control points.
- Quality indicators are used to show how quality is checked using variables or attributes.
- QC makes use of standard sizes, dimensions and tolerances to enable quality to be checked.

needs to be done first; how long each activity will take; when each activity has to be done by; which activities can be done at the same time; which activities depend on the completion of other activities and if any activities are more essential or critical than others.

The critical path is the one that takes the shortest time. It is often helpful to work backwards from the deadline, plotting when an activity has to start in order for you to finish on time. You could use a Gantt chart to plot the critical path of your product. Map each task against the time available, so you can prioritise the critical activities. Remember to use your Gantt chart or any other planning tool as a working document in which you record any subsequent changes you might make, if, for example, delays occur. It is worth remembering that you will only be awarded marks for a production plan made before you manufacture your product.

> *To be successful you will:*
> - Demonstrate effective management of time and resources, appropriate to the scale of production.

2. Take account of time and resource management and scale of production when planning manufacture

> **SIGNPOST**
> 'Take account of time and resource management and scale of production when planning manufacture' Unit 2 page 55

Many of the activities that you undertake during your A2 project will overlap. This is inevitable because designing and making is a complex process. Planning a project is not easy, because it involves estimating how long each activity will take. Use your AS experience of project planning to help you plan your A2 project. Did you have to amend your time plan? Did some activities take longer than you expected? How did you overcome these problems? Did you find it helpful to use a Gantt chart to plan your project? Did you use one to plan your manufacture?

In industry, project managers often use **critical path analysis** to plan the successful outcome of a project. This involves working out all the critical activities that must be undertaken and how one activity relates to another. Decisions have to be taken as to which activity

3. Use ICT appropriately for planning and data handling

> **SIGNPOST**
> 'Use ICT appropriately for planning and data handling' Unit 2 page 56

The aim of using information and communications technology (ICT) for planning and data handling is to enhance your design and technology capability. Although you will not be penalised for non-use of ICT, you should use it where appropriate and available. If you do have access to ICT, you can use it for a range of activities. For example:

- find out costs of materials using e-mail or the Internet
- use spreadsheets to work out quantities and costs of materials and components
- plan the critical path of your product using a colour-coded Gantt chart.

> *To be successful you will:*
> - Demonstrate good use of ICT.

E Product manufacture (40 marks)

1. Demonstrate understanding of a range of materials, components and processes appropriate to the specification and scale of production

SIGNPOST
'Demonstrate understanding of a range of materials, components and processes appropriate to the specification and scale of production' Unit 2 page 58

'Working properties of materials and components' Unit 3A page 80

'Selection of materials' Unit 4A page 174

Your AS coursework experience should enable you to select and use materials and components with growing confidence. By this stage of your course, you should have developed a better knowledge of their working characteristics and a wider variety of skills. Your modelling, prototyping and testing of materials, components and processes should also give you confidence in making your product. Most of the problems related to the product assembly should, hopefully, have been ironed out.

Your aim now should be to manufacture a well-made product (see Figure 5.12). This may not necessarily be made to a very much higher quality than the product that you made at AS. However, your A2 product should fulfil the design and manufacturing specifications more closely than your AS product did.

Scale of production

Your 2D and 2D elements will be made as a one-off, whichever scale of production you design it for. You should make and finish them to the best of your ability and explain how they could be manufactured in high volume.

> ### To be successful you will:
> - Demonstrate clear understanding of a wide range of materials, components and processes.

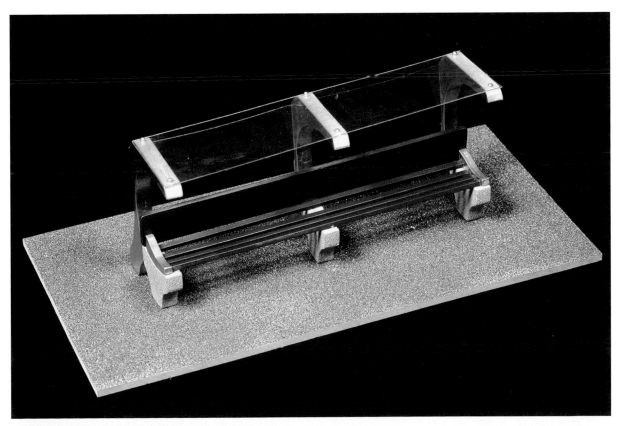

Figure 5.12 *Manufacturing a product*

2. Demonstrate imagination and flair in the use of materials, components and processes

SIGNPOST
'Demonstrate imagination and flair in the use of materials, components, processes and techniques' Unit 2 page 59

You are expected to work creatively, innovatively and imaginatively with materials, components and processes. You can only do this if you have a good understanding of the materials and processes you use. During the modelling and prototyping stages you may have had the opportunity to explore a range of interesting materials and trial new ways of using them. Your ability to use materials and processes with flair should result in the production of a quality product that:

- meets the specification and is well finished
- is easy to use and well designed.

3. Demonstrate high-level making skills, precision and attention to detail in the manufacture of high-quality products

SIGNPOST
'Demonstrate high-level making skills' Unit 2 page 59

At A2 you should place a greater emphasis on the prevention of faults through your use of quality assurance and quality control. This will involve using your production plan to monitor your product manufacture:

- Plan where you will check for quality during manufacture.
- Use tolerances and dimensions to check the accuracy of component parts of your product.
- Check the accuracy of machines prior to cutting.
- Test components parts prior to assembly.

Your making skills should result in the production of a high-quality product (see Figure 5.13). The product should be capable of being tested against the specification and used by the client or users for its intended job.

> **To be successful you will:**
> - Demonstrate demanding and high-level making skills that show precision and attention to detail.

Figure 5.13 *Evidence of high level making skills can be shown through the use of photographs*

4. Use ICT appropriately for communicating, modelling, control and manufacture

SIGNPOST
'Use ICT appropriately' Unit 2 page 60

You should use ICT to help your product manufacture where it is appropriate and available, but you will not be penalised for its non-use. You are not expected to know how to use specific hardware or programs, but you should understand the benefits of using ICT for manufacture.

> **To be successful you will:**
> - Demonstrate good use of ICT.

5. Demonstrate a high level of safety awareness in the working environment and beyond

SIGNPOST
'Demonstrate a high level of safety awareness in the working environment and beyond' Unit 2 page 60

Safety should be a high priority in your work at all times. Safe production means identifying all areas of possible risk and documenting safety procedures to manage and monitor this risk. This means that you should:

- identify hazards and use risk assessment procedures to ensure safe use of materials, tools, equipment and processes

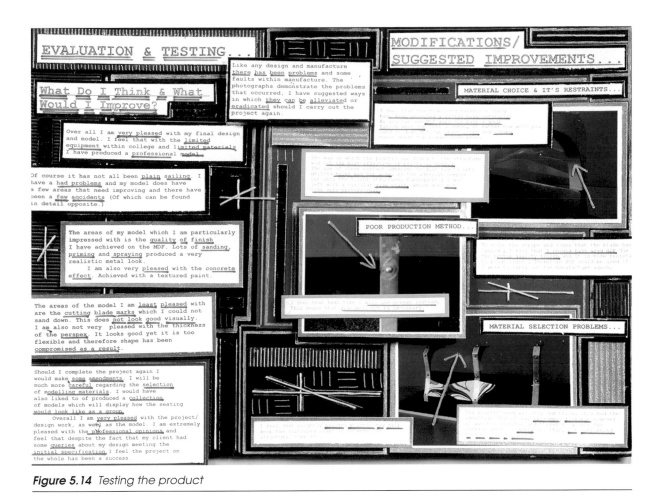

Figure 5.14 *Testing the product*

G Appropriate project (10 marks)

The G criterion has been included to reflect that your project meets all the coursework requirements (see 'What is a Graphics project?' on page 267). It is very important to check the appropriateness of your project with your teacher or tutor at the start of your project, to make sure that it will enable you to address all of the assessment criteria by which your project will be marked. After your teacher or tutor confirms that your project is appropriate, you will still need to keep an eye on the assessment criteria in order to achieve all the available marks. You will also need to take account of feedback from your teacher or tutor in order to improve your work as it progresses. Remember to include photographic evidence of your modelling and prototyping and the product manufacture, especially to highlight difficult techniques or hidden details.

Student checklist

1. Project management

- Take responsibility for planning, organising, managing and evaluating your own project.
- Include photographic evidence to show hidden details, or to demonstrate the processes you used at each stage of manufacture.
- Include only the work related to the assessment requirements.

2. A successful A2 coursework project will:

- Identify the needs of a specified user, client or users in a target market group.
- Develop a designer/client relationship that enables the use of feedback when making decisions.
- Select and use relevant research that targets more closely the brief and specification.
- Make closer connections between research, feedback and the development of ideas.
- Demonstrate high-level communication and presentation skills and appropriate use of ICT.
- Be a manageable size so you can finish it on time.
- Include a 2D element developed using graphics media.
- Include a 3D model using at least one resistant material.
- Include an increased emphasis on industrial practices, including the use of feedback.
- Include clear photographs of modelling, prototyping, testing and manufacture.
- Detail the manufacture of one product and show how it could be manufactured in quantity.
- Use a better understanding about materials, components and manufacturing processes.
- Demonstrate a wide variety of skills, a broad use of materials and increasing knowledge of A2 technologies.
- Manufacture a product that matches specifications more closely than at AS level.
- Allow time to evaluate your work as it progresses and modify it if necessary.
- Be well planned so you can meet your deadlines.

3. Evidencing industrial practices in coursework

- Use industrial-type terminology and technical terms.

- Include a range of designing activities that are similar to those used in industry.
- Include a range of manufacturing activities that are similar to those used in industry.

4. Using ICT in coursework

- Develop the use of ICT for research, designing, modelling, communicating and testing.
- Develop the use of ICT for planning, data handling and manufacturing.

5. Producing a bibliography

- Reference all secondary sources of information in a bibliography, e.g. from textbooks, newspapers, magazines, electronic media, CD-ROMs, the Internet, etc.
- Reference scanned, photocopied or digitised images. Do use clip art at this level.
- Do not expect marks for any work copied directly from textbooks, the Internet or from other students.

6. Submitting your coursework project folder

- Have your coursework ready for submission by mid-May in the year of your examination.
- Include a title page with the Specification name and number, candidate name and number, centre name and number, title of project and date.
- Include a contents page and numbering system to help organise your coursework folder.
- Ensure that your work is clear and easy to understand, with titles for each section.

7. Using the Coursework Assessment Booklet (CAB)

- Complete the student summary in the CAB and *remember to sign it!* The summary should include your design brief and a short description of your coursework project.
- Ensure that the CAB contains a minimum of three clear photographs of the whole product, with alternative views and details.
- Write your candidate name and number, centre name and number and 6301/01 in the CAB by the product photographs and on the *back of each photograph.*

Design and technology capability (G6)

Summary of expectations

1. What to expect

Unit 6 brings together all the knowledge, understanding and skills that you have gained during your Advanced GCE course. Although no new learning is expected during the unit itself, it is essential that you prepare fully for the Unit 6 exam.

2. How will it be assessed?

Unit 6 is assessed through a three-hour Design Exam, in which you are asked to produce a solution to a given design problem and describe how the solution can be manufactured. Your design solution should reflect a study of the 'technologies' involved in the Graphics with Materials Technology Specification. In this kind of exam you should demonstrate your ability to think on your feet, not to recall information.

The assessment criteria shown in Table 6.1 cover knowledge and understanding related to designing and manufacturing. You will be assessed on your ability to organise and present ideas and information clearly and logically. The style of assessment will remain the same each year, but there will be a different design problem each time the unit is assessed.

Table 6.1 *Design Exam assessment criteria*

Assessment criteria	Marks
a) Analyse the design problem and develop a product design specification, identifying appropriate constraints	15
b) Generate and evaluate a range of ideas	15
c) Develop, describe and justify a final solution, identifying appropriate materials and components	15
d) Represent and illustrate your final solution	20
e) Draw up a production plan for your final solution	15
f) Evaluate the final solution against the product design specification and suggest improvements	10
Total marks	90

3. The Design Exam

Your centre will be sent a Design Research Paper at least six weeks before the exam. The paper will give you a context for design, together with bullet points that give you direction about what to research. The Design Paper will have one design problem that is based on the research context. In the assessment criteria a) to f) you are asked to:

- analyse the design problem, making connections between it and your research
- develop a product design specification in response to the problem and your research
- generate and evaluate a range of ideas based on your product design specification
- develop, describe and justify a final solution, identifying appropriate materials and components
- illustrate your final solution using dimensioned drawings, with details and quantity of materials and components
- draw up a production plan for your final solution, including manufacturing processes, the sequence of assembly and quality checks
- evaluate your final solution against your product design specification and suggest how it could be improved.

You may take *all* your research material into the exam and use it as reference throughout, but this is *not* submitted for assessment. The pasting of pre-prepared or photocopied sheets is *not* permitted. You will *not* be allowed to use ICT facilities during the examination.

You will be provided with answer sheets that are all you will need to answer the Design Paper. More answer sheets may be used if absolutely necessary. You may separate the answer sheets, but must secure them with a treasury tag at the end of the exam. Suggested times for each section of the exam are given at the foot of each answer sheet.

4. How much is it worth?

The Design Exam is worth 15 per cent of the full Advanced GCE.

Unit 6	Weighting
A2 level (full GCE)	15%

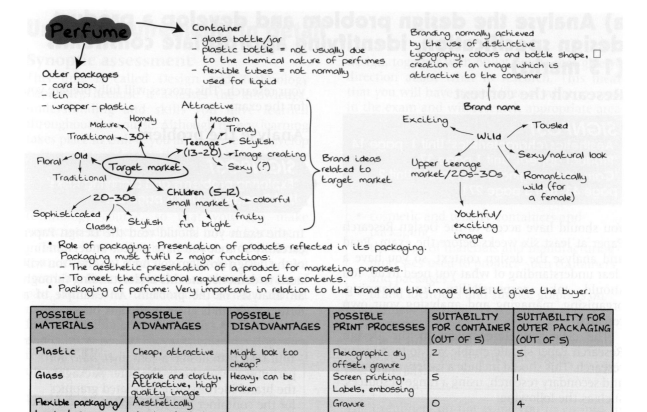

Figure 6.1 *Analysis needs to ensure that six relevant aspects of the design problem are highlighted, together with at least three points of expansion/justification for each aspect*

Making connections

Your analysis of the problem should be a short task. It should enable you to make clear connections between the design problem and what you have researched. This will make it easier to pinpoint relevant information that will help you develop a product design specification. In the exam you should *not* copy out any research materials. Your ability to research the context and to analyse your research material will be assessed through the quality of your analysis and product design specification.

Although you can use your research material in the exam, you are not allowed to paste, stick or staple any pre-prepared or photocopied work onto your answer sheets. Although it may be appropriate to use ICT for research, you will not be allowed to use ICT facilities during the examination.

Develop a product design specification, identifying appropriate constraints

SIGNPOST
'Develop a design specification' Unit 2 page 45, Unit 5 page 273

Your product design specification is important because it forms the basis for generating and evaluating your design ideas *and* for evaluating your final solution. The specification headings are printed at the top of your answer sheet and you should use them to provide a framework for your written product design specification.

Your specification criteria should be *specific* to the requirements of the set design problem and should demonstrate how your research and analysis have helped you make decisions about what your product will be like. Your specification criteria should be written as short, reasoned sentences that explain the requirements of the product that you will design in response to the design problem. (See the example of a design specification in Figure 6.2.)

The printed specification headings ask you to develop the following criteria:

- purpose: what the product is for
- function: what the product needs to do
- aesthetics: how the product should look, its style, form, aesthetic characteristics
- market and user requirements: market trends, user needs and preferences, ergonomic constraints
- performance requirements of the product, materials, components and systems: mechanical properties and working characteristics
- processes, technology and scale of production: manufacturing processes and technology required to make the product at an appropriate level of production – either batch, mass or continuous production

- quality control: the quality requirements of your product (and target market) and how you will achieve quality using quality control and quality standards
- legal requirements and external standards: how the product will meet safety requirements, referencing external standards where appropriate (British Standards are *not* to be copied out)
- cultural, social, moral and environmental issues that may influence your design ideas.

Helpful hints

There is a suggested time of 30 minutes available for section a) which is worth 15 marks. Check out how many marks are available for analysing the design problem and how many marks are available for developing a product design specification. Spend an appropriate amount of time on each activity so you have the opportunity to achieve the available marks.

Design specification

Purpose: The perfume bottle must fulfil packaging requirements by presenting the perfume aesthetically well for marketing purposes; it must also meet the functional requirements of the contents and create a barrier to prevent contamination of product.

Function: The design of the bottle must be based on the brand name 'Wild', and this could also be used as the perfume name. The brand identity and associated graphics should also be produced for the packaging.

Aesthetics: The perfume should be attractive to its target market and have a 'wild' style.

Market and user requirements: Target market should be able to use the product easily.

It should be different from other products on the market to give it an element of competitive edge, perhaps in its shape. Needs to contain perfume for easy dispensation.

Performance requirements: Should be made to a high quality and must function efficiently in the way the perfume comes out of the bottle, e.g. by an atomiser, etc.

Processes, technology and scale of production: Should be batch produced to test the market.

Quality control: Must be made to a good packaging quality in line with British Standards, and be able to display a kitemark.

Legal requirements and external standards: Must have correct labelling such as a unique bar code. It doesn't have to be on the bottle but must be on the outer packaging.

Value issues: Must not exploit third world workers. Should consider environmental concerns such as the environmental impact of the perfume packaging in its manufacture, use and disposal.

Figure 6.2 A product design specification

b) Generate and evaluate a range of design ideas. Use appropriate communication techniques and justify decisions made (15 marks)

Generate a range of design ideas

SIGNPOST
'Generating ideas' Unit 2 page 46, Unit 5 page 274

You are expected to generate a minimum of three feasible design ideas for each of the main aspects of the given design problem. In the example of the perfume or aftershave, three ideas for each of the 'container, outer package and brand name' would be expected. Initial ideas should be in the form of good quality sketches, with annotation in sufficient detail to demonstrate your understanding of commercial methods of manufacture. Make sure that your annotation is *specific* to your chosen materials and avoid the use of terms such as 'plastic' or 'metal'. You should instead be referencing specific materials such as acrylic, aluminium or MDF, and specific processes such as blow moulding or lithography. This is your opportunity to make use of your research information and to demonstrate what you have learned during your course.

SIGNPOST
'Use knowledge and understanding gained through research to develop and refine ideas' Unit 2 page 48, Unit 5 page 275

You are expected to generate your own ideas; drawing and developing an existing product will not gain the available marks. Your choice of scale of production is also very important, because you need to evidence your understanding of commercial methods of production. You should therefore be designing for batch, mass or continuous production, not a one-off product.

Making use of the information you acquired during your research should enable you to develop a number of alternative design ideas and select one (or possibly two) that are the most promising. Figure 6.3 shows how one student generated ideas.

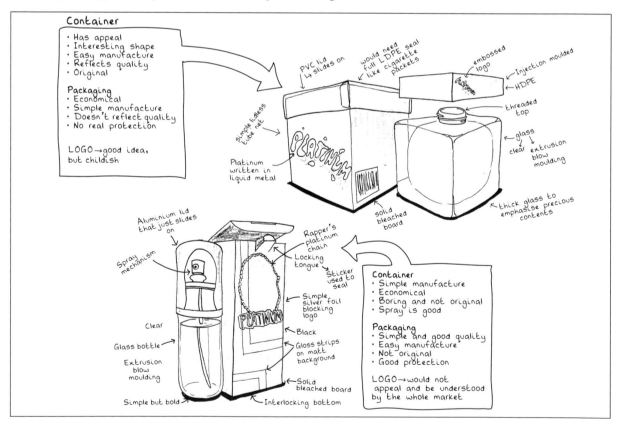

Figure 6.3 *Generating and evaluating ideas for a container and outer packaging including brand identity and graphics*

At this stage the examiner is looking for evidence of your design thinking in response to the requirements of your product design specification and your knowledge and understanding of materials and manufacturing processes. You will also gain marks for quality of communication, so make sure that you practise your drawing skills in advance of the examination.

Evaluate design ideas and justify decisions made

> **SIGNPOST**
> 'Evaluate and test the feasibility of ideas against specification criteria' Unit 2 page 49, Unit 5 page 277

All first ideas should be evaluated against the key points of your product design specification.

The design or designs chosen for development should be explained and justified, using written notes. Refer back to your product design specification – how will your ideas meet the requirements you set out to achieve? Which are the best ideas and why?

You should, at the very least, provide a sensible justification of each of your ideas by explaining the pros and cons of each. Figure 6.3 shows how one student evaluated ideas.

> ### Helpful hints
> There is a suggested time of 30 minutes available for section b) which is worth 15 marks. Make sure you allow sufficient time to annotate your sketches and to justify which ideas are worth developing in section c).

c) Develop, describe and justify a final solution, identifying appropriate materials and components (15 marks)

Develop a final solution, identifying appropriate materials and components

> **SIGNPOST**
> 'Developing and communicating design proposals' Unit 2 page 49, Unit 5 page 278

In this section you are asked to produce 2D and/or 3D sketches that develop and refine your initial design idea(s). It is not appropriate to simply redraw your best idea from section b). The examiner is looking for clear evidence of how you have adapted and changed the characteristics of your initial ideas, so the final solution will better meet the needs of your product design specification criteria. You should use a good graphic style to communicate the aesthetic and functional aspects of your design. It is often helpful to use a backing sheet of isometric paper as a guide for 3D sketches.

> **SIGNPOST**
> 'Demonstrate a wide variety of communication skills' Unit 2 page 50, Unit 5 page 279

You should annotate your 2D and/or 3D sketches to explain any hidden details, to evidence aspects of the product function and to explain the aesthetic, functional and mechanical properties

of your chosen materials and components. Demonstrate your understanding of materials and commercial manufacturing processes through the use of appropriate technical terms. Your sketches should provide sufficient detail to explain how the product will work or function. Figure 6.4 shows how one student developed a final solution.

> **SIGNPOST**
> 'Working characteristics of a range of materials and components' Unit 1 page 16
> 'Demonstrate understanding of materials/components' Unit 2 page 51, Unit 5 page 281
> 'Working properties of materials' Unit 3A page 80
> 'Selection of materials' Unit 4A page 174

Describe and justify your final solution

In the second part of section c) you are asked to describe and justify your final solution. This is *not* a re-run of your design specification in a) so you should not be describing what the product *should* do, but explaining *why* and *how* the design you have developed is *the best solution* to the problem.

The most effective way to do this is to list and respond to the headings printed at the top of

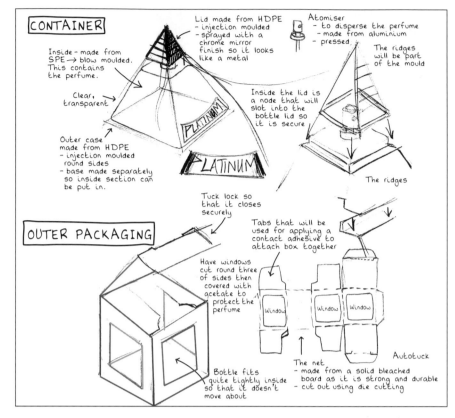

Figure 6.4 *Developing a final solution for a container and outer packaging, including brand identity and graphics*

your answer sheet. You should use short, reasoned sentences to describe and justify your design solution in relation to:

- function – how your product design meets the functional/mechanical aspects of your product design specification
- appearance – how the aesthetic characteristics of your product design meets the look, style, form and characteristics outlined in your specification
- performance – how your product design meets the quality and safety requirements of the market and users outlined in your specification
- materials, components and systems – how the working characteristics of your chosen materials, components and/or systems meet the requirements of your specification
- processes – how suitable your chosen manufacturing processes are for the scale of production stated in your specification
- technological features – how modern technology such as the use of CAD/CAM and ICT could enable efficient manufacture of your final design solution.

Once again you will be making *connections* between what you set out to achieve (your product design specification) and how you hope to achieve it (your final design solution). It would be helpful to have your specification in front of you when you describe and justify your final solution, so that the points you make are relevant to the design problem. Figure 6.5 shows how one student described and justified a final solution.

SIGNPOST
'Evaluate design proposals against specification criteria' Unit 2 page 52, Unit 5 page 281

Helpful hints

There is a suggested time of 30 minutes available for section c) which is worth 15 marks. Check out how many marks are available for your final design solution (you will lose marks if there is no development) and how many marks are available for its justification (you will gain marks if you use accurate and appropriate technical terms). Spend an appropriate amount of time on each activity so you have the opportunity to achieve the available marks.

FUNCTION AND APPEARANCE

Container: Holds the perfume, dispenses it via an atomiser. Includes the name 'Platinum' but only discretely.

Packaging: Holds the perfume container, has windows so that product can be seen and advertised, the tuck lock closes so perfume is secure. Surface graphics include - name, brand identity logo, bar code, amount of perfume, standard e mark and that it is a fragrance for women.

PERFORMANCE

Container: Perfume is held in the inner container, then this is protected by an outer layer. Dispenses the perfume and lid secures the perfume inside when not in use.

Packaging: To attract attention there are windows so the buyers know what they are purchasing. Secure locks so that bottle cannot fall out. Surface graphics provide the buyer with sufficient information.

MATERIALS

Container:

· Inner container made from PE which is tough and hardwearing.

· Outer container HDPE, which is strong and durable and will protect the inner container

· Lid made from HDPE as well to make sure it's durable

Packaging:

· Box made from solid bleached board which is a strong, hard wearing board and will hold the weight of the perfume bottle

· Windows covered with acetate so that the bottle can be seen - transparent

PROCESSES

Container:

· Inner - blow moulded, producing little waste. Outer sides injection moulded first then layer can be inserted and a separate base attached so it's safely concealed

· Atomiser - pressed aluminium - includes the pump mechanism

· Lid - injection moulded then sprayed with a chrome mirror finish so it carries on the futuristic theme and looks like metal

· Platinum - logo printed onto self-adhesive label by flexography

Packaging:

· Net cut by fast die cutting

· Acetate and net assembled together using a strong contact adhesive

· Graphics applied using offset lithography, a quick and cheap process.

Figure 6.5 *Describing and justifying the final solution*

d) Represent and illustrate your final solution (20 marks)

Represent and illustrate your final solution

SIGNPOST
'Demonstrate a wide variety of communication skills' Unit 2 page 50, Unit 5 page 279

This section asks for you to 'represent and illustrate' your final design solution, using:

- clear construction/making details
- dimensions/sizes
- details and quantity of materials/components
- clear and appropriate communication techniques.

Do not be confused about this, as it is *not* a repeat of the previous section c), or a catalogue-style illustration in full colour! You should instead produce good-quality dimensioned drawings of your final design solution, using clear and appropriate drawing techniques, that show how to manufacture the product, together with details and quantity of your chosen materials and components.

Orthographic 3rd angle drawings are not required in the Design Exam. Your ability to produce fully dimensioned working drawings is tested in your coursework. However, in the exam your drawings should provide sufficient detail for the examiner to understand how your final design solution is to be manufactured in an appropriate scale of production.

Appropriate graphic techniques could include dimensioned drawings that show different views of the product and exploded drawings that show any hidden details. Make sure that your dimensions conform to British Standards. You should also include a conventional style cutting list that shows details and quantity of the materials and components required to manufacture your final design solution.

Remember that marks are available in this section for the use of clear and appropriate communication techniques and for good use of technical terms that demonstrate your understanding of industrial manufacture. Figure 6.6 shows how one student represented and illustrated the final design solution.

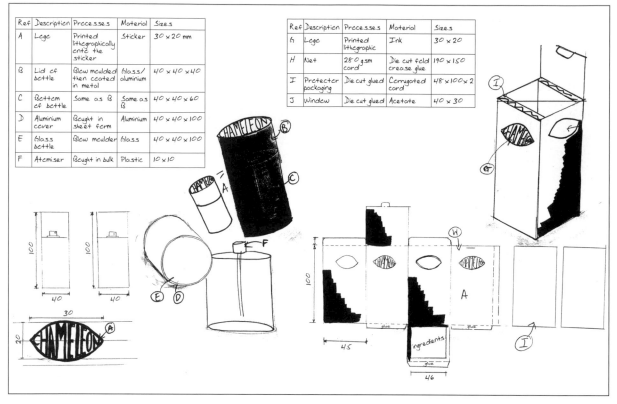

Ref	Description	Processes	Material	Sizes
A	Logo	Printed lithographically onto the sticker	Sticker	30 × 20 mm
B	Lid of bottle	Blow moulded then coated in metal	Glass/ aluminium	40 × 40 × 40
C	Bottom of bottle	Same as B	Same as B	40 × 40 × 60
D	Aluminium cover	Bought in sheet form	Aluminium	40 × 40 × 100
E	Glass bottle	Blow moulded	Glass	40 × 40 × 100
F	Atomiser	Bought in bulk	Plastic	10 × 10

Ref	Description	Processes	Material	Sizes
G	Logo	Printed lithographic	Ink	30 × 20
H	Net	280 gsm card	Die cut fold crease glue	190 × 150
I	Protector packaging	Die cut glued	Corrugated card	48 × 100 × 2
J	Window	Die cut glued	Acetate	40 × 30

Figure 6.6 *Representing and illustrating the final solution*

Helpful hints

There is a suggested time of 40 minutes available for section d) which is worth 20 marks, making this the highest scoring section in the Design Exam. Check out how many marks are available in this section for:

- clear construction/making details
- dimensioned drawings of your final design solution
- a cutting list with details and quantities of the materials and components required
- clear and appropriate communication techniques.

Spend an appropriate amount of time on your drawings and cutting list so you have the opportunity to achieve the available marks.

e) Draw up a production plan for your final solution (15 marks)

SIGNPOST
'Stages of production' Unit 1 page 27
'Produce a clear production plan' Unit 2 page 54, Unit 5 page 282

Draw up a production plan

This section of the Design Exam you to draw up a production plan that describes the production requirements of the solution to include, where appropriate:

- assembly processes/unit operations
- sequence of assembly
- quality checks.

In your coursework units you produced very detailed plans for the manufacture of your product. In the Design Exam, however, you should produce a simple production plan to show:

- the main industrial/commercial manufacturing processes you intend to use to produce the elements of your assembly
- clear sequencing of your proposed assembly with specific commercial assembly techniques
- quality checks performed at specific stages of manufacture and assembly.

The type of processes and assembly required will depend on your chosen materials and the scale of production identified in your product design specification. For example, a high-volume product with thermoplastic component parts may require the use of injection or blow moulding and limited finish.

On the other hand, a batch-produced product with a number of identical component parts may require the use of simplified manufacturing processes and assembly methods and a high-quality finish.

It is not advisable to combine one-off and mass-production methods in your production plan, because it will signal to the examiner your lack of manufacturing understanding. Instead you should concentrate your efforts on the use of batch, high volume or continuous production techniques and the use of appropriate technical terms. These may include references to technological features; for example, how the use of modern technology such as CAD/CAM could enable efficient manufacture of your final design solution.

Quality control

SIGNPOST
'Quality control in production' Unit 1 page 33
'Devise quality assurance procedures' Unit 3 page 63, Unit 5 page 287

Quality control is an important aspect of your production planning. Refer back to your quality control criteria in your specification. It may help to remember an industry saying that 'quality cannot be manufactured into a product, it has to be designed into it'. Make sure that any comments you make about quality control refer to your specific product manufacture, rather than being generalised comments about how 'checks should be made at certain stages'. You need to state exactly where and how you will check for quality by identifying:

- the critical control points (CCPs) *where* you will check for quality and accuracy
- quality indicators that describe *how* you will check for quality. For example, will you use dimensions and tolerances to check the accuracy or fit of joints or mechanisms? Will you check quality against specific aspects of the product design specification? Will you use sensory checks of touch and vision to check for finish?

Your production plan should be clearly organised if you are to demonstrate your manufacturing understanding to the examiner. You are more likely to achieve higher marks in section e) if you use a flow diagram to show the sequencing of your proposed assembly, with details of specific materials, commercial manufacturing processes and quality control checks. Figure 6.7 shows how one student drew up a production plan.

Helpful hints

There is a suggested time of 30 minutes available for section e) which is worth 15 marks. In order to achieve the available marks you must address all the headings printed at the top of your answer sheet, including the manufacturing processes, sequence of assembly and quality checks.

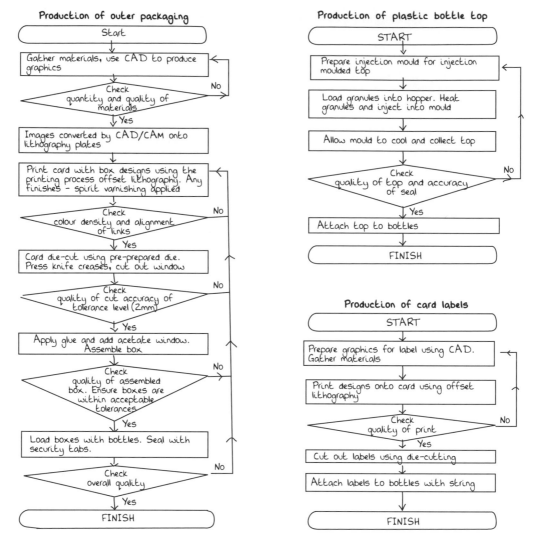

Figure 6.7 *Part of a production plan*

f) Evaluate your final solution against the product design specification and suggest improvements (10 marks)

Evaluate your final solution against the product design specification

SIGNPOST
'Objectively evaluate the outcome against specifications and suggest improvements'
Unit 2 page 62, Unit 5 page 288

The last section of the Design Exam asks you to evaluate your final solution against your product design specification and suggest improvements. Efficient time management throughout the exam will ensure that you have sufficient time to complete this section.

Your evaluative comments should be objective and justified in order to achieve the available marks. If you evaluate your final solution against your product design specification you will have something to measure your product design against. How could you test the product to ensure it performs in use? What kind of tests would be appropriate? How well will it function? How safe will it be to use? Are your chosen materials sufficiently durable to enable the product to meet its quality requirements? Could you make identical products? How and why? What are the product's best features in relation to the design specification? In what way is it a

Design spec point	Evaluation of final solution against the design specification	Improvements	Other
Hold, promote and protect	• The container holds the aftershave well due to the properties of the glass – non-corrosive and resistant. The blow-moulding allowed for a strong seal – a screw thread. The packaging promotes as it is colourful and seen as a modern and trendy image (simple typeface). The corrugated card protects the contents well. The polypropylene cap also has properties which allow for it to be resistant to wear.	• Promote – the possible introduction of more colours on the packaging or an image to support the corporate identity – branding.	• In reality – a test market may be established to monitor reactions of consumers towards the product.
Performance	• The graphics of the Platinum brand name are modern in appearance and eye-catching to the target market. The brand name is in text style, which makes you look closely at the product. With a corrugated card package containing a glass bottle, safety should be considered (see improvements).	• Safety – within the box there may need to be more card (in sections) to support and secure the bottle inside, to stop it from moving around, as glass is delicate, it needs to be protected.	• In reality, many tests would be done to test for durability and protection of the packaging.
Legal requirements	• Will contain all legal requirements, screen printed in silver onto the reverse of the packaging. Encourage recycling – PP, glass and card can all be recycled.	• A clearer incorporation of the recycling symbol on the front of the packaging to appeal to the target market; a young market may be recycling-conscious, and this might act as a USP also.	• In reality, test against set British Standards and research the necessary legal requirements which are to be included.
Culture	• This design is not offensive to any party and, as stated above, clearly incorporates all necessary environmental and legal requirements and symbols.		• This would mean either higher profits can be made or prices can be reduced in order to compete with competitors.
Process	• The item will be mass-produced as it will be sold to a wide market and all products are to be identical. This will also reduce costs of production.	• An injected-moulded PET outer packaging container may offer more protection to the glass bottle and offer better aesthetic qualities.	

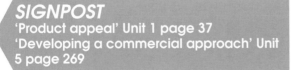

Marketing issues

The product would be sold direct to the target market, so a mixture of soft and hard sell promotion would appeal to the target market. It is to be sold in specialist outlets, so the image is portrayed as desirable and quality is high. Therefore higher prices can also be justified, as people will pay for this image.

SWOT

Strengths
– image
– aesthetics
– branding

Weaknesses
– protection
– materials choice (packaging)
– costs of materials

Opps
– product range
– inclusion of images to either appeal to a wider market or to a niche

Threats
– competitors if price is set high
– demand if quality is poor, or poor market research is conducted

• General further improvements involve the use of market research. From this a corporate image can be produced, along with a container and packaging that appeal to the target market. Further research into properties of materials and testing should also be undertaken. Consideration should be given to the inclusion of images also.

Figure 6.8 Part of an evaluation of the final solution

marketable product? Figure 6.8 shows how one student evaluated the final design solution.

Suggest improvements

SIGNPOST
'Product appeal' Unit 1 page 37
'Developing a commercial approach' Unit 5 page 269

You should suggest realistic improvements that would enable your product to better meet the needs of your product design specification. This may include improvements in response to the performance, quality, safety, market and user needs that you identified. How could any of these aspects be improved? What changes could you make to your product design to make its manufacture more efficient in relation to your chosen scale of production? How could you make use of modern materials and/or modern technology, including the use of CAD/CAM or automation to enable more efficient manufacture of your product?

Helpful hints
There is a suggested time of 20 minutes available for section f), which is worth 10 marks. Check how many marks are available for evaluating the final solution and how many marks are available for suggesting improvements. Spend an appropriate amount of time on each activity so you have the opportunity to achieve the available marks.

Exam practice

Revise what you have learned

Revising and reviewing what you have learned during your advanced level course will help you prepare for the Unit 6 exam. In Unit 6 you will find 'Signposts' that refer you back to other AS and A2 units. It would also be helpful to re-read your Unit 1 notes about industrial practices, such as specification development, scale of production, manufacturing processes, quality, and health and safety.

Practise doing timed design tasks

You should be familiar with all of the design processes that you will follow during the exam. They are similar to those that you used in your coursework.

Read the whole of Unit 6, so you are absolutely clear about the exam requirements. Practise doing timed design tasks that correspond to the design processes you will follow in the Design Exam.

a. Analyse

- Analyse a design problem set by your teacher or tutor.
- Analyse a body of research information related to the design problem – perhaps downloaded from the Internet, from CD-ROMs or from textbooks.
- Develop a product design specification based on the design problem and your research.

Exam tip

Analyse your own research before the exam, so you are familiar with it. Make a list of the product design specification criteria and organise your research information under the same headings. Write a conclusion to each section of your research. This process will fully prepare you for the exam. Practise developing product design specifications.

b. Generate ideas

- Generate a range of design ideas based on your product design specification and research.
- Evaluate the feasibility of each idea against your product design specification.
- Justify the design you have selected for development.

Exam tip

Practise producing quick design ideas in a set time. Always annotate your drawings and give reasons for your design decisions. Always make use of your product design specification when evaluating design ideas.

c. Develop ideas

- Develop a chosen idea into a workable solution.
- Describe and justify the solution in terms of its purpose, function, appearance, performance, quality, safety, the materials, components, systems, processes and technological features to be used.

Exam tip

Show how your ideas develop by annotating your drawings. Explain your thoughts about your chosen idea. How does it meet the specification requirements? How might it be made? Will it be easy and practical to make? Are the manufacturing processes appropriate to the scale of production? Are any parts or components the same size or shape? Can their manufacture be easily repeated? Can you use any bought-in parts or components?

d. Represent and illustrate the solution

- Illustrate and annotate the solution using 2D or 3D drawings.
- Use exploded drawings to show hidden details.
- Include dimensions and sizes of component parts.
- Provide details and quantities of materials and components.

Exam tip

Annotate your drawings so an examiner can understand how you intend the product to be made. You are not required to produce a full working drawing in the exam.

e. Production plan

Describe how to make the product. Include the following:

- commercial manufacturing processes
- the sequence of assembly
- quality checks.

Exam tip

Produce a flow diagram to show the sequence of assembly for the product. Describe the manufacturing processes and show where and how you would check for quality.

f. Evaluate

Evaluate your solution against your product design specification, suggest how your product could be tested after manufacture and suggest realistic improvements.

Exam tip

Suggest improvements to the design and

manufacture of your product so it better meets the needs of your specification.

Practise doing a complete timed exam

Practise a complete exam, working to the same time and under the same conditions as the Unit 6 Design Exam. Choose from the following four Design Research Papers and corresponding Design Papers. Use the assessment criteria and marks below to help you.

Table 6.2 *Assessment criteria*

Assessment criteria	Marks
a) Analyse the design problem and develop a product design specification, identifying appropriate constraints	15
b) Generate and evaluate a range of ideas	15
c) Develop, describe and justify a final solution, identifying appropriate materials and components	15
d) Represent and illustrate your final solution	20
e) Draw up a production plan for your final solution	15
f) Evaluate your final solution against the product design specification and suggest improvements	10
Total marks	90

Design Research Paper 1

A kite manufacturer wants to increase sales of a new box kite by basing its design on a cartoon character. The kite is to be advertised on the company web site. Investigate the following:

- existing box kite designs
- styling, colour and images used in the creation of cartoon characters
- web site advertising.

Design Paper 1

Your task is to design a new 'sky diver' box kite, together with advertising materials to be used on the company web site. The kite needs to be based on a cartoon character and should be suitable for 7–12-year-old children.

Design Research Paper 2

International sporting events have to deal with the problems of language and communication. Investigate the following:

- the types of signage used at past international sporting events
- the materials and manufacturing processes used to produce sporting event signage and their methods of support
- the aesthetic and functional requirements of signage in general.

Design Paper 2

Your task is to design suitable signage and its support structure for an international youth indoor sporting event. Incorporate all of the following events in your design: volleyball, gymnastics, badminton, tennis, swimming and cycling.

Design Research Paper 3

Communicating a company's image is essential in today's competitive market, especially for food manufacturers, where branding is an essential marketing tool. Investigate the following:

- signs, symbols and brand marketing used in graphics for breakfast cereals
- the use of colour, typography, shape, form and layout
- promotional gifts used in marketing products for children.

Design Paper 3

Your task is to design the brand identity, packaging and a promotional gift for 'Crackle Pops' breakfast cereal. The promotional gift must incorporate at least one resistant material and be capable of being assembled into a 3D toy.

Design Research Paper 4

Cosmetic and toiletry products are presented for sale in a variety of containers and outer packages. Investigate the following:

- cosmetic and toiletry containers and packaging
- the range of materials and manufacturing processes used to produce such containers and packages
- brand identities and surface graphics on cosmetic and toiletry containers and packages
- the aesthetic and functional requirements of cosmetic and toiletry packaging materials.

Design Paper 4

Your task is to design a container *either* for perfume *or* aftershave; the outer packaging; the brand identity and associated graphics for the container and its package. Your designs must be based on one of the following brand names:
Salsa; Chameleon; Wild; Platinum; Odyssey; Frantic.

Student checklist

1. Understanding the research context

- Make sure that you read the Design Research Paper carefully, so you have a clear understanding of what you need to research.
- Take responsibility for planning, organising, managing and analysing your own research.
- Make sure that you analyse your research *before* the exam.

2. Preparing for the Design Exam

- Review what you have learned about designing and making products.
- Practise doing timed design tasks that correspond to the different sections of the exam paper.
- Practise doing a complete timed exam.
- Make sure that you understand the assessment requirements, so you will know how much time to spend on each section of the exam paper.

3. Evidencing your capability in the Design Exam

- Produce a short analysis of the problem you are asked to solve.
- Show the influence of your research in your product design specification.
- Clearly show how your ideas unfold. Annotate your drawings.
- Demonstrate a variety of communication skills, including graphics, diagrams, flowcharts and written work.
- Allow enough time to evaluate your work as it progresses.
- Use industrial terminology and technical terms to demonstrate your understanding of commercial manufacture.

4. Exam hints

- Collect together all drawing and writing materials and practise any drawing techniques you may need to use.
- Prepare a backing sheet with heavy lines as a guide for written work.
- Prepare a backing sheet of isometric paper as a guide for 3D sketches.
- Collect together all your folios, sketch books, notebooks and technical data compiled as part of your exam research. You may also refer to any relevant British Standards. Be sure to analyse your research *before* the exam.
- During the exam you will be provided with headed answer sheets. Extra sheets may be used if *absolutely* necessary. Be sure to write your name and number and centre name and number on any extra sheets you use. Remember to fasten together all your answer sheets *in the correct sequence* with a treasury tag at the end of the examination.

5. Exam day

- Arrive half an hour before the exam to get yourself organised.
- Read the exam question carefully and identify key points in the design problem. Do not copy out the design problem.
- Use a pencil or pen for writing and sketching. Try not to rub things out, as its better if the examiner can see your thinking. Take a pencil sharpener with you.
- Only use colour after you have completed a section, or spend the last ten minutes adding colour.
- If you make a mistake or go completely wrong, state it and attempt to get back on the right track. Do not throw away any work, as it will be helpful for the examiner to see your 'design thinking'.
- You will be assessed on your ability to organise and present information, ideas, descriptions and arguments clearly and logically. You will get marks for quality of communication.

Help! What if my project goes wrong?

- If your Unit 6 Design Exam doesn't meet your expectations, don't worry! Your teacher or tutor will be able to advise you on the best way forward.

Glossary

A

Adjustable sensing inputs – transducers configured in potential dividers to sense changes in the surrounding environment and produce an electrical signal accordingly.

Advertising – any type of paid-for media that is designed to inform and influence existing or potential customers.

Advertising Standards Authority (ASA) – regulates all British advertising in non-broadcast media.

Aesthetics – the sensory responses that people make to external stimuli such as colour, shape, sound, touch, smell and taste; individuals respond differently to these stimuli, finding them pleasing or disagreeable; 'visual aesthetics' refers specifically to those that are concerned with appearance; people respond to 2D and 3D products using a range of senses.

Agitation – the movement of film and photographic solutions in development to ensure an even coating.

American National Standards Institute (ANSI) – administrator and coordinator of the United States voluntary standardisation system founded in 1918.

Analogue signals – type of electronic signal that can exist at any level between the two extreme points of high and low

Annealing – the process that relieves internal stresses within a material as a result of cold working.

Artificial intelligence (AI) – a branch of computer science concerned with developing computers that think/act like humans.

Axonometric – drawings showing an object in 3D including isometric, planometric and oblique.

B

Bandwidth – the amount of HYPERLINK 'data.html' data that can be transmitted or received in a fixed amount of time; expressed in bits per second (bps) for digital devices. The higher the bandwidth the faster the rate of data transfer. See 'ISDN'.

Bar code (data communication tag) – machine-readable pattern of stripes printed on a component part or a finished product to identify it for production processes, stock control, pricing or retail sales.

Binary signal – a digital signal with only two values: high ('on' or '1') or low ('off' or '0').

Bitmaps – images of a surface or solid model produced by an intensity of points called picture elements or pixels.

Biochips – in biochips, semi-conducting molecules are inserted into a protein framework, in which the proteins grow to take up the shape required for an electronic circuit. The manufacture of biochips will lead to the further miniaturisation of electronic products.

Blow moulding – the process of using air, under pressure, to give form to products by 'blowing' plasticised material into moulds.

Brand – a product or service with a marketing identity that belongs to one producer.

Brand loyalty – involves buying a chosen brand of product, because it provides a perceived level of reliability, quality or lifesyle.

Brand name – (also trade mark or trade name) protects and promotes the identity of a product or process, so that it can't be copied by a competitor. A branded product usually has additional features or added value over other generic products – making it 'special' in the eyes of consumers.

Break-even point – the point at which products sold at a certain price will equal in value the cost of their manufacture.

British Standards Institute (BSI) – an independent, non-profit-making and impartial body that serves both the private and public sectors. BSI works with manufacturing and service industries to develop British, European and International standards.

Buy in – components or sub-assemblies that are 'bought in' or purchased from another manufacturer for use within a product, e.g. a gearbox may be bought in by a car manufacturer.

Buying behaviour – establishes how people make buying decisions and factors that influence these decisions.

C

CAD modelling – involves a number of representation techniques such as vector or raster graphics. Models are used for variety of purposes ranging from a simple record of a design with critical manufacturing dimensions through to a fully interactive simulated product or system model that can be viewed in operation.

Caliper – used to describe the thickness of card material; measured in microns (mic) which are equal to one thousandth of a millimetre.

Camera ready artwork – artwork that has been pasted into position to be photographed for final film; usually a page containing all the necessary image texts and tints in position needed to produce a printing plate.

CEN – the European Committee for Standardisation. It implements the voluntary technical harmonisation of standards in Europe, in conjunction with worldwide bodies and European partners.

CENELEC – the European Committee for Electro-technical Standardisation.

Cleaner design – aimed at reducing the overall environmental impact of a product from 'cradle to grave'.

Cleaner technology – the use of equipment or techniques that produce less waste or emissions than conventional methods. It reduces the consumption of raw materials, water and energy and lowers costs for waste treatment and disposal.

Closed question – provides a limited number of possible answers to choose from.

Closed-loop control system – a system that has a degree of feedback built into it that enables a degree of checking to see whether actions or processes have been carried out.

Clutch – type of shaft coupling that allows a rotating shaft to be easily connected or disconnected from a second shaft.

Comparative testing – comparing the characteristics and properties of different materials, often in relation to specification requirements; simple comparative testing makes use of standard tests under controlled conditions, so each material is tested in exactly the same way; results can be compared to see which material is the most appropriate.

Compatibility – hardware and software applications capable of being used in combination without any technical problems. If systems are compatible, computers can 'talk' to other computers and to equipment such as CNC machines.

Competitive edge – the reason why a customer might choose a certain product rather than its competitors.

Components – the parts of a product that go to make up the whole.

Composite image – a photo or other graphic image that is made up of a number of images

Compound gear train – a series of gears connected together, some on common shafts, which allow rotational speeds to be increased or decreased.

Computer integrated manufacture (CIM) – a system of manufacturing that uses computers to integrate the processing of production, business and manufacturing information in order to create more efficient production lines.

Computer simulation – a method of modelling manufacturing processes, designs or other product characteristics using computer software.

Computer-aided engineering (CAE) – the use of computer systems to analyse and simulate engineering designs under a variety of conditions to see if they will work.

Conductivity – the measure of a material's ability to have heat or electricity passed through it.

Consumer demand – the potential for the demand to 'pull' products through the distribution system; often stimulated by marketing and promotion.

Consumer goods – products purchased by consumers for their own consumption.

Consumer society – a social culture in which consumers are encouraged by advertising to buy consumer goods.

Continuous Improvement (CI) – (Kaizen) involves companies systematically looking for opportunities to make manufacturing operations leaner and more cost effective whilst also getting the most from their workforce by involving employees in all aspects of their operation. It aims to develop a culture where all employees are communicating better, working proactively to meet the company's business objectives and where implementing improvements is a matter of course.

Conversion – in timber production, the process in which a felled tree is turned into a useable source of timber.

Co-polymerisation – the process in which two or more monomers combine to form a new material.

Corporate Identity Systems – these define and describe the way in which an organisation presents itself both internally and externally; the system expresses and reflects the values of the company to employees, suppliers and customers; it commonly deals with the permitted variations of use of logotypes and symbols applied in different ways to its products, stationary, delivery vans, uniforms, promotional materials, etc.

Corporate image – the identity or 'personality' of a company or organisation, created through the use of different graphic images.

Crating – a way of drawing a product by imagining it being made up of a number of boxes joined together.

Critical control points (CCPs) – points during the production cycle at which a product is monitored to ensure that it is successfully manufactured to specification; ensures that any faulty components are rejected before they are processed further or built into the final assembly.

Critical path analysis – the breakdown of the whole manufacturing process into an ordered sequence of simple activities.

Critical technologies – technologies that need to be in place for a product to develop, e.g. the availability of well-established sensor technologies is critical in the development of industrial robots.

Cupping – a fault in timber where the board is hollowed along its length.

Current – the flow of an electric charge through a conductor.

Customer profile – a profile of potential customers, such as gender, age, family group, income, education, beliefs and attitudes, taste, lifestyle and perception of products.

D

Darlington pair – two smaller transistors connected together, increasing the overall gain and sensitivity.

Data storage devices – devices such as CD-ROMs that are used for storing digital electronic information known as data.

Demographics – patterns and trends of population and society, such as age, gender or income bracket.

Dendrites – a crystal that has branched during its growth and has a tree-like look.

Design management – the planning of a product to include organisational, economic, legal and marketing considerations, as well as decisions about form and function.

Design specification – sets out the criteria that the product aims to achieve.

Designing for manufacture (DFM) – aims to minimise costs of components, assembly and product development cycles and to enable higher quality products to be made.

Desktop publishing (DTP) – the use of a computer program to design the arrangement of text, images and graphic devices on a printed page; the electronic data included in the resulting files can be passed directly to computerised printing systems.

Desktop videoconferencing (DTVC) – videoconferencing applications that use video cameras mounted on standard desktop computer system such as an Intel-based PC, Apple Macintosh or Unix workstation.

Die cutting – the process of cutting and creasing sheet materials, so that the material can be formed into 3D products such as cartons or packages.

Die cutting tools – these are required for the cutting and creasing of cardboard or plastic sheet components that make up cartons or boxes; the design for die cutting tools are normally produced on a CAD machine, but the tool is made by hand.

Digital imaging – the process of creating a digital copy of an illustrated or photographic image.

Digital photography – the process of recording images using a digital camera.

Digital printing – this involves linking printing presses with computers, bypassing the need for making printing plates.

309

Digital signal – a signal that can only take fixed values between two points.

Digital system – an electronic system that can exist in one of only two states: on or off.

Diode – an electrical component that will only allow a current to pass through it in one direction.

Direct costs – see 'Variable costs'.

Downtime – unproductive period when a computer system or machine is not operational usually because of technical problems, maintenance or, in the case of machines, setting up new tools or reprogramming tools on a CNC machine.

Download – to get electronic information, software, files and documents from the Internet on to a computer system.

Dynamic images – these give a sense of movement to a graphic layout; their shape directs the viewer's eye to different parts of the page and adds impact.

E

Economies of scale – these occur when the cost of producing each product falls as the total volume of products produced increases.

Efficiency – a measure of what you get out in relation to what you put in.

Elastic deformation – when a material returns to its original shape and length once a deforming force has been removed, it is said to have been elastically deformed.

Electromotive force (emf) – a source of energy that can cause a current to flow in an electrical circuit or device.

Electronic Data Interchange/Exchange (EDI or EDE) – the electronic transfer of commercial or organisational information from one computer application to another; it is also known as paperless trading.

Electronic Product Definition (EPD) – this makes use of CAD/CAM systems in which all of the product and processing data is generated and stored electronically in a database. The whole production team has access to the database, which evolves as the new product is developed. EPD enables the use of computer integrated manufacture (CIM).

Enabling technologies – these provide the useful technologies, for example drive motors and power control systems that make critical technologies effective.

Environmentally friendly plastics – plastics capable of bio-degrading naturally.

Enzymes – naturally occurring proteins used to create industrial products and processes. These enzymes are the same kind that help us digest food, compost garden rubbish and clean clothes.

ETSI – the European Telecommunications Standards Institute.

European Standards Organisation – joint standards organisation called CEN/CENELEC/ETSI.

Evaluation matrix – this is used to compare and evaluate a number of ideas against specification criteria. Each idea is given a score showing its strengths and weaknesses. Very weak ideas are eliminated, resulting in the emergence of strong ideas, which can be developed individually or combined in some way.

Expert systems – part of a general category of computer applications known as artificial intelligence (AI). They either perform a task that would normally be done by a human expert or they support the less expert to complete a task.

Exploded – (drawings) show a product or component pulled apart, laid out in an ordered and linear form.

External failure costs – these occur when products fail to reach the designed quality standards and are not detected until after being sold to the customer.

Extranet – an intranet that is partially accessible to authorised outsiders.

F

Fabrication – the joining and fixing together of various materials and components to form a new product.

Feedback – information generated within a system or process to enable modifications to be made to maintain the operation of a system or to ensure a consistent level of production.

Field effect transistors (FET) – an electronic voltage amplifier.

Figure – the natural decorative pattern of the timber's grain.

File server – a computer with data that can be accessed by other computers.

File Transfer Protocol (FTP) – the method of transferring information files from Internet libraries directly to a computer.

Final film – the intermediate print production step between artwork and plate making.

Finishing – in commercial printing, this is the way a document is collated, bound, folded or glued.

Finite resources – see 'Non-renewable resources'.

Firewall – a hardware or software system designed to prevent unauthorised access to or from a private computer network (see 'Intranet').

Fitness-for-purpose – a product's fitness-for-purpose can be evaluated through its performance, price and aesthetic appeal.

Fixed costs – (indirect or overhead costs) fixed costs remain the same for one product or hundreds as they are not directly related to the number of products made. They include design and marketing, administration, maintenance, management, rent and rates, storage, lighting and heating, transport costs.

Force field analysis – this maps the forces for and against an idea or concept and the forces for and against changing it.

Form – created when a 'shape' becomes three dimensional, e.g. a circle becomes a sphere or cylinder.

Fourdrinier machine – a machine that releases and processes pulp to form a continuous paper roll.

Function – the means by which a product fulfils its purpose.

G

Gantt chart – a simple chart that maps each task against the time available, together with an order of priority.

Geometric modelling – using computer programs for representing or modelling the shapes of three-dimensional components and assemblies. Geometric models are the basis of all CAD/CAM systems.

Glass Reinforced Plastic (GRP) – a matting of glass strands held rigid in a polyester resin; also called fibreglass.

Global manufacturing – the manufacture, by multinational companies, of products that may be designed in one country and manufactured in another.

Global market place – the marketing of products, such as washing machines and cars, across the world. To be successful in this global market place, a company has to have a product that appeals to people in different countries and cultures.

Graphic devices – lines or shapes (usually coloured) used in a layout to add visual interest and/or to help identify different sections of information within the design.

Gravity die casting – the process by which molten metal is poured into metal or graphite moulds.

H

Hard sell – a hard sell advertisement has a simple and direct message, which projects a product's Unique Selling Points (USPs).

Hazard – source of or situation with potential harm or damage. Hazard control incorporates the manufacture of a product and its safe use by the consumer.

Heat treatment – the changing of a material's properties and characteristics due to the application of an external heat source.

High-technology production – the production of 'high tech' products, which emphasise technological appearance and modern industrial materials.

HTML (Hyper Text Mark Up Language) – text-based coding system and scripting language used when writing web pages.

HTTP (Hyper Text Transfer Protocol) – a transport protocol used when transmitting hypertext documents across the Internet.

Hyperlink – an electronic connection that allows links between different web pages to be made, usually shown in a different colour and/or underlined.

I

Indirect costs – see 'Fixed costs'.

Industrial terminology – this includes the use of technical terms such as critical control point or production plan to demonstrate an understanding of industrial practices.

Injection moulding – a highly automated manufacturing process in which a plasticised material is injected into a mould cavity under high pressure.

Interface – a device that allows electrical signals into and out of a computer.

Internal failure costs – these occur when products fail to reach the designed quality standards and are detected before being sold to the consumer.

Intranet – a network based on the Internet belonging to an organisation, which is accessible only by authorised users with user names and passwords.

Inventory – a company's merchandise, raw materials, and finished and unfinished products that have not yet been sold.

ISO (International Organisation for Standardisation) – worldwide federation of national standards bodies from some 130 countries, one from each country. ISO is a non-governmental organisation established in 1947.

ISDN (Integrated Systems Digital Network) – a high-speed, wide-bandwidth electronic communications service for carrying digital data, digitised voice or video across digital phone lines.

ISO 9000 – a set of management processes and quality standards to ensure that a product meets the customer's requirements.

Isometric paper – paper that has vertical lines, with all other lines drawn at 30° to the horizontal; useful as a backing for sketching an isometric view, in which the product is drawn at an angle with one corner nearest to view. In this type of drawing all vertical lines on a product remain as vertical, while all horizontal lines are drawn at 30° to the horizontal on the paper. No vanishing points are used and the height, width and length are shown as parallel sets of lines.

ISP (Internet Service Provider) – a company providing a connection to the Internet.

J

Jidoka – (autonomation) a Japanese term for the automatic control of defect; a machine finds a problem, finds a solution, implements it without outside assistance and then carries on.

K

Kaizen – See 'Continuous improvement'.

Kanban – a Japanese term for a card signal or visual record, 'Kan' meaning card, 'Ban' meaning signal.

Kerning – the space between characters, which can be adjusted so that parts of the characters overlap; the purpose of this is to make words fit on a line without affecting readability; LY LY and AT AT are called 'kerning pairs' as they can overlap.

Kitemark – a seal of approval by the British Standards Institute, awarded to any product that meets a British Standard as long as the manufacturer has quality systems in place to ensure that every product is made to the same standard.

L

Laminating – the process of sticking sheets of laminate or veneers together in either flat sheets or over curved formers.

Lash-ups – quick, rough models used to work out and test the relationship between different parts of a design.

Lattice structure – the pattern adopted when the atoms of a liquid solidify.

Layout – the arrangement of images and text in relation to each other.

Layout paper – this allows images to be traced and is commonly used for sketching.

Level/scale of production – the size of production, e.g. a one-off such as a bridge or thousands such as chocolate bars.

Life cycle assessment (LCA) – this evaluates the materials, energy and waste used in a product through design, manufacture, distribution, use and end-of-life, which could be disposal, reuse or recycling.

Lifestyle marketing – the targeting of potential market groups and matching their needs with products.

Light-dependent resistor – a semiconductor whose resistance changes as the amount of light falling upon it changes.

Linkages – a series of levers connected together to change the direction of motion.

Liquid crystal display (LCD) – a numerical and alphanumerical display system used in calculators, digital watches, etc.

Local Area Network (LAN) – a collection of computers connected together to share information and other computer resources such as a printer.

Logic – a structured way of thinking or a set of operating principles applied to a manufacturing system or product to allow it to perform a specified task.

Logic gate – a series of electronic switches that give a known output when a certain configuration on the input pins exists.

Logistics – the detailed organisation and implementation of a plan or operation such as supplying and moving parts, components and finished products within and from a manufacturing system.

Logotype – a logotype is the use of a distinctive typeface to identify the goods or services of a particular organisation, or the brand name of a particular product; often used in conjunction with a symbol.

M

Malleability – the ability of a material to be beaten or pressed into a shape without breaking or fracturing when cold.

Manufacturing specification – clear details of product manufacture such as accurate drawings, clear construction details, dimensions, sizes, tolerances, finishing details, colour tolerances in printing/reproduction processes, quantities and cost of materials and components.

Market driven – the concept that promotional activity and marketing pushes products through the market group to pull products through the distribution system.

Market led – the concept that promotional activity and marketing stimulates demand in customers in a distribution system.

Market potential – the potential for a product to sell into a specific target market group.

Market research – this identifies the buying behaviour, taste and lifestyle of potential customers and establishes the amount of money they have to spend, their age group and the types of products they like to buy.

Market segment – a group of people with similar needs who wish to buy a certain type of product or service.

Market segmentation – a marketing technique that targets a group of customers with specific characteristics.

Market timing – attempting to predict future market directions, usually by examining recent price and volume data or economic data, and investing based on those predictions.

Marketing – anticipating and satisfying consumer needs while ensuring a company remains profitable.

Marketing plan – a set of marketing activities developed to match a company's products to selling opportunities; involves developing a competitive edge by providing reliable, high-quality products at a price customers can afford, combined with the image they want the product to give them.

Mechanical Advantage (MA) – a mathematical relationship which exists between the load and effort in relationship to levers. The greater the MA, the easier it becomes to move the object.

Media – agencies such as the press, television or posters that carry advertising.

Metal crystals – basic unit cells which make up the lattice structure of a metal.

Metal grains – small crystals that form between dendrites on cooling.

Micro-structure – the structure of material as observed under a microscope.

Milestone planning – project management process involving identification of key points or milestones that needed to be reached in a production process if a product is to be successfully completed on time and to budget.

Miniaturisation – this came about through developments in microchip technology, resulting in ever smaller products.

Modelling – visualising design ideas using hand or computer techniques in two dimensions (2D) or three dimensions (3D).

Monomers – a compound whose molecules can join together to form a polymer.

Moodboard – boards used by a professional designer to explore moods or themes and to give a product an identity; moodboards communicate ideas on design, illustrating themes, trends, form, colours, texture and styling details; these product 'stories' are inspiration for generating design ideas.

Multimeter – a measuring device used to measure current, voltage, and resistance, and also used as a continuity tester.

Multinational companies – companies that operate in more than one country and used to be mainly associated with mineral exploitation or plantations, such as cotton or food.

N

Niche markets – target market groups for whom products are designed and marketed.

Noise – unwanted electrical signals, e.g. fuzzy TV pictures caused by electrical interference.

Non-renewable resources – finite resources, such as oil or coal, which will eventually be exhausted unless action is taken.

O

Offset printing/lithography – currently the most common commercial printing method; ink is offset from a printing plate to a rubber roller and then to paper.

One-point perspective – the simplest form of perspective drawing in which the front view is drawn as a flat two-dimensional image. All receding lines are then taken back to a single vanishing point, to give a three-dimensional view.

Opacity – how paper is judged in its degree of transparency.

Open-loop system – a control system that incorporates no feedback, being a pure linear progression from the input to the output.

Optical character recognition (OCR) – where an electronic device recognises written letters or numbers.

Orthographic views – see 'Working drawing'.

Output transducers – devices such as bulbs, motors and alarms.

Overhead costs – see 'Fixed costs'.

P

Parallax error – the mis-alignment of the image when taking a photo; this occurs when using a camera that does not allow the user to view the image through the lens.

Parametric designing – this involves establishing the mathematical relationships between the various parts that make up a shape or a product. Once the parameters are determined a designer can model exactly what would happen if particular sizes were redefined because if one measurement is changed all the others are changed in the correct proportion.

Patents – issued by government authority, these documents grant the sole right to make, use or sell a design, making it both unique and protected.

Performance modelling – working prototypes that enable the designer to test the function of a design against the design specification.

Perspective – this allows an object to be drawn as it is viewed by the human eye, with parallel lines converging at a vanishing point; only the vertical edge closest to the viewer is in scale.

PEST – part of the basic structure of a marketing plan. It involves analysing values such as political, economic, social and technological issues related to marketing a product.

Photovoltaic cell – a semiconductor that generates a small voltage when exposed to bright light.

Pictorial – shows the most realistic view of a product, sometimes called an 'artistic impression'.

Piezo-electric actuators/transducers – an electronic device capable of generating a small voltage when pressure is applied, or a small movement if voltage is applied.

Planning horizon – how far to plan forward, determined by how far ahead demand is known and by the times required to pass through the manufacturing operation.

Plasticity – the ability of a material to be moulded.

Plates – sheets of treated aluminium alloy on which a print image is chemically etched; one plate is required for each colour; used in the printing process to transfer and ink image onto board or paper.

Poka-yoke – a Japanese term meaning a device or procedure to prevent a defect during order taking or manufacture (also called baka-yoke). The nearest translation is 'foolproofing' or 'mistake-proofing'.

Polluter pays – the concept that those generating, handling and treating wastes should pay large fines if they allow potentially harmful materials to enter the environment.

Polymerisation – a chemical reaction that occurs when a polymer is formed.

Post-production – the manipulation and addition of elements – usually computer generated – of filmed scenes in studio and on location.

Potential divider – two resistors connected in series, set up to divide the potential in the ratio of resistor one to resistor two.

Potentiometer – a three-legged device that can be configured to work either as a potential divider or rheostat.

Presentation drawings – drawings to communicate ideas about a product or environment using a variety of suitable drawing techniques; a children's pop-up book could be presented in a simple, colourful style, whereas a high-tech interior may benefit from a more technical style of presentation.

Prevention costs – the costs of 'making it right first time'. Prevention costs include those relating to the creation of and conformance to a quality assurance system and the management of quality.

Primary processing – the conversion of raw materials into usable stock for production, e.g. steel making.

Primary research – facts and figures that are collected specifically to provide information and help achieve the research objectives.

Primary sector – this is concerned with the extraction of natural resources such as mining and quarrying.

Product Data Management (PDM) software – this integrates the use of computer systems, including CAD/CAM and computer integrated manufacturing (CIM). PDM software enables the design and development of virtual products on screen. The software organises and communicates accurate, up-to-date information in a database, monitors production and enables fast, efficient and cost-effective manufacturing on a global scale.

Product design cycle – the process leading to the design and manufacture of a product involving design, make, redesign and remake, which starts with a perception of need and includes many influences throughout the process such as government policy, manufacturers, advertisers, retailers and consumers.

Product viability – essential to the existence of a manufacturing company and to the employment of its workforce; relates to the cost of manufacture, the product's market potential and the potential profit from manufacturing the product.

Production capacity – the maximum number of products that can be made in a specified time.

Production chain – the sequence of activities required to turn raw materials into finished products for the consumer.

Production plan – this shows how to manufacture a product, based on the breakdown of the whole process into an ordered sequence of simple activities; includes all specifications, the stages of production, resource requirements and the production schedule.

Production schedule – an ordered sequence of processes that are required to manufacture a product.

Production team – flexible, organised, skilled, versatile people, who work collectively, make joint decisions and share the responsibility for the design and manufacture of products.

Productivity – a measurement of the efficiency with which raw materials (production inputs) are turned into products (manufactured outputs). High productivity results in lower labour costs per unit of production and a higher potential profit.

Profit – the amount left of the selling price of a product, after all costs of manufacture have been paid.

Programmable logic controller (PLC) – small but complex systems containing timers, counters and many other special functions capable of almost any type of control application, including motion control, data manipulation and advanced computing functions such as manufacturing plant management.

Protocol – an agreed standard or set of rules. See 'File Transfer Protocol' and 'HTTP'.

Prototype – a detailed 3D model made from inexpensive materials to test a product before manufacture.

Pulley – a circular disc normally with a V-shaped groove cut around its circumference.

Q

Qualitative research – an investigation to find out how people think and feel about issues and why they behave as they do.

Quality – conformity to specifications and ensuring fitness-for-purpose. Making products right first time, every time, with zero faults, to ensure customer satisfaction.

Quality assurance (QA) – a system applied to every stage of design and manufacture; ensures conformance to specifications to make identical products with zero faults.

Quality control (QC) – checking at critical control points against specifications for accuracy and safety, so that a product meets consumer and environmental expectations.

Quality indicators – quality control techniques, such as inspection, testing and sampling, that are applied at critical control points during manufacture to ensure the product meets specifications. Quality indicators may be attributes that can only be right or wrong, such as using the correct type of wood, or variables that can vary between specified limits, such as meeting a tolerance of +/–1.0 mm.

Quality Management System (QMS) – this uses structured procedures to manage the quality of the designing and making process.

Quality of design – a product that is well-designed and attractive to the target market, meets specifications, uses suitable materials, is easy to manufacture and maintain, and is safe for the user and the environment.

Quality of manufacture – this refers to a well-made product that uses suitable materials, meets specifications and performance requirements, is manufactured by a suitable, safe method, is made within budget limits to sell at an attractive selling price, and is manufactured for safe use and disposal.

Quantitative research – an investigation to find out how many people hold similar views or display particular characteristics.

Questionnaire – a standardised set of questions designed to collect data that is relevant to the research objectives.

Quick Response Manufacturing (QRM) – a manufacturing system able to respond quickly at all levels of business or production processes in response to market trends and changing demand patterns.

R

Rapid Prototyping (RPT) – a CNC application that creates 3D objects using laser technology to solidify liquid polymers in a process called stereo-lithography.

Rectification – the process of converting an alternating current into a direct current.

Recycling waste materials – a form of waste management in which waste materials from the production process are used in a different manufacturing process.

Relay – a device used to interface two separate circuits that operate at two different supply voltages.

Renewable resources – resources that flow naturally in nature or are living things which can be regrown and used again. They include wind, tides, waves, water power, solar energy, geothermal, biomass, ocean thermal energy and forests.

Resistor – an electronic component used to control the current flowing in an electrical circuit.

Reusing waste materials – see 'Recycling waste materials'.

Right first time – the aim of quality assurance, to make sure the product is right first time, every time. It involves making products that meet the specification, on time and to budget.

Risk assessment – identifying risks to the health and safety of people and the environment.

S

Seasoning – the process of reducing the moisture content in timber.

Secondary processing – the working of a material using engineering processes such as turning or milling.

Secondary research – facts and figures that are already available, having been collected for another purpose by a range of organisations.

Secondary sector – this is concerned with the processing of primary raw materials and the manufacture of products.

Selling price (SP) – the price at which a product can be sold in order to make a profit. It generally includes variable costs, fixed costs and a realistic profit.

Semiconductor transducers – semi-conductors whose resistance changes depending on the surrounding environment.

Server – a host computer that distributes and stores data on a network.

Shape memory alloys – plastics that revert to their original form when heated.

Sketch model – a quick model produced in the early stages of product development, using inexpensive materials such as card, paper, expanded polystyrene, styrofoam and wood.

Sleeve – a versatile packaging accessory with applications ranging from the protection of a delicate item to food such as a ready meal; also used as an alternative for a lid; made from card or PVC.

Smart materials – the properties of smart materials can change in response to an input, such as piezo-electric actuators; provide opportunities for the development of new types of sensors, actuators and structural components, which can reduce the overall size and complexity of a device.

Soft sell – promotes a product's image with which consumers can identify and is often associated with brand advertising.

Solar panels – these panels normally have water pumped through them that gets heated by the sun's energy.

Specialised components – components that are manufactured specifically for a particular product application.

Standard components – components such as nuts and bolts that are supplied ready to use.

Standards – documented agreements with technical specifications or other precise criteria to be used consistently as rules, guidelines or definitions of characteristics, to ensure that materials, products, processes and services are fit-for-purpose.

Statutory rights – what consumers should reasonably expect when buying or hiring products and services. Statutory rights are enforced and regulated by a wide range of legislation relating to consumer protection and fair trading.

Strategic technologies – ways of thinking and operating, e.g. artificial intelligence.

Sub-assemblies – component parts of a product that are already made up of smaller components.

Supply chain – companies and organisations that collaborate to produce raw materials, components and end-products for specific end-uses aimed at specific target market groups.

Survey – a way of collecting quantitative data, often about behaviour, attitudes and opinions, of a sample in a target market group.

Sustainable development – a concept that puts forward the idea that the environment should be seen as an asset, a stock of available wealth. If each generation spends this wealth without investing in the future, then the world will one day run out of resources.

SWOT – part of the basic structure of a marketing plan. It involves analysing a product's strengths, weaknesses, opportunities and the threats from competition.

Symbol – a graphic device used to identify the goods or services of a particular organisation, or the brand name of a particular product; often used in conjunction with a specific logotype; as they are easily understandable by most cultures, common symbols are also often used to communicate information such as locations, directions and safety or hazard warnings.

Synoptic assessment – the drawing together of skills, knowledge and understanding acquired in different parts of the whole A level course.

T

Target market group – all the customers of all the companies supplying a specific product.

Target marketing – the process of identifying market groups and developing products for it.

Technical drawing – this contains factual information relating to appearance and dimension and is based on British Standard BS7308. See 'Working drawing'.

Telematics – a new technology that allows a product to be managed electronically from receipt of the customer order through development, manufacturing, delivery and after-sales support.

Tempering – the process of removing the brittleness caused as a result of hardening.

Tertiary sector – this is concerned with industries that provide a service; employs the most people in developed countries and includes education, retailing, advertising, marketing, banking and finance.

Test marketing – this involves introducing a product in a small sector of a target market to test its viability before incurring the expense of a full-scale product launch.

Test models – these models used to test different parts of a design and are often built from kits to test mechanical, structural or control problems.

Thermistor – a semiconductor whose resistance can change depending on the temperature around it.

3D CAD systems – a computer-aided design system that can produce virtual images in three dimensions to present more realistic representations of products and assemblies.

Thumbnail – a small rough sketch showing the main parts of a design in the form of simple diagrams.

Thyristor – a three-legged electronic semi-conductive switch that can be used as a latch.

Time bucket – the unit of time on which a production schedule is constructed and is typically daily or weekly.

Time delay circuit – a capacitor/resistor network that is capable of producing an electronic time delay.

Tolerance – the degree by which a component's dimensions may vary from the norm and still be able to fulfil its function.

Total design concept – design using multimedia 'toolkits' to access an integrated on-screen design modelling environment that includes systems linked to production databases to analyse and plan for manufacture.

Total Quality Control (TQC) – the system that Japan has developed to implement Kaizen for the complete life cycle of a product.

Transducer – a device for converting physical signals into electrical signals. An input transducer responds to a physical change by producing an electrical signal to represent the change. An output transducer takes an electrical signal from a system and produces a physical change as an output.

Transistor – a semiconductor device that can exist as an insulator or conductor. It can be used as an electronic switch and amplifier.

2D CAD drawings – drawings that have length and width but no depth.

Two-point perspective – the most common form of perspective drawing in which the vertical lines stay vertical, while all other lines recede to two vanishing points. These points are placed on a horizontal line called the eye level. If a product is drawn below eye level, the top will be visible. If it is above eye level, the underside is visible.

Typography – the design and application of different letter (and number) forms to printed text to ensure appropriate legibility and ease of reading.

U

Unique selling proposition (USP) – a product's unique features and advantages over a competitor's products.

Unix – a computer operating system that was developed by AT&T in the 1960s. The system was used extensively during the establishment of the Internet.

Upload – this means to send electronic information from your computer to another location via the Internet or other types of network (the opposite of download).

URL (Uniform Resource Locator) – the convention used when naming pages on the World Wide Web.

V

Vacuum forming – a plastic processing method in which a softened plastic sheet is pushed down by atmospheric pressure on to a mould to make products such as baths.

Value analysis – the process of close study of a product in order to reduce manufacturing costs and/or increase the product's perceived value.

Variable costs – (direct costs) variable costs increase with the number of products made. They include depreciation of plant and equipment.

Vector graphics – (object-oriented graphics) these are images comprised of a collection of lines rather than dots as in bitmap graphics.

Videoconferencing – a conference conducted between two or more participants at different sites using computer networks to transmit audio and video data.

Virtual reality – this combines computer modelling with simulations to enable the development of an artificial 3D product or sensory environment; 3D virtual products can be created and viewed from different angles and perspectives.

Virtual Reality Modelling Language (VRML) – a specification for displaying and interracting with 3D objects on the World Wide Web.

Voltage – a difference in potential between two points in a circuit.

W

Waste minimisation – this involves reducing, reusing or recycling materials used in manufacture.

Web browser – software such as Netscape Navigator or Microsoft Internet Explorer that provides the interface between a computer and the Internet; for example, it allows the capture and display of web pages.

Work order – see 'Production schedule'.

Working drawing – this is drawn full size or to scale and contains factual information relating to appearance and dimension. An orthographic drawing is produced to BS7308, which forms the basis of the international standard to which all technical drawings are made. It should include all the necessary information for you or anyone else to make the product.

Work schedule – see 'Production schedule'.

World Wide Web (WWW) – part of the Internet consisting of millions of pages of electronically stored information and graphics, complete with hyperlinks. The web has now become a gigantic global marketplace for products, services and self-promotion.

Index

Building exam confidence

Revise for Product Design: Graphics with Materials Technology

NEW

REVISE FOR
Product Design:
Graphics with
Materials Technology

John Halliwell
Barry Lambert
Consultant: Peter Neal

Heinemann
Inspiring generations

This new revision guide has been written to match the latest specification. It is the only one available for A Level Product Design: Graphics with Materials Technology.

- Written specifically to prepare students for the Edexcel assessment so students know exactly what to expect.

- Advice on answering the different types of question students will encounter and guidance on what examiners are looking for.

- Clear summaries of key information help students identify the main points of the course.

- A focus on application of knowledge to industry builds students' confidence about this aspect of the exam.

- Clarification of the recent changes to the coursework criteria.

Endorsed by
edexcel

Why not order a copy today?

Contact our Customer Services Department for more details:

(t) 01865 888068 (f) 01865 314029 (e) orders@heinemann.co.uk (w) www.heinemann.co.uk

Heinemann
Inspiring generations

H966